Home Is Where the Wind Blows

CHAPTERS FROM A COSMOLOGIST'S LIFE

Home Is Where the Wind Blows

CHAPTERS FROM
A COSMOLOGIST'S LIFE

Fred Hoyle

University Science Books
Mill Valley, California

University Science Books
20 Edgehill Road
Mill Valley, CA 94941
Fax (415) 383-3167

Production manager: *Susanna Tadlock*
Manuscript and copy editor: *Gunder Hefta*
Editorial coordinator: *Jane Ellis*
Text and jacket designer: *Robert Ishi*
Map illustrator: *Bill Nelson*
Indexer: *Andrew Joron*
Compositor: *Wilsted & Taylor*
Printer and binder: *Maple-Vail Book Manufacturing Group*

This book is printed on acid-free paper.

Library of Congress Cataloging-in-Publication Data

Hoyle, Fred, Sir.
 Home is where the wind blows : chapters from a
cosmologist's life / Fred Hoyle.
  p.   cm.
 Includes bibliographical references and index.
 ISBN 0-935702-27-X : $28.50
 1. Hoyle, Fred, Sir. 2. Astrophysics—History.
3. Astrophysicists—Great Britain—Biography. I. Title.
QB460.72.H69A3 1994
520'.92—dc20
 [B]   93-48080
                                                          CIP

Printed in the United States of America
10 9 8 7 6 5 4 3 2 1

Contents

PART THREE

Home Is Where the Wind Blows: 1959– 307

Foreword

A S ONE LOOKS BACK over the advances in astronomy and astrophysics during the past 50 or 60 years, one sees that the pathway is marked by sudden leaps forward, both by new observational opportunities and by striking instances of perceptive theoretical insight. These discontinuities are followed and spaced by stolid observational or theoretical research that breaks no new ground but tracks along the newly opened pathways, as on a plateau following a major upward step. Outstanding examples on the observational side are the opening up of previously unobserved regions of the electromagnetic spectrum (for example, the radio spectrum in the 1930s and 1940s) and the discoveries, enabled by observations made in new wavelength regions and by advances in instrumentation, of previously unsuspected components of the Universe. Obvious examples of these are the quasars and neutron stars—pulsars— although the existence of the latter had been predicted decades earlier.

Examples of similar major advances on the theoretical side concern the structure, energy sources, and evolution of stars. The vision of nuclear processes as the energy sources responsible for the energy output of the Sun and stars, from the conversion of hydrogen to helium, led to the consequent development of the theory of inhomogenous stellar structure and an understanding of stellar evolution from main sequence ("original" stellar structure), to red giants, to post–red-giant structure, to the degenerate-matter configurations of white dwarfs and neutron stars.

Fred Hoyle has been a key figure in these major, and usually controversial, upward steps on the theoretical side, including some that appeared to lead nowhere in particular at the time but that have become "sacred cows" in today's astronomical scene. (An example is the accretion through gravity of diffuse matter by an existing object, which was first examined by Hoyle and Lyttleton for capture of interstellar matter by stars and is now almost universally invoked for processes in active galactic nuclei.)

Friends, admirers, and even critics of Fred Hoyle will welcome this full autobiography, an account of Sir Fred's personal as well as his scientific life. Those who enjoyed his earlier book *The Small World of Fred Hoyle* may have been tantalized by the ending of that book at the point of his marriage to Barbara Clark and will have looked forward to a continuation of it in Part Two and Part Three. They may have been tantalized also by the single picture, on the paper cover of *The Small World*, of Fred as a cheeky-looking schoolboy in short trousers, appearing indeed as if he had many much more worthwhile things to do than to listen to an apparently incompetent schoolteacher, and they would have welcomed more pictures.

Home Is Where the Wind Blows now satisfies these expectations. The earlier book showed how the innate talents of a very gifted person can surface in an environment not conducive to eventual arrival at the top of one's field, be it science, music, art, politics, or any field of human endeavor. In Fred's case, one can attribute this to genes, to a supportive family, and to that spark that comes one knows not whence.

Fred did not have an easy road to the top. His struggles through childhood and grammar school to achieve acceptance as a student at Cambridge University were described in the earlier book and appear again here in Part One—fortunately, because the earlier book is out of print. And there are many delightful pictures here of Fred's family and of the young Fred, as well as of physicists and astronomers with whom he had important interactions—for example, Paul Dirac and Arthur Eddington, among many others.

Part Two begins with a chapter that gives a very interesting account of Fred's interactions with Sir Arthur Eddington, who—quiet, withdrawn man that he was—clearly influenced Fred greatly. But Part Two then takes Fred away from his early career in research, as the Second World War plunges him, like so many scientists in the United States and the United Kingdom, into war-related research. Fred's lot was to be assigned to radar research, at the Admiralty Signal Establishment (ASE) in a place called Nutbourne.

Historians of the Second World War may find here many new insights into this world of the "boffins," while astronomers can enjoy the stories involving his ASE colleagues Tommy Gold and Hermann Bondi. The image of Bondi banished by hay fever to the top of Mount Snowdon for his part in a radar experiment, and of Gold, in charge of a horse box full of secret equipment, being rescued from utter boredom in a remote spot in Cornwall by a roster of visitors dispatched there to entertain him, will surely linger in astronomers' minds.

Chapter 16, "The Origin of the Chemical Elements," should lay to rest once and for all a curious misapprehension that has arisen among some younger astronomers, that work on nucleosynthesis in stars was inspired by Fred Hoyle's interest in the 1948–1949 steady-state cosmological model. The facts are the other way around. Fred's seminal work, begun in 1945 and published in 1946, on building all the elements by nuclear processes as stellar evolution progresses, arose out of his profound physical insight and influences from Ralph H. Fowler and others in Cambridge (such as Dirac and Eddington), and the history is told in this book.

There existed the accurate mass-spectrograph measurements (by Francis William Aston) of the isotopes of the elements and the advances in nuclear physics that had led to acceptance of the carbon–nitrogen cycle for nuclear processes in stars. Two timely and very fortunate opportunities now came to Fred: The first was an official visit to Washington, D.C., and to the U.S. Naval headquarters in San Diego, in late 1944. During the latter, Fred made an unauthorized visit to Pasadena, where he met and was enchanted by Walter Baade, through whom he visited Mount Wilson Observatory. Baade, as remembered by those of us privileged to have known him, was a wonderful source of sparkling conversation, as well as a fount of astronomical knowledge and experience. Fred was charmed, and his creative instincts were captured by what Baade told him about novae and supernovae. Supernovae! Implosion to precede explosion! Implosion would produce the high temperatures and densities he needed in stars for conditions to be right for rapid interactions among nuclei of the heavier elements to occur, to which the statistical mechanics he had learned from R. H. Fowler could be applied.

The second piece of luck was that, upon Fred's return to Cambridge in 1945, he met Otto Frisch, who fortuitously possessed Mattauch's recently available tables of nuclear masses. Thus, Fred had the idea, the physical know-how, and the experimental data with which to calculate what would happen to nuclei under conditions of statistical equilibrium at very high densities and temperatures in stars about to explode as supernovae, which was published as the marvelous 1946 paper on nucleosynthesis.

It would seem natural, if the elements much heavier than carbon, nitrogen, and oxygen could be built in stars and dispersed into the interstellar medium by supernovae, to make a following intuitive leap—that such processes could represent continuity in the Universe, thus a key to an alternative cosmology that would avoid an unimaginable creative event followed by unexciting dissipation to infinity of such creation. This plunged Fred into controversy, which has beset much of his scientific life.

Some of Fred's theoretical work that was received with scepticism by colleagues was later dramatically proved correct by the experimentalists. An example is his prediction that there should be an excited level in the nucleus of carbon-12, at 7.65 MeV, that would speed up the interaction of the unstable beryllium-8 nucleus (formed by two alpha particles) with a third alpha particle. If there were not this excited level in carbon-12, the carbon nucleus would be formed slowly enough for rapid interaction with another alpha particle to form oxygen-16, and the cosmic abundance of carbon would be *much* lower than observed.

Fred's suggestion to the experimental nuclear physics group at Caltech was first greeted wth scepticism—if such a level existed, it would already have been detected. Fred persisted, in his gentle but persuasive way, and he succeeded in persuading the group to design a new, more sensitive experiment. Sure enough, the excited level in carbon-12, at 7.65 MeV, was found!

This autobiography reveals many threads besides science that have been a part of the fabric of Fred's life. His family life is one; his love of and skill in mountaineering is another. Descriptions of his mountaineering experiences appear throughout the book. Fred's love of animals is another. Pervading the entire book is Fred's sense of humor. There are amusing and entertaining stories and anecdotes that are dotted like golden nuggets throughout the book, together with many perceptive observations on the quirks ("fads and fetishes") of astronomers. Chapter 22, "Droll Stories," contains a pungent account of the saga of the Isaac Newton Telescope (INT), conceived in 1945 and then, when it was finally constructed, placed at a site 300 feet above sea level in an area named Pevensey Marshes. Those old enough to have experienced the famous lecture on "Time" by former Astronomer Royal Harold Spencer-Jones will greet Fred's account of this, and indeed of the whole INT story, with delighted hilarity.

The account of the mighty effort Fred had to make to achieve the funding, building, and success of the Institute of Astronomy reveals the dark side of the Cambridge scene. Names are not spared here; the machinations of two *eminences noires* and of the Science Research Council in the events that led to Fred's early retirement from Cambridge University are fully described. (Ray Lyttleton is quoted as having described one of these people as "rusted in" to what should have been a rotating position.)

The history of the Anglo-Australian Telescope and its successful outcome is described rather briefly, although Fred Hoyle played a big role in its planning and eventual success, because it has been fully told in *The Creation of the Anglo-Australian Observatory* by Gascoigne. I can add one additional gem as an illustration of the way Fred's sense of humor could always be called upon in an awkward situation. Early in the negotiations between the U.K. team and the Australian National University, which sought at that time to take over the whole project and remove the U.K. astronomers from it, a formal luncheon took place at which the Australian Minister for Education and Science, Malcolm Fraser (later Prime Minister of Australia), and Fred Hoyle were to give brief speeches. Fred had prepared an elegant response to what we expected to be a friendly welcoming and upbeat speech from Fraser. In the face of Fraser's vitriolic attack on the Brits, what could Fred say? The other two members of the U.K. team sat in ghastly suspense. Fred rose slowly to his feet, made a brief formal "thank you" for the ANU luncheon, and then pulled out one of his traditional funny stories (I have a poor memory for funny stories, but I have never forgotten this one). It had to do with "time" and the fact that a sleepy country railway station in Ireland had two clocks, one on each side of the line, and they told different times. The assembled luncheon gathering relaxed, its embarrassment stroked away.

The autobiography ends in an upbeat way, with some philosophical insights that have links to some of the writings of William James and Carl Jung. And Fred's many fans will be pleased to know that he is deeply involved in exciting research on a new cosmological model that makes predictions that can be tested observationally. I hope the die-hard big-bang aficionados will fail in their expected attempts to block its public presentation.

—Margaret Burbidge

Home Is Where the Wind Blows

CHAPTERS FROM A COSMOLOGIST'S LIFE

PART ONE

34, Primrose Lane
1939–1958

The British Isles

[3]

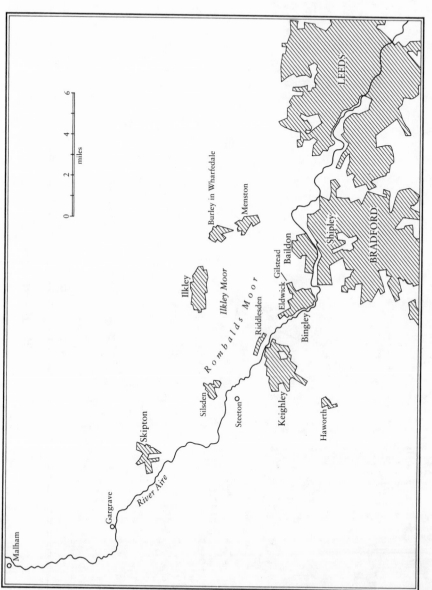

Southern Yorkshire

[4]

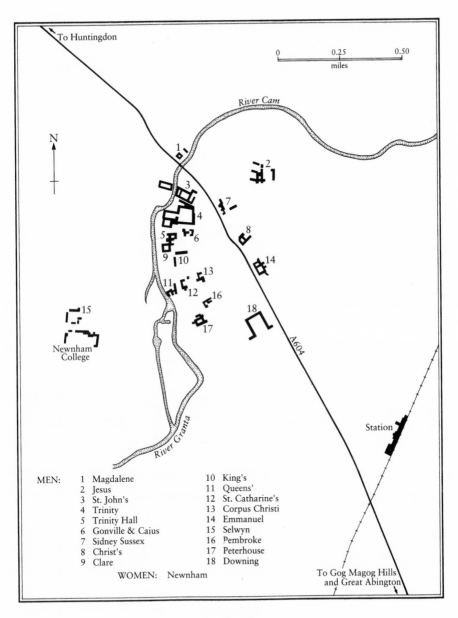

To Huntingdon

River Cam

N

1 Magdalene
2 Jesus
3 St. John's
4 Trinity
5 Trinity Hall
6 Gonville & Caius
7 Sidney Sussex
8 Christ's
9 Clare
10 King's
11 Queens'
12 St. Catharine's
13 Corpus Christi
14 Emmanuel
15 Selwyn
16 Pembroke
17 Peterhouse
18 Downing

Newnham
College

River Granta

A604

Station

0   0.25   0.50
        miles

MEN: 1 Magdalene   10 King's
         2 Jesus   11 Queens'
         3 St. John's   12 St. Catharine's
         4 Trinity   13 Corpus Christi
         5 Trinity Hall   14 Emmanuel
         6 Gonville & Caius 15 Selwyn
         7 Sidney Sussex   16 Pembroke
         8 Christ's   17 Peterhouse
         9 Clare   18 Downing
         WOMEN: Newnham

To Gog Magog Hills
and Great Abington

Cambridge

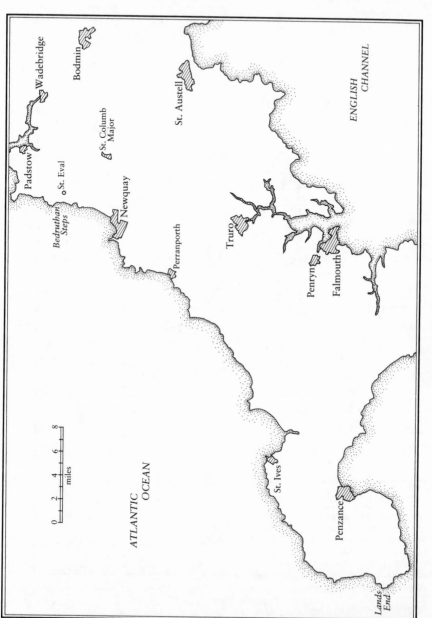

Western Cornwall

[6]

Chapter I

The First World War

MY FATHER always wore a trilby hat. It wasn't anything to do with the weather, because, until his later years, he worked bareheaded in his garden. Yet for the short step into the local village of Gilstead, for a walk to the town of Bingley a mile away, and certainly for business in the bigger city of Bradford, in the county of Yorkshire, he always wore a trilby hat.

This male fashion would be only a minor shift of custom from the 1920s to the present day, were it not for the large size, to the modern eye, of those hats. The effect of wearing a brimmed hat rather than a close-fitting bonnet is to make the head look bigger. If you were suddenly transported back to the 1920s, the difference that would surely hit you with the force of an executioner's axe would be that nearly every man walking the streets would appear to have a head twice as large as those we see today. Here, in one move, we have the reason why it is so difficult to notice the circumstances and events in one's own day that will appear remarkable, important, and even amazing in the future. No commentator in the 1920s would have thought to record the size of trilby hats. If it hadn't been for the existence of the camera, this really quite big shift in human psychology would be well on its way to becoming lost. Why did men in the 1920s want to make their heads seem so big? Or, to put it from their point of view, why do men today want to make their heads seem so small?

The river Aire rises in limpid pools in limestone hills near the village of Malham, in what has become known in recent years as "Herriott country." After passing Skipton (which, I suppose, means sheeptown), the Aire opens into a wide valley filled with sediments that extend as far as Keighley. Continuing downriver a further three or four miles, the valley narrows into a bottleneck at the town of Bingley. Thereafter, it opens out again in a flattish bottom at Shipley, where it continues nearly eastward to Leeds, and thence eventually out via the river Humber to the sea.

People commanding a strategic bottleneck traditionally live by exacting tribute from travelers and traders, and so it seems to have been with Bingley as far back as the Domesday Book of William the Conqueror. In industrial times, traders plying their wares between the agricultural communities of the Wharfe Valley around Ilkley and Otley and the early manufacturing center of the Halifax district took care to steer clear of both Bingley and Shipley—the latter probably because of swamps, the former because of the rapacity of its inhabitants. Their route is being imitated nowadays by modern motorists. It went over Ilkley Moor and down through the village of Eldwick to Gilstead. Here it turned into Primrose Lane to avoid Bingley, passing immediately before the very door of the house where I would be born. The community of Bingley responded to this scheme by doing its worst to convert Primrose Lane into a quagmire. The tug-of-war from Domesday down the centuries culminated in July 1758, when an indictment was made against the inhabitants of Bingley, for the solid reasons that the King's subjects could not "go, return, pass, ride and labour without great danger of their lives and loss of their goods [wherefore] the inhabitants of the parish of Bingley shall repair and amend when and so often as it shall be necessary the Common Highway aforesaid [Primrose Lane] so as aforesaid being in decay."

The bottleneck of Bingley had more valid commercial advantages during the second half of the nineteenth century. Its importance continued, but with the bang gradually fading to a whimper, through the years of my boyhood. The steep sides of the Aire Valley offered sources of waterpower for textile mills, which provided the main employment for the outlying villages of Morton on my side of the valley and of Harden, Cullingworth, and Haworth on the western side. There were sheep in plenty on the hill farms on both sides of the valley, with spinning mills close by, a way of life that formed the background to the novels of the Brontë sisters. You could see their village of Haworth from my village of Gilstead. For that matter, you could see Brontë country from my bedroom window.

Gilstead wasn't a place of manufacture, however. In my boyhood, there couldn't have been more than two or three hundred village people, earning their livings in diverse ways. There was a big quarry supplying stone to Bingley, of which my maternal grandfather, William Pickard, had been the foreman. He died in his early thirties, long before I was born, apparently from silicosis, although it wasn't called that in those days. In addition to rising from stonemason to quarry foreman, William Pickard, by the time of his early death, had built a house for himself and his young wife, my maternal grandmother. Had he lived, my mother would surely not have had to work for so long in the local mill to earn the money that took her eventually to the Royal Academy of Music in London.

There were two farms, Walsh's and Robinson's, around which my youth revolved. There were a few substantial properties and the more modest houses of folk, like my father, who worked down in the Aire Valley itself. One of the substantial properties had a determining effect on my life—the Milnerfield Estate. A mystery surrounded this estate, a mystery that still persists, of why, in an industrial area, a large, valuable tract of land should never have been built upon.

I never had sight nor sound of any member of the Salt family, which was the nearest thing to divinity I ever heard of in my youth. The Salt family owned the Milnerfield Estate. The Salt family had a big textile mill down where the Aire Valley broadens into the Shipley region. The family built houses close to the mill for its employees, and the conglomerate ultimately became so large that it formed a township in its own right, known as Saltaire. The Milnerfield Estate stretched all the way up the valley from the lowlands around Saltaire to the highlands around Gilstead. Where the estate abutted the village, a stone wall fully ten feet high was built. The remarkable thing about this stone wall was that its construction was so precise, stone fitting exactly on stone in a manner the ancient Incas would have approved of, that at no point was even the most agile village boy able to climb it. In my youth, we made hundreds and hundreds of attempts, but never once did anyone manage to climb on top of the big, overhanging flagstones that capped the wall.

Nearest to the village was a narrow lane about six feet wide, with the Great Wall of Milnerfield on one side and a dark strip of woodland on the other, the woodland also being enclosed by walls, so that you went along the lane between high walls on both sides. We called it the "Sparable," but later in life I learned that its proper name was the Sparrow's Bill. Even on a bright day, it was dark along the Sparrow's Bill, and, since there were many twists and turns to it, you had the impression that an ogre was waiting around the next corner, an ogre from which it would be impossible to escape because of the walls on either side. It was by forcing myself to go alone along the Sparrow's Bill at night that, somewhere about the age of seven, I learned not to be afraid of the dark.

The perimeter of the Milnerfield Estate remote from the village was more readily negotiated, even though the wall was everywhere well constructed and fairly smooth. The estate was no small affair, by the way. Before the outlying farms were sold off, it was an extensive block of land that lay strategically athwart the ancient traders' route between the Wharfe and Aire valleys. The trouble from our point of view was that low wages permitted a truly great number of gardeners and gamekeepers to be employed, so that, except by using real guile, you could hardly make a hundred yards inside the place without having a brawny adult seize you by the ear and march you at the double to one of the exits. In retrospect, the geomorphological cunning of us village boys fills me with admiration, for really it couldn't have been better done. We realized that, since the estate was on a general downslope, water had inevitably to flow through the wall. There was one particular place along the Sparrow's Bill where the masons had left a low hole in the wall to permit the passage of a stream, and, provided the stream wasn't in flood, an agile young boy or girl could just squeeze through without becoming too seriously wet. This was our point of attack in exploring the Milnerfield Estate. One day we came on a partridge's nest with perhaps fifteen brown eggs in it. We hadn't been admiring the nest for more than a minute or so before a game-keeper came roaring down on us with a threatening stick. I can't vouch for the intention of the other lads, but I know I had no particular inten-tion of seizing the eggs. It was the sheer beauty of the nest that impressed me, to a point where I can still conjure a vision of it to this day.

I was never a great one for taking eggs from birds' nests, for the reason that I could see no interest in them once they were taken. If you blew them, there was only an empty shell that did nothing, and, if you didn't blow them, they soon started to smell terribly. I found it far more inter-esting to watch what happened if you left the eggs where they were. On one occasion when I was scouting alone I found a kingfisher's nest along the banks of the same stream by which we entered the Milnerfield Estate. I told one of the lads about it; whether he took the eggs or someone else did I never knew. When I found the nest empty, it seemed an absurd waste. Here had been something that would have made marvelously col-ored birds that were then not to be.

Strange events were in train at the Milnerfield Estate, an almost literal enactment of the first part of the story of *The Sleeping Beauty*. The prac-tice was for the managing director of the Salts' textile mill—in my earliest days, a certain Sir John Roberts—to live at Milnerfield House. He drove daily to and from the mill in a horse-drawn carriage along a mile of drive-way, on the sides of which the grass was mown meticulously by hand. But the house itself was a grim pile indeed, designed by some more than usu-ally insane architect in a fashion that would have done credit to the stories of Edgar Allan Poe. And grim stories as to what went on in the house were rife in the village, stories that delighted my young ears, I can tell you. Deaths and disasters, ghosts and ghouls—no good ever came to anybody who lived there. When I was about nine years old, it fell out just as the old ladies of the village said it would. Just as the princess in *The Sleeping Beauty* pricked herself on a spindle, so the managing director of the Salts' textile mill pricked himself on a thorn bush. Within hours, blood poison-

ing had set in, and, within a week, a hearse was carrying his coffin down
the mile-long driveway to the cemetery in Saltaire. The gardeners and
gamekeepers melted away. Nobody ever came to live there again. Grass
grew thick down the carriageway, the splendidly cultivated gardens and
orchards fell more and more into disuse, more and more of the green-
house windows were broken owing to chance events, and, above all, the
house simply fell down, to be carted away stone by stone, until today there
is literally nothing but a place left for sixty years to become a jungle—
just like the environs of the castle in *The Sleeping Beauty*.

There was a time early in the century when it was very much otherwise,
when the estate looked as if it might swallow the village itself. Four houses
known as "Milnerfield Villas" were built outside the high wall to accommodate the highest-ranking servants—the butler, the head gardener, and
so on. They were substantially built, as estate agents say, a step ahead in
quality of most of the other houses in the village. In 1910, a decision was
taken to reverse this process. The houses, including the four known as
Milnerfield Villas, were sold, probably because it was found that estate
folk didn't mix with village folk. I have a fair idea why. When I was about
eight, apparently during some crisis at the estate, a footman erupted from
the big iron gates that guarded the entrance to its carriageway, and then
rushed with a bundle of letters to the village post office. I remember nothing of the man's appearance, but I do remember, as if it were yesterday,
that he was wearing white gloves. In a derisive gang, we followed him to
and from the post office, dogging his heels as closely as we dared. Since
white gloves and earthy village life obviously didn't mix, the estate decided to sell its scattered properties outside the big, flagstone-capped wall.
And my mother decided that—somehow, someplace—she would raise
the money to buy the residence at the lower end with the high-sounding
address of 4, Milnerfield Villas, changed later by the post office to 34,
Primrose Lane. It was there, on 24 June 1915, that I was born.

It wasn't the high-sounding address my parents had wanted. It was the
view. From the house you can see over the flagstone-capped wall into the
kitchen garden and orchard of the estate. More relevant, however, the eye
lifts to the summit of a moor two miles away—Rombald's Moor, to give
its official name, although nobody local ever called it anything but Bail-
don Moor. The point is that, with astute use of Ordnance Survey maps,
you can walk from the summit of Rombald's Moor to the outskirts of
Edinburgh without ever descending even into a village, sleeping at night
only in remote farmsteads. If at Edinburgh you take David Balfour's route
across the Firth of Forth and thence beyond the Highland line, you can
continue through remote places to the end of Britain at Cape Wrath. So,

throughout my youth at 4, Milnerfield Villas, I had half-a-thousand miles of open country beckoning me everlastingly towards adventure.

Owen Glendower, in Shakespeare's *Henry IV, Part I*, says to Hotspur:

> . . . at my nativity
> The front of heaven was full of fiery shapes,
> Of burning cressets; and at my birth
> The frame and huge foundation of the Earth
> Shaked like a coward.

So it was at my own birth, the fiery shapes and the shaking of the Earth being caused by the thundering guns of armies locked in the battles of the First World War. My father was already in his mid-thirties when the First World War broke out in 1914. Shortly after I was born, he was conscripted into the British Army, opting for the Machine Gun Corps, which choice he said in after years was dictated by his dislike of "bull." The life expectancy of machine gunners was so short that "bull" was not demanded of them. Nor were machine gunners likely to be accused of insubordination, a risk that an infantryman of thirty-five years holding strong views about the intelligence of his senior commanders might easily have been exposed to.

So it came about that my mother and I were left alone to live on five-pence a day, the government allowance to the wives of soldiers serving a country that needed them so much, according to General Kitchener. Had it not been for me, my mother could have returned to her former job of schoolteaching; but with me so very young, and of a somewhat frail disposition, I believe, she felt a job that did not require her to leave me during the daytime was to be preferred, if it could be found. During the first years of the century, my mother had worked in a Bingley mill. With the money she saved, and with help from her own mother and elder sister, she had then studied music at the Royal Academy. The outcome was a training and a qualification that permitted her to embark on a career as a professional singer. Times were hard, as they always seem to be for young musicians, so my mother was tempted out of full-time professional work to teach music in schools. However, in 1911, when she married, her career was blocked, because in those days married women were not employed as teachers, although, during the war years, the rule was held temporarily in abeyance.

Even before 1911, my mother had begun transferring her main interest from singing to the piano. From my earliest memories to her death forty years later, she hardly ever passed a day without spending two or three hours at the piano. The solution to the problem of a job in 1916, therefore, lay literally to hand—to go out of an evening and play the musical accompaniment to silent films at a cinema in the town of Bingley. I was put to bed in my mother's bed, and off she would go, leaving me to fall asleep, which I did without difficulty until I reached the age of two. My earliest memory is of lying awake for a while and wondering what I would do if my mother never came back. Then I cried for a bit and at last went off to sleep. There were two teenage girls next door, Ethel and Mary Clark, but it seems that when they came round to sit with me I said I was all right alone.

Eventually, my mother lost her job at the local cinema because her musical tastes did not suit those of its manager. Her idea of accompanying films was to play parts of Beethoven sonatas. Imagine whooping Indians charging soundlessly in jerky black and white, done with the execrable cinematographic technique of 1917, accompanied by the thunderous roll of the third movement of the *Moonlight* Sonata. About a week after my mother's dismissal, however, there came a tap on the front door of the house. It was the manager come to ask my mother to return to her job. After she'd left there had been a decline in attendance. On inquiry around the town, the man had been told, "We didn't come to see your films, we came to hear Mrs. Hoyle play."

The early development of musical appreciation made it impossible for me to learn to play a musical instrument myself. To a young child of two who knew how the *Waldstein* Sonata goes, the trite pieces one must play to learn a musical instrument seemed indescribably boring. The gap between appreciation and achievement had become too large to be bridged. Learning comes best if a child does not become too sophisticated too early on.

It turned out better where numbers were concerned. With much time on her hands during the day, my mother taught me the numbers, and almost immediately I began to set little problems for myself. I was coming up to three years old when one morning I asked my mother if two sixes made twelve. She answered yes, and then asked me how I knew. I can't claim I remember my answer, but, from its reported nature, I think it has to be true, since an adult would hardly have thought of arguing like this: "One and six make seven, so two and six make eight, so three and six make nine, so four and six make ten, so five and six make eleven, so six and six make twelve." By remembering my previous results and using the same slow but sure method, I managed between the ages of three and four to construct a good deal of the multiplication tables for myself. The reason for the plodding method was simply that I worked at it in my head after being put to bed at night.

The life expectancy of a British machine gunner in the First World War was only three or four months. It was inevitable, then, that every day my mother watched in dread for the arrival of the postman bearing the government's heartfelt regret that my father could serve as cannon fodder no longer. As each day passed without the arrival of a letter expressing the government's grief, it could only have seemed to my mother that the inevitable had been postponed for just a little longer.

As the days, weeks, and months passed inexorably away with their grim toll on the western front, it was said with macabre humor by the men of the British Machine Gun Corps that there were two whom death could not touch: the two aitches, Holmes and Hoyle. Random chance—or luck, as we say—always works to make it seem like that, and I daresay something of the same sort happened with fighter and bomber pilots in the Second World War. If the life expectancy of the average machine gunner was four months, there would be a lucky one among twenty who would survive for a whole year, a lucky one among four hundred who would survive for two years, and a lucky one among eight thousand who would survive for three years. As well as being lucky, my father had characteristics that must have aided his survival. He was compact in build with a bubbling sense of humor, very quick in sprint races, and with exceptionally keen eyesight. Since exceptionally keen eyesight must surely have been one of the really important advantages that a man in the trenches of the First World War could have had, quite likely Holmes was similarly endowed. But I never learned about Holmes, because I don't think he and my father knew each other. And a day eventually came when their meeting became impossible, the day Holmes was killed. It was from then onwards that my father said he began to feel really bad, and suffered from the immovable conviction that, for him too, the time must surely be coming, just as after a long partnership in cricket both batsmen often fall to the bowlers in quick succession.

If my father had been lucky until March 1918, he was unlucky on the twenty-first of that month—unlucky to be exposed in the trenches on the day the German chief of staff, Erich Ludendorff, launched what is generally conceded to have been the fiercest assault of the whole war—"Operation Michael," to use its code name. The *Encyclopaedia Britannica* entry reads as follows:

> The launching of "Michael" took the British by surprise, but the ensuing Second Battle of the Somme did not develop as Ludendorff had expected. While the German Army south of the Somme achieved a complete breakthrough, the big attack to the north was held up by a heavy concentration of forces near Arras. For a whole week Lu-

dendorff mistakenly persisted in trying to carry through his original plan in the north, instead of exploiting the unexpected success in the south. . . .

Here, in only a few words, you have an overall distant view of the situation. But what actually happened there on the ground?

I never heard my father say exactly where his post was located, and I doubt that he really knew, because almost surely he wasn't issued a map, although he had long been the leader of a machine-gun crew of eight men. But, from his description, never put together in one piece but coming out on odd occasions over the years, it is likely my father's post was in the Somme Valley, in the south, where Ludendorff made his unexpected breakthrough.

There is at least one semiofficial history of the British Machine Gun Corps. The writer says that very little is known of what happened to machine gunners in the Somme Valley on 21 March 1918, because few of them survived the day. According to the writer, there was a thick morning mist. At first light, the German field command sent patrols, amply equipped with stick grenades, through the mist. Because of the mist, the machine gunners were caught unawares and were almost completely taken out, mostly without firing a round in their own defense. In just a few spots, the writer says, a capricious swirl of mist happened to clear the ground, permitting the machine gunners to see ahead to open fire effectively, and so to save themselves. If I believed this story, I would have to suppose that my father's life, which as a result of his survival made a big difference to my own life, turned on a capricious swirl of heavy morning mist. But there was more to the matter, issues that either the writer of the semiofficial history did not know about or did not wish to discuss. The question this history actually raises is how the German patrols made their way through thick mist and managed to take out the British machine-gun positions so unerringly. I happen to know the likely answer to this question. It was because the leaders of British machine-gun posts were under orders at all times while in the front line to fire bursts at ten-minute intervals—to fire mostly, therefore, at nothing at all. Consequently, the Germans knew, from simple observation, exactly where the British posts were located. Presumably, the German patrols came through the mist on compass bearings. Knowing the exact direction and more or less the exact distance they had to go, they would perhaps walk at first and would then crawl the last yards until they heard voices.

The machine gun was the single most important ground-based weapon of the First World War, a crucial fact the British High Command never seems to have understood from the first day of the war to the last. By

March 1918, Douglas Haig was commander-in-chief of the British forces in France. Haig had written in a report dated 1915 that he thought a ration of two or three machine guns to a battalion would be quite sufficient, which showed that, despite an already long career as a professional soldier, Haig had not understood the basic strategic lesson of the American Civil War of 1861–1865—namely, the astonishing extent to which a few posts equipped with even the crudest of early automatic weapons could hold up large numbers of infantrymen, always provided the posts were carefully concealed. Even riflemen in concealment—snipers, as we call them nowadays—were effective in the field to a degree that officers in staff colleges could not believe, until during the First World War the fact was rediscovered in the hardest conceivable way—or, as in Haig's case, not rediscovered at all.

From the earliest days when he had become the leader of an eight-man machine-gun crew, my father had ignored the order to fire random bursts at ten-minute intervals, telling his crew never to fire except on critical occasions. Although his ignoring of orders was probably dictated at first by thoughts of survival, practical observations in the field soon convinced him that his policy had to be correct. The Germans always went for machine guns first, which proved, as my father told me in later years, that the positioning of the guns was critical, at any rate in German eyes. Of course it was, because the machine gun more than anything else defined the nature of trench warfare, a truth it was possible to stumble on in practice, but a truth that remained perpetually hidden at British headquarters. Time and again British commanders sought to break German defenses with heavy artillery barrages followed by advances of large numbers of infantry, only for the infantry to be mown down by German machine guns—thereby leading to those many letters that came daily through the British post expressing the heartfelt grief of a government that nevertheless maintained those who made these disastrous misjudgments in their posts, largely, I suppose, because summary dismissal would have been "bad form."

So it came about on 21 March 1918, when the German ground attack eventually came an hour or so after the activities of the stick bombers, that my father's post was still operating. Now indeed he had unprecedented luck. In an ordinary engagement, where every hundred yards of ground was bitterly contested, the loss of support on either side would have been disastrous. But the staggering magnitude of the German advance in the Somme Valley meant that the German infantry simply poured through the holes on either side of my father's position. Doubtless it would be the intention of some German officer to clear out the few still-active British machine guns in due course, but why bother for the moment when the way ahead was clear? So hell-for-leather the Germans went, at first in hundreds of yards, then in quarter miles, and at last in whole miles, leaving a crew of eight men still alive, one a dazed, half-shell-shocked man approaching forty and the others recent recruits on the right side of twenty. Through his mental mist, my father told me that he almost became convinced, because of the uncanny silence now fallen over the shattered British line, that he had indeed been killed and had passed into some grotesque new form of existence.

Towards evening, his judgment of the situation gradually returned. After more than two years of practical experience, and with the keen sight I mentioned before, he had learned to judge battlefield situations, to judge where men were fighting, where guns were firing, and the lie of the land generally. At last he understood that the unbelievable had happened. As if aided by powers never seen before, the Germans had done what our official historian says they did: "The German Army south of the Somme achieved a complete breakthrough." Seeing how it was, my father explained things to his young crew, telling them there were two possibilities. They could stay put, with the prospect of being taken prisoner, or they could move back and try to penetrate the now greatly advanced German front from its rear. My father then said that, in his physically shocked condition, he didn't feel he could manage to crawl great distances across vile, crater-strewn ground. So he told each member of the crew to make their own decision. One of the crew, a young fresh-faced farmboy from Somerset, decided to stay with him. The others elected to try their luck.

It was a long, distressed night that followed, but, in the way the human body sometimes does, my father began to feel stronger towards dawn on the twenty-second, so that eventually he was able to move. He and the farmboy from Somerset set off, not walking openly, of course, but edging from one depression to another. The problem was whether to try crawling through the German position or to stay behind it, moving parallel to what had become an entirely new front, in the hope of there being a gap somewhere along it. Not being an infantryman himself, he thought better to trust to observation and geography rather than to risk a tussle with German infantry. So it was a hole in the front or nothing. After three days of crawling and dodging, a possible gap was found. A gap it turned out to be, and, after a week, by 28 March or thereabouts, my father and the farmboy from Somerset rejoined the British concentration near Arras. He never saw the other six members of the crew again.

The war over, the British government showed considerable generosity to young officers, sending many of them free of charge to university. I never detected a comparable generosity in the treatment of ordinary war widows. In all cases known definitely to me, the sons of war widows left school at fourteen to earn what they could to support mothers in straitened circumstances. I cannot say this would necessarily have been my own fate had luck, sharp eyes, and common sense not aided my father's return from France. My mother might, with her qualifications, have obtained a teaching job again, so that events for me could have fallen out not too differently from the way they actually were.

By the late summer of 1918, German prisoners were coming over in droves. They were ragged, half starved, and desperately hungry. Rations were distributed to them in groups of ten, rations that were by no means generous. One day my father watched a group distributing its combined ration. One of the group was first elected by lot. The chosen man then divided the ration into ten portions as equally as he could manage. Thereafter the others drew lots among themselves as to how to distribute nine of the portions, leaving the last and tenth portion for the man who had done the division.

The German government put out peace feelers in 1916, but, owing to the political confusion, from which no Allied leader emerged with credit, the German initiative came to nothing. This moment in 1916 was perhaps the crucial turning point of the twentieth century. Had there been peace in 1916, needless slaughter would have been stopped and the 1917 communist revolution in Russia would most likely not have happened.

The year 1916 is not very far back in time, in the sense that it lies within the span of a single human lifetime. A visit to any major library would permit you to examine the issue of *The Times* for any day in 1916 you pleased. Yet, so far at least as some aspects of technology are concerned, a single human lifetime has covered greater change than occurred in the whole of the preceding thousand years. The biggest social change, I believe, has been in communication. No young person today could, I think, conceive of a society in which there was an almost impenetrable barrier of communication on a day-to-day basis between government and the upper levels of society on the one hand, and the mass of ordinary people on the other. Notably, we have radio and television in most households today telling us of happenings, not just in Britain and other developed countries, but in every corner of the world. We have newspapers available at a cost that can be sustained by most households. The cost of subscribing to *The Times* in 1916 amounted to half an ordinary man's wage. Aside from possibly at the Milnerfield Estate, there would be no daily copy of *The Times* in our village, and probably not more than a dozen copies throughout the whole Bingley area. There was no radio or television, of course, so that news traveled literally by word of mouth. In the 1920s, my father would bring a copy of the Bradford daily paper back home when he returned from business in the evening, but, unless the skies fell in on a national or international basis, our sources of news were intensely local. There were no opinion polls either, so governments were just as out of touch with ordinary people as ordinary people were out of touch with governments.

The almost instant formation of public opinion today, over issues the public would scarcely have heard of in 1916, constitutes one of the greatest changes the world has ever seen. Almost everyone today, except those in the media, agrees that the media are a blessed nuisance; but the media have changed the world from one in which people could be trivially manipulated to one in which a catastrophe like the First World War has surely become impossible. It was a catastrophe that was to have repercussions all through my life and of which we are not free even to this day.

Chapter 2

Forgotten Times

WHENEVER I SURVEY the modern scene, with its special shops for the youngest children, playschools and TV programs for somewhat older children, and free medical and dental care for all, the circumstances of my youth seem as far remote from the world in which a child grows up today as one might expect for two different planets revolving around quite different stars. I have to doubt, therefore, that much of what I can say of my own times will seem meaningful in a modern context, although it may conceivably have the virtue that comes with things long gone by, like, it is hoped, a bottle of old wine one comes on unexpectedly in the corner of some cellar.

For balance and sophistication, for understanding how the world works, young people today are a decade ahead of where we were. Equate fifteen-year-olds today with twenty-five-year-olds in the 1920s and the comparison would be about right. Necessarily so, for without television, radio, or newspapers in most households, we knew nothing of what was going on outside an area that literally was not more than a few miles across. You might think our deficiencies would have been put right at school, but the same parochial fog also hung over the schools, or at least over the schools I attended. Until I reached the upper forms of the local grammar school, I do not think any teacher ever mentioned events on a national scale, still less on an international scale.

In its review of the year 1925, *The Times* began as follows:

> The year is likely to be remembered as that of Locarno, and the name Locarno may possibly be blessed by many generations. Such at least was the intention of the representatives of those nations which took part in the Pact initiated there and afterwards signed in London. The consummation of this agreement had involved much anxious and delicate negotiation, as susceptibilities had to be consulted and apprehensions allayed; and it had necessitated the retracing of imprac-

ticable but well-meant steps taken the year before. While it was recognized to be no more than a beginning it was confidently believed to have laid lasting foundations. In effect, the Pact bound France, Belgium and Germany, under guarantee of Great Britain and Italy, not to make war on one another, and ensured the inviolability of the frontiers concerned, subject to the reference of any outbreak of hostilities to the League of Nations, of which Germany was to become a member. The Pact also foreshadowed international arbitration and concerted measures towards disarmament.

Six weeks or so after this was published, I took my junior scholarship examination, in curious circumstances I will describe later. If I'd been asked by the examiners to describe the Pact of Locarno, I would have known as little about it as I did about the Diet of Worms. Imagine a similar situation today with an apparently major pact signed, sealed, and delivered between the world powers. The immense emphasis in the media would acquaint every child with what was going on. Yet the Pact of Locarno proved a load of rubbish, with exactly the opposite to what was intended eventually happening in the 1930s. Human nature does not change to the extent that it would be any different today. Unfortunately, nations act conspicuously in their own interest, and the devil take the hindmost. This is such a clear lesson of history that it is surprising how leaders, writers, and commentators of all kinds go on, generation after generation, chasing the same old chimeras. By not knowing anything about it in my younger years, we were spared from traps like this, into which the responsible section of society fell so completely.

The modern generation is much better fed and clothed than we were. The feeding shows itself to the eye in stature. I eventually grew to a height of five feet nine, which in my time was a bit above average, but which today would be on the short side. There were no such things as vitamin pills in my youth, and fresh fruit was unknown for about four months every winter. Improved clothing, such as the parka and the rubber boot, came partly with greater affluence and partly through manufacturing innovations.

Mine was quite possibly the worst-shod generation there has been since the inception of the industrial age. In poorer times, people had worn wooden clogs, often without stockings, a trick I tried out once or twice but could never master. You needed a curious shuffling motion to be able to make progress at anything like speed. In my early years, I often watched older people loping about in clogs, and the process never failed to astonish me. In rain, you inevitably got your feet wet with clogs but, once indoors with bare feet, the circulation soon returned and within half an hour clogged persons would be dry again. With the prosperity engendered by the industrial revolution, the aim became to shut out the weather entirely from one's footgear by wearing whacking great boots. In my first two years at school, everybody wore boots, summer and winter alike. Then, in a flash, it seemed boots were out and shoes were in, perhaps because some manufacturer had discovered how to produce shoes at a significantly lower cost than boots. The shoes were leather-soled and they didn't last long, especially with the sort of wear we gave them. So it came about that most children had holes in their shoes, which let in the rain like so many sieves, with the consequence that, in bad weather, a considerable fraction of us sat in school for hours on end with cold feet. Nobody among my acquaintances had several pairs of shoes, which necessarily made the resoling process irregular, even if one's parents could afford it. After resoling, you had a month or so of bliss with dry feet, then another month or more while the holes got worse and worse to the point where something just had to be done . . . and so on, round and round the repair cycle, until the uppers disintegrated, a situation that was not unknown even by the time I reached university. I'll just bet that, in those days, as many as a third of undergraduates had a hole in the sole of one or other of their shoes.

I was never conscious of being grievously bothered by the shoe situation, but I was indeed conscious of the need for some garment like an anorak. There were no special school buses in those days. Between the ages of nine and eleven, I had a walk to school of about a mile and a quarter, which, in normal weather, I enjoyed much better than riding in a bus. But there were icy cold days in winter when it was otherwise. The first mile of my journey lay exposed along a north-going road that afforded not the slightest shelter from the wind. Then you turned sharply at right angles into the shelter of a row of cottages—Brick Kiln Row, if I remember aright—and beside a high wall, which protected the garden of one of the better-off families in the Eldwick district. I can remember how my ears and temples used to thaw out when I reached this welcome protection. For many years afterwards, I blamed these icy experiences for a hugely painful infection of the middle ear that I contracted at this time.

With the exception of whooping cough and scarlet fever, I suffered more or less all the standard diseases of childhood. Such things were generally as much an advantage as a disadvantage. You had three or four days of high temperature when the world seemed to be coming to an end, then a few days of pampering in bed, followed by a week off school, or perhaps by two glorious weeks, if you could manage to swing it on credulous adults. The infection of the middle ear was something else altogether. It was a whole week of acute and unmitigated pain, day and night alike.

Our doctor, Dr. Crocker, came from time to time and looked sadly at me with his dark, spaniel-like eyes, but he might just as well have stayed away, for all the good he did me. Today, given a suitable antibiotic, the thing would have been over and done with in only a few hours, probably even before the pain became acute. Eventually, the eardrum burst and the pain was at last relieved. Dr. Crocker said I'd been lucky the abscess had burst outwards.

There was a form of spelling test in schools known as dictation, which quite likely is still in use. The teacher simply read out some passage that you had to write down. In the years after the middle-ear infection, I always listened with great care in dictation tests, which was probably necessitated by a hearing defect that either followed the middle-ear infection or had an alternative cause that I will mention in a later chapter. I was not overtly aware of any considerable disability until my thirties, when the trouble gradually grew into a progressive deafness of the left ear that became a handicap at just the stage in my life when I was involved in science politics, with the unwelcome correlation that the higher the politics in which I was embroiled, the worse the deafness became. There may be some who have succeeded in affairs of state despite deafness, but the disadvantage is so severe that such cases must be few and far between. My own disability was not in those days serious enough for me to fail to understand what was being said. The trouble in a committee is that, if you have to listen carefully, thinking all the time to check you have the discussion right, you miss the sudden interventions on which swaying fellow committee members often depends. Just as you need to be born with keen eyes to be a good ball games player, with fine balance to be a gymnast, with sound heart and lungs and muscles to be a runner, so you need to be born with sharp ears to be a successful politician. I never met one myself who couldn't hear a pin drop on the other side of London.

Unless you had a taste for the cut and thrust of surgery, the medical profession tended to be an unsatisfactory vocation in the distant days of my childhood, because there was so little that could be done effectively about most afflictions. The doctor had to wait for nature to take its course, hoping for abscesses to burst in the right direction, as in the case of Dr. Crocker. Even as late as the early 1940s, a medical friend who graduated from Cambridge in my year, and who became a public health officer in an inner-city area, expressed his dissatisfaction at what he regarded as the well-nigh hopeless struggle against tuberculosis. The fact that tuberculosis, a scourge throughout the nineteenth century and in my youth, is today much reduced demonstrates another of the major revolutions of the present century. It seems, however, that no good ever comes in human affairs without accompanying ill effects. Dramatically increased medical skills have inevitably led to exploding world populations, a critical problem we are somehow managing to duck in our own time, but one that is likely to prove one of the grimmer issues of the next century.

Although people in my youth lived under shadows that have since been lifted, they were not apprehensive about it. I recall a considerable outbreak of smallpox, in about 1927, I think. Few among my friends and acquaintances, young or old, rushed to be vaccinated. With modern media pressure, doctors in such a situation would be overwhelmed. There is little doubt that the media take a pernicious delight in whipping whole populations into terror over dangers much less than this smallpox outbreak of around 1927, dangers that often prove quite illusory. The young today will hardly be able to appreciate what a profound relief it would be to them if they could only shrug off the modern atmosphere of whipped-up hysteria over things that never happen. There was a bit of hysteria in my youth over spiritualism and ghosts, which the press did its best to hammer us with. This made something of an issue out of walking along the Sparrow's Bill alone at night, but it was exciting stuff, and you could stand up to the ghosts for yourself.

Occasionally, the shadows thickened. I remember walking into a mean house built on a slope halfway between the village of Gilstead and the town of Bingley. It was a back-to-back affair in a row fifty yards or more long, facing across a narrow street towards an exactly similar back-to-back row. A boy I'd known more as an acquaintance than as a friend had lived there, but he had died of cerebral meningitis. Together with other of his acquaintances, I filed past the coffin, with the inevitable thought of "there but for the grace of God . . ." and noting how little light came in through the two downstairs windows of the house.

My earliest medical memories predate by about a decade the popular television series *Dr. Finlay's Casebook*, based on the stories of A. J. Cronin. Dr. Crocker did not come to our house in the fabulous bull-nosed Morris car in which Dr. Finlay and his senior colleague rode around their practice, because the bull-nosed Morris was still a vision for the future. Dr. Crocker must have come by pony and trap, just as we had deliveries of fresh milk from Walsh's farm twice daily by pony and trap. A bull-nosed Morris would indeed have been well-nigh useless, because the roads were unpaved and full of potholes. I have a vision, again from around 1927, of a Sunbeam car lurching from pothole to pothole through the village. It must have been about this time that a private bus service was started. It went from Bingley through Gilstead as far as the nearby village of Eldwick. It always traveled with an open tailgate, where there were two or three steps, which passengers climbed to a height of about four feet above the road. What with the hills and potholes, there were sections where the bus traveled so slowly that you could easily slide yourself onto this step affair, which, being at the back, was hidden from the driver's view. Then the bus picked up speed, and you hung on for dear life until you came to another slow spot close to your destination, when you dropped off again for dear life onto terra firma. We village boys appreciated this free service, and we always had a pleasant word for the owner driver. "Aye, Mr. Murgatroyd," we would say.

The potholed road through Gilstead must have been pretty bad ten years earlier in the wartime years because, in late 1918, roadmen appeared with a big steamroller. I know this was so because somebody from the village came one afternoon and told my mother I had been standing and watching the steamroller for upwards of two hours, and that I couldn't be shifted away from it. Doubtless the visitor thought me peculiar, but, for myself, I like to believe that, already at the age of two and a half, I'd demonstrated what my métier in life was going to be: an observer of the world and a ponderer on its problems. I cannot claim to remember this incident of the steamroller. I am simply reporting what was said to me in after years. Many people do claim, however, to have memories from their earliest years. It has never seemed to me wise social policy to challenge claims of this sort, but I am very doubtful of their accuracy. I doubt that any of us remembers with any clarity at all what happened to us before the age of five, except for traumatic experiences. The misapprehension comes from confusing later memories, which are real, with earlier events. This indeed is an issue over which the poorer one's memory happens to be, the easier it is to be deceived. The brain seems at the age of five to more or less wipe itself clean of earlier memories, "clearing store," in computer language, in preparation, presumably, for sterner times ahead. I suspect that only the memory of things one continues to do by rote on a day-to-day basis passes unimpaired through the age of five. Probably this is the real reason why education begins formally at five.

I have precise memories of things that happened to me at the age of six. On my sixth birthday, I went for a walk with my parents and said: "In another year, I'll be *seven!*" I even know the exact bit of road we were walking when I said it, and I remember comparing unfavorably the boyishness of my bare knees with the manly long trousers my father was wearing, the itch to grow up quickly becoming almost ungovernable.

In contrast to this sharp clarity, such early memories as I believe I have float insubstantially in my mind. My earliest vague recollection is of standing outside the village church and of a bell ringing. Since the village church never had a bell of its own, the bell would have had to have been brought there for the occasion, which probably fixes the incident as occurring on 11 November 1918, the occasion of the armistice at the end of the First World War. It was probably announced by a sort of town crier who had come up from Bingley. From out of a distant fog, I also seem to remember being given a small trinket to mark the occasion.

One has to be cautious not to confuse true memories with what we are told later by parents and friends of the family. An incident that I feel sure my parents would hardly have thought to talk about in later years occurred when I was about four. I always got up first in the mornings, as young children mostly do. For some reason, I sat down one morning on the carpet. Perhaps it was to look at stamps on letters that the postman had just delivered. I put down a hand on the carpet, to receive the shock of my young life. This shock happened to repeat itself a year or so ago in the garden of a pub where I was eating an evening meal in failing light. I had put my hand on a crawling wasp and had been stung for my carelessness. The recent case verified my impression that the experience was fairly traumatic, which, I suppose, is why I remember it.

My parents told me in later years that I discovered how to tell the time when coming up to the age of four. Here I have to be careful not to confuse the later information from my parents with the impression I have of the actual moment of discovery itself. There is a detail, however, in my apparent memory that only I would have known. So probably there is some substance to the memory. Since the incident was my first bit of research, there would also be a reason for it to have stuck. It happened like this.

One of the things my father did immediately after being demobilized early in 1919 was to fix an old grandfather clock, which ticked away boldly for years thereafter in a corner of our "sitting room," as we called it in unashamedly middle-class terminology. For a while, the grandfather clock was a talking point between my parents and between my father and others who came into the house to help him with it. I became more and more intrigued—and frustrated, I suppose—by this thing that everybody around me called "time." Where *was* "time," I asked myself. I hunted around trying to find it. Eventually, when the clock began to work, the mystery partially resolved itself. "Time" had to do with the hands of the clock, which, it being a grandfather clock, were quite obvious to the eye. Yet, as one mystery became a little clearer, others took its place. "Time" was never the same twice running—which made you think a bit, didn't it?

Whatever "time" was, it had to do with the motion of the hands. I knew this to be true, an easy deduction, because one of my parents would ask, "What's the time?" Then the other would look at the hands of the clock and give the answer. Not to be outdone, I got into the way of asking,

"What's the time?" My repeating of the question must have seemed inane, and it is to my parents' credit that they kept on answering it, for, if they hadn't, I would never have made my little discovery.

I'd been put to bed one night, but even then I contrived to shout downstairs, audibly in our house: "What's the time?" One of my parents answered: "Twenty past seven, and that's the last time." If it was to be the last question for the day, there was nothing left but to think a bit before I went off to sleep. An idea suddenly occurred to me. Could it be that "time," instead of being a mysterious number unknown to me called "twenty past seven," was really two separate numbers, twenty and seven? Discoveries mostly need two steps, just as a tune needs an answering refrain. A second idea hit me almost immediately. There were two hands on the clock. Perhaps one number belonged to one hand and the other number belonged to the other hand. A few more repetitions of the question "What's the time?" the following day showed that this was indeed so. Because the numbers on the clock face were big and clear, it was easy now to see there were two sets of them. One hand went with one set and the other hand went with the other set. Refinements remained, like the meaning of "past" and "to," but, to all intents and purposes, the problem was solved and I could turn to other puzzling things, like what made the wind blow.

The other early memory I think I can claim to be genuine was also connected with the weather. I have a memory of being taken out sledging (or, as Americans say, sledding) in winter by an older boy whose parents were friends of my parents. This must have been before I reached the age of five, since then I was allowed to go out by myself. We sledged on the roads, which provided a harder base for a few inches of snow than open country would have done, permitting our sledges to have quite narrow steel runners. In a good year, you could sledge down Primrose Lane, the ancient trading route that ran past our house, for about a mile until it reached a humpbacked bridge over the Leeds and Liverpool canal. This was about halfway down into the Aire Valley. Indeed, you could go further if you could be bothered to pull your sledge two or three hundred yards over level ground from the canal bridge to the railway bridge, but we preferred to walk back up to the village again. We developed what we called a "track," a band of hard-packed snow that you could go down at a fair old lick, the sort of thing modern skiers might call a "piste." To a modern eye, the road was narrow, and, since it had quite high stone walls on either hand, our track bore some similarity to the Cresta Run, had anybody in the village known of the Cresta Run. We lads had our own individual sledges, and we went on them headfirst. You went with your cap turned around with its peak to the back, in the fashion of a catcher in baseball. Nothing of this would have been possible, of course, if there had been cars on the road.

The principal hazard was the other sledgers coming up the track, and we were perpetually shouting cries of warning to them. Going headfirst at a fairish speed between high stone walls must have looked a little dangerous, and, at a distance in time, I'm surprised my parents allowed it. They did so, I suppose, because this was the way things had been for as long as they could remember. Actually, the procedure was safer than it looked. The head projected six inches at the front, and the whole of the legs projected at the back. With the toes, you could brake a bit, and, by combining hand pressure on the front of the sledge with the toes pushing on the ground, it was possible to steer with commendably high precision (another process, by the way, that didn't do our shoes much good). I only came to grief once, and then because I was unwise enough to tackle a hill when it was almost completely iced. Soon seeing the danger of getting out of control, I edged to where there was snow near one of the walls and managed to stop with nothing worse than a bit of scraping against the wall. My sight, when I was seven or eight, was still good, and I liked sledging because it was the one physical activity where I was on a par with the village boys. I was always slower at running, less strong, and not too sparkling at ball games either.

These sledging episodes come into the area of clear memory. So too does an event that, from outside evidence, must have occurred when I was coming up to eight—which is to say, in the spring of 1923. I lived not much more than a mile from Shipley Glen, a V-shaped ravine cut by a stream that drains from the slopes of Ilkley Moor, where (as an old song has it) you do not go without a hat if you are wise. Many hundreds of hours of my youth were spent playing all manner of games in Shipley Glen. Every Easter Monday, a fair was held on a wide stretch of open ground immediately to the north of the ravine, attended, wet or fine, by folk who mostly walked from the chain of small industrial towns lying between Bingley and Bradford.

Because my father had exceptionally good sight, combined with quick eye-to-hand control, he could play precision games like billiards and darts with great facility. It was his wont, whenever we attended either the fair in Shipley Glen or any other in the district, to win prizes at the darts stalls. Eventually their owners drove him off with a fury born of despair. "Tha's 'ad *enough*," they would shout after him. If my memory serves me right, the prizes were mostly cheap crockery, which thereafter my mother had to suffer against her natural inclinations.

What was exceptional about the fair in 1923 was that I also won two prizes. My game was a ball-and-bucket affair. The balls were small, hard, and highly elastic in their properties, while the buckets were wide-mouthed and shallow. The buckets, moreover, were carefully tilted to make nearly sure that a ball thrown into any one of them would immediately bounce out again, in which case you won no prize. However, if you were lucky enough to throw a ball so that it skimmed the inside of a bucket, going round and round like a rider on the Wall of Death, there was a chance it would fall gradually to the bottom and stay there, and then you won a prize. This happened twice for me. I won two metal badges, each about an inch across and colored bright blue. They were singularly worthless, in the terms in which the world accounts worth, but how do you reckon true worth for a boy who normally never won anything but was now treading on air as he wore the badges proudly, one on each side of his jersey?

I remember the other incident because it was epoch-making. At the edge of the fair, away from the babble of the crowd, was a mysterious enclosed booth. At first, nobody went nigh nor by it, so that one felt the owner must be sunk deep below the breadline—times were hard in 1923, harder than we can conceive of today. But, as the afternoon wore on, a miracle happened. A queue to the booth began to form, growing ever longer as time passed by. At last, my father could restrain his curiosity no longer, and, even though he ascertained the cost of entrance to be no less than sixpence, he stuck it out in the queue until his turn came. Reader, do not think this was a topless affair or some other pornography that was being offered. Such things were not permitted in my youth. The best you could then manage were those seaside postcards, colored in bright red, which offered jokes so thin as to be without the slightest interest to the sophisticated industrial population of my native district.

My father reeled, glassy-eyed, from the covered booth. "They call it wireless," he said. I regret my father did not cough up another sixpence for me to experience the miracle he had just witnessed. In retrospect, however, it was easily possible to reconstruct the situation, for pretty soon we had bundles of copper wire, cardboard tubes, and bits of insulating ebonite all over the house; and there were similar scenes in a dozen other houses in our little village. These were the days in which low-power radio stations were springing up all over Britain, and there was intense competition among the villagers as to who could receive the call signs of the most distant transmitters. Grown men would sit up late at night, hunched over their little crystal-rectifying receivers, listening with headphones for those whisper-faint call signs.

Everybody knew about wavelengths, because each transmitting station had a different wavelength, and you had to make up a different tuning coil for each one. This was why the houses were forever strewn with insulated copper wire. I tried my hand at winding coils, but, at the age of eight, I made such a mess of it, getting the wire into a tangle like an unruly ball of wool, that eventually both of my parents balked—my mother at the mess and my father at the cost of the wire I was rendering useless.

Perhaps it was because of my lack of success at winding coils that I turned my thoughts to the problem of what "wavelength" meant, rather as I had done earlier with telling the time. But "wavelength" was a much harder nut to crack, and, somewhat naturally, I had to drop it eventually. As things turned out, I didn't find myself in a position to understand the real meaning of a radio wavelength until I was approaching twenty, so the solution to this problem had a long wait—though not so long as others were to take in later life. Although I was immediately disappointed in this particular regard, perhaps the sight of all those coils and of the huge span of wire—the aerial—running from the house to the very top of a tall, neighboring tree influenced me to find out eventually what it all meant.

I began this chapter by comparing the world as youngsters see it today with the world as we saw it. Our world would have seemed incredibly mean to a modern child, mean with hovels, with cheap, leaky shoes, with undernourishment, and without cars to ride in or television to watch. Yet, to us, it was a world of perpetual wonderment, a world of unicorns that might just perhaps turn into real beasts, a world in which there might even be a ghost around the next corner of a dark lane. What we actually found around the next corner, to our perpetual disappointment, was a couple hugging each other, which seemed the silliest possible waste of time.

A considerable fraction of my own time between the ages of five and ten was spent in the invention of games, as was that of all the boys around me. I do not recall ever being provided, in those years, with a game by adults—there were none at the schools I attended. We made everything up for ourselves. This was a piece of real luck, for I had caught the tail end of one of the greatest periods in human history for the invention of games—football, rugby, cricket, tennis, golf, baseball, winter sports—almost every game you can think of had its ideas and rules developed in that period of which I saw the closing moments. By contrast, the inventiveness of the modern era is as near zero as makes no difference. Today it needs an international revolution to change the rules of football by even an inch. So far as sport is concerned, we do not live today in an inventive era. For other activities, such as science, the situation isn't quite so obvious, but it is a pertinent question whether the state of affairs in sport isn't typical of all society, science included. Science magazines and science programs on radio and television maintain strenuously that it is otherwise, of course, but most of the stuff one hears has to be taken with a pinch of salt. It is expensive gloss, just as new-style boots, and rackets, and golf clubs are only glossy details that tend to conceal the fact that nothing fundamental is happening within the games themselves. I rather suspect that historians of the future will see our present age as one of low inventiveness all round, to be explained as a combination of too much affluence and the ubiquitousness of communications and media pressure.

Chapter 3

Coming to Grips with Who You Are

A CHILD SOON learns it has a special relationship with its mother, and with its father a little while later, but it is not until the age of about three that the child perceives a similar relationship for others. Then it does so mostly for children whom it meets and plays with, who can also be seen to have mothers and fathers. Grandparents are quite a problem to fit into place, especially as there may be two sets of them. The concept that one's own parents also have parents is an awkward one. In my own case, this critical aspect of life was made still more awkward by the fact that, while both my grandmothers lived close by in Gilstead, neither of my grandfathers was still alive, so there were no obvious pairings to be spotted. George Hoyle, my paternal grandfather, came from Rochdale in Lancashire. Like so many menfolk in the district, he earned his living in the textile business, but his real interests, according to older relatives who knew him, were in mathematics and chess.

I was never able in my youth to sort out the larger family relationships that, from listening to my parents, I realized I possessed. My paternal grandparents had both been married before, so, as well as my own direct line, there were children from two sets of former marriages. I had great-grandparents with a family of thirteen, some lines of which had proliferated further. The details seemed too amazing to be true. I was somehow related to the Hammond family of Bradford, which owned Hammond's Brewery and counted itself among the plutocracy of the district. I do not recall seeing a Hammond, although every day in my teens I passed billboards on my way to school that bore big posters that read "Hammond Ales," displaying a hefty fellow drinking from a large tankard, posters that were plastered throughout the Bradford area. I would pause in front of one of them to search my empty jacket in the hope that a coin might have slipped through a hole into the inner lining, and would think to myself how fine it must be to own a brewery. The Hammonds were elusive people. I learned in later life that they had sent their carriage to my great-

32

grandfather's funeral but didn't attend themselves, despite the relationship and despite his being regarded as one of the outstanding Yorkshire poets of his day.

Better known to me was my Uncle Harry (actually my great-uncle), who visited my parents from time to time. I always liked Uncle Harry, even though he dressed in rags. The rags seemed less important to me than his invariably complimentary remark: "Now *there's* a grand lad." On one of his visits, my mother was having trouble in shutting a warped door. Knowing Uncle Harry was supposed to have received training as a joiner, my mother permitted him to fix it, which he rapidly did by slicing away at the door with an axe, much to my father's grief when he came home in the evening. Like me, Uncle Harry didn't have two penny pieces to rub together, because, if he was ever so fortunate, he immediately made his way, at the double, to the nearest pub. A time came when the family decided Uncle Harry's rags were not helpful to its own social standing. So they fitted him out with a resplendent new suit, only for the suit to disappear within the week, doubtless into the hands of a pawnbroker. To the family's frenzied demand to know what had happened to the suit, Uncle Harry declared that it had blown away in a strong wind on Saltaire Bridge, and from this view of what had happened, he could not be budged; the seeds of imaginative greatness must have lain tragically dormant in my Uncle Harry.

Then there was Black Uncle Jack (actually my great-uncle), who my father always insisted was much the strongest man in the district. I never saw Uncle Jack, but stories of his prowess lit up my young life as the nighttime tipping of molten iron from a furnace lights up the clouds overhead. After a day's work, Black Uncle Jack would become blind drunk. He was not welcome in the local pubs because, in such situations, he would flay alive anybody who should contradict him in the smallest degree. The character in world literature most like my Uncle Jack was Brandy Bottle Bates in the stories of Damon Runyon. There was, I believe, general relief in the district when, one dark night, following just such a combative evening, my Uncle Jack fell, on his way home, into the local canal, and became "drowned dead," as Charles Dickens's Mr. Peggotty would have said. This was the Leeds and Liverpool canal, the "cut," as it was known in the local vernacular.

There is no way a person could judge today from visiting a modern pub how beer drinking used to be in the good old days. There was a middle-aged man in my village—not a relative, this time—of whom most drinkers stood in some awe. He heard one evening of a pub some four miles away that was opening that very evening under new management and, in the interests of goodwill, was dispensing free beer. With a thunderous cry of, "Why didn't anybody tell'm?" he raced out of the village and downhill pell-mell into the town of Bingley, where he set off, still in full cry, in the direction of Crossflatts. He was said to have reached the pub at half past nine. Half an hour later, at the statutory closing time of ten o'clock, he rolled out of the pub into the street, ten pints of beer to the good—or to the bad, as the temperance folk saw it.

Life was so closed in by what we would regard today as severe poverty and by a lack of communications—closed in except for the local cinema, for which one paid 1*d* (or 0.4p) to gain entrance into what was popularly known as the "bug hole"—that a large majority of the menfolk spent their evenings in the pubs. Beer in those days was priced at only a penny or two to the pint; at such low cost, drinking was necessarily excessive, and the tendency was for the male population to be formed into two disparate groups—either you drank to excess or you didn't drink at all. I do not recall a commercially produced alcoholic beverage ever being brought into our house. This wasn't because my parents were doctrinaire about alcohol, the way some people were, but because they saw it as an all-or-nothing situation.

A Mr. Bartle came to live in our village. He was an extreme temperance advocate. Between the ages of eight and twelve, I saw quite a bit of Mr. Bartle, a man with a cherry-red nose, because his house was close by, and partly because, in co-operation with the local church, he made temperance propaganda among us young fry. The organization in question was called the Band of Hope, a title more frank in its honesty than most. The main consideration was that, if we joined the Band of Hope, we obtained the use of a large warm church hut in winter on a particular night of the week—Tuesday, I think. There were clear bribes, from time to time, of buns and cakes, which were consumed with much speed and relish. Also from time to time there were harangues from visiting personages, which, even in my tender years, I couldn't help viewing with a certain morbid fascination. A singer in a stiff collar stands out sharply in my memory, a singer who rendered the toreador's song from *Carmen* accompanied by piano and trumpet, an occasion not to be missed. More routinely, we would be shown lantern slides of drops of water and alcohol, the water teeming with ugly-looking creatures, the alcohol devoid of such things. "There," the lecturer would say, "nothing can live in alcohol," whereon Mr. Bartle would intone in a sombre voice, "Never let a drop of it pass your lips." Meanwhile, some of us would have realized there were other very different interpretations of what we had just seen.

When I was eleven or so, Mr. Bartle persuaded me to enter the annual Band of Hope examination, the syllabus for which, in that particular year, included an extensive section dealing with the brewing of alcoholic beverages. Whether somebody had thought it a joke or clergymen were themselves looking for instruction in the art of home brewing, I don't know, but, astonishingly, there it was, and, since by now I had become keenly interested in chemistry, the examination seemed like money for old rope. More precisely, the first prize was £10 in book tokens. My itch to instruct the clergy in the practice of home brewing was such that I did indeed win first prize, not only for the dioceses of Ripon and Bradford but even nationally, a meteoric situation indeed for our little village. I had to journey to the city of Bradford to receive my prize and to be patted on the head by the Bishop. I would have preferred it if I had been patted by the winner of the second prize, a good-looking farmer's daughter from Gargrave in the upper Aire Valley. Gargrave was twenty-five miles from Bingley, so it seemed impossible I would ever see the girl again, and I didn't.

Drinking fermented fruit wines was thought to be in order by all but the more rigorous teetotalers, although with poorly controlled fermentation, such as then was practiced, fruit wines must have contained esters and higher alcohols, which really are damaging to the human body. The whiz woman at making fruit wines in my village was my maternal grandmother, Mary Ellen Preston, known familiarly as Polly. In my time, Polly was a white-haired, good-looking woman in her sixties, the terror of the district, although to me she was never anything but my kind old grandmother. People said I was like her, and, in a stubborn obstinacy to get ourselves into trouble, this seems to have been true. In recent years, my wife asked a hundred-year-old relative what Polly was like in her youth. The old lady got as far as "What was Polly like?" and then broke into a cackle of laughter, after which we couldn't get a further word out of her, so shocking, apparently, had been her recollection.

I was distantly aware that my mother's sister, my Aunt Leila (girls in those days still suffered from Byronic names), was somehow different. She was tall, always beautifully dressed in clothes she made herself, and she took me, from time to time, to Lingard's shop in Bradford, where she bought ice cream and I was able to watch the amazing mechanics of containers that shop assistants kept shooting along wires to a central cash desk. In the early years of the century, Aunt Leila had earned money from sewing to help my mother with her musical training in London. What was different about Aunt Leila was that she was illegitimate. I have never learned who the father was, so all I can say is that, unless he died or Polly had given him his marching orders, he was a fool not to have claimed such a daughter.

It is impossible today to understand the intensity of the stigma of illegitimacy as it existed in the late nineteenth century, and as it existed even in the nineteen-twenties, during my boyhood. This situation for Polly must have been made worse by the peculiar Yorkshire illusion that, by rubbing hard enough on a wound, it will somehow be healed. If a man struggled ill to work, he would be greeted by, "Ee, lad, th'art not looking thizen,* art tha?" Or to a girl sensitive about a little adolescent acne, "Ee, lass, th'art real spotty, arn't tha?" Bernard Miles once told me a story, due I think to Harry Secombe, who had finished a turn at the Alhambra Theatre in Bradford, leaving his audience rolling in the aisles. On Secombe's way out of the theater, a stage hand called after him: "Ee, 'arry lad, tha a'most made mi laugh." Bernard Miles also told me, unaware that I was Ben Preston's great-grandson, that Preston was one of the three surviving poets in the Yorkshire dialect. Unfortunately, the real dialect is hard to understand today, even for Yorkshire people. The real dialect was already pretty attenuated in my boyhood, and now, a half-century later, it is essentially gone.

Ben Preston conveyed the popular excitement about railways in a racy jingle he wrote for a Bradford newspaper on the occasion of the opening of the railway through the Aire Valley. From our present-day, technologically sophisticated standpoint, it is hardly possible to appreciate the emotional impact of the railway as it appeared to those who actually witnessed it. An important aspect, easily overlooked, was that an age-old class structure based on the horse was largely destroyed, a class structure whose remnants we still see today in the Enclosure at Ascot. The aristocracy and plutocracy had their own horses and carriages, while, by clubbing together, the middle class gave financial support to the stagecoach system. Poorer folk walked. With the coming of the railway, all this was changed in a moment. Anyone with a few pence could suddenly travel faster and farther than any aristocrat on his horse. If we take literally the ebullience of Ben Preston's poem "The Locomotive," we come close to a sense of how things must really have felt to those who experienced the historic coming of the railway:

> The neigh of the dragon, a terrible cry,
> Wild, piercing, and shrill, has gone up to the sky;
> He pants for the start, and he snorts in his ire,
> His life-blood is boiling, his heart is on fire.
>
> O man, O my brother, how stubborn thy will,
> How dauntless thy courage, how Godlike thy skill:

---

*The word *thizen* is a corruption of *thyself.*

> The Earth, with its elements, yields to the brave,
> The fire is a bondsman, the vapour a slave,
> The vales are uplifted, the mountains are riven,
> And the way of the dragon is shining and even.
> Let us gaze and admire, and declare, if we can,
> How mighty the God that created the man.

From Ben Preston's other writings, it is clear that "the man" was not used generically, but was an explicit reference to the great railway pioneer George Stephenson, whom he greatly admired.

Ironically, Ben's best-known bit of writing, the hymn *Onward Christian Soldiers*, is not attributed to him at all. At least Ben's younger daughters, who were alive in my time, were always emphatic about his authorship of the hymn, and the actual words themselves, together with the circumstances in which they were written, would seem to bear the family out. The time was around 1864, and the occasion the annual Easter parade of schoolchildren in Bradford, who, by tradition, carried a large number of banners. There really was a "royal banner" and there really was a "cross of Jesus" carried at the head of the parade. S. Baring Gould, to whom the hymn is usually attributed, was, as I understand it, a young clergyman newly arrived in the Bradford district who wanted a special hymn to celebrate the occasion. Hearing of Ben Preston's reputation as a poet, Baring Gould asked my great-grandfather for words appropriate to the circumstances of the parade. No attribution was given at the time because, by 1864, Ben Preston had changed from the early proreligious stance of "The Locomotive," perhaps under the influence of Charles Darwin's book *On the Origin of Species*, which appeared in 1859. Ben Preston was also, at the time, turning from being a liberal writer to becoming a socialist one. Eventually, the hymn was attributed to Baring Gould by default.

The notion that things are frequently not the way they are supposed to be entered my head at an early age, a favorable situation for a prospective scientist. Catholics were anathema in our district. The local Catholic school was cleverly placed so that its pupils were unlikely to encounter lads of the much more numerous Protestant population. Even so, the Protestant youth would occasionally converge, with evil intent, on the Catholic school, stoning the Catholic pupils in true biblical fashion as they emerged into a veritable hellbrew, scattering them shrieking in all directions like birds under a hawk. I took no part in these activities because two of my best friends in the village were Catholics, a boy of about my own age and an older girl who was good at organizing games. They were outsiders who stayed only for a few years.

Then there were the Jews, a considerable number of whom had settled in and around the city of Leeds, having escaped from persecutions in Poland and the Ukraine. They set up as tailors, often enough, which brought some of them into a business connection with my father. We had no such words in our local speech as "cheat" or "swindle." The standard local expression, if you were cheated, was to say you'd been "jewed." I used the word many times myself before its connotation occurred to me, but I ceased to use it after the following little episode, which requires an introduction describing my father's business, and indeed the general business practices at the time in the city of Bradford.

Bradford was then a world leader in the weaving of high-quality cloth. It exported its products internationally, as well as supplying Bond Street tailors in London. Occasionally, looms would go slightly wrong and would turn out cloth with some very minor flaw, which only the eye of an expert could detect. Such high-grade "seconds" were sold off at small fractions of the price of flawless cloth. My father was friendly with the foremen and managers of many of the Bradford mills, and so was privy to where seconds were to be obtained. His business was to build a stock of seconds, which he then passed on to a wide range of customers, exporting the cloth to places as far away as China, and selling it to the Jewish tailors from Leeds.

Payments for "deals" were always made, as far as I could see, by check. As soon as a purchaser handed his check to a vendor, the vendor would immediately "ship" the goods. Shipping the goods consisted of calling in one of the flat horse-drawn carts that hired themselves out almost like taxis. Any day you went into Bradford during business hours you would see a score of these flat carts rattling along cobbled streets. The whole center of Bradford was a mass of horse dung, and in winter you could detect a dozen piles of the stuff smoking in the cool air.

The morality of cashing checks was a little curious. If a vendor went immediately to the purchaser's bank and the check bounced, this was distinctly bad form. But if the vendor permitted more than an understood length of time to elapse before he presented the check, the more fool he, if it bounced. Friday afternoon was the usual deadline for settling deals.

On the occasion in question, my father had forgotten, by Friday afternoon, to cash a check he'd received from a Jewish tailor, presumably due to the pressure of other business. Over and over through the following weekend, my mother and I had to agonize over my father's frenzied cries that he would be "jewed." Since the sum was of an order that seemed considerable in those days, we were naturally appalled at the situation— I was about ten at the time, just the sort of age to be appalled. My father went off to business as usual on the Monday morning, returning as usual in the afternoon at about five o'clock. My mother gave him a cup of tea and then waited with mounting impatience as he drank it. "Well, Ben?" she said, unable to contain herself any longer. "He paid," said my father at last, and then he added, as if to defend the local prejudice: "But they tell me he's religious." These were the seeds that, flowering in later life, have caused me to entertain reservations about everything I hear, unless it is based either on observed facts or on mathematical calculations.

The first person known to me with whom I have genes in common was a certain John Preston, who, in the late eighteenth century, amassed a fortune through dealing in wool, owning, at one time, a considerable fraction of the property in central Bradford. John Preston had the enviable gift of selling at the top of the market, declaring, "Ah'll sell, even if ah repent it." He became such a byword for this tactic that his fame even crossed the Atlantic. A cartoon sketch of him is said to have hung for many years at the entrance to the Merchant's Club in New York.

John Preston had no children, but he had a favorite nephew, William Preston, who was directly in my own ancestral line. William Preston was given a good schooling at John's expense, and all seemed set fair for my ancestral line. I might thus have been born with a silver spoon in my mouth, as my father was fond of remarking about the well-to-do. But, unfortunately, John Preston did what other rich and clever businessmen have been known to do. Doubtless imagining he would never die at all, John Preston died intestate, a dreaded word in our family. A sister or sisters of John Preston won his riches, and young William Preston was left to sample life in a woolsorter's job in a Bradford mill.

William Preston still had one shot left in his locker. He was educated in a day when educated young men in Bradford were not to be found on every tree. With this shot he won the affections of Anne Hammond, from the wealthy Hammond family. For marrying below her station, Anne Hammond appears to have been cut off with the proverbial shilling. The shilling sufficed, however, to educate her two sons, Ben and John Preston, the one becoming the poet and the other an artist. Ben Preston escaped from the woolsorters' shed at the age of eighteen by writing a political poem that was said to have won a local parliamentary election for the Liberal candidate, the poem being a lampoon on the Tories. Thereafter, Ben eked out a somewhat precarious existence as a poet, as a journalist with the local Bradford paper, and by poetry readings from Shakespeare and Milton—industrial England was first becoming literate at this time— which sometimes took place in St. George's Hall in Bradford, where my mother would perform professionally as a singer two generations later. Both for Ben Preston and his younger brother John, the Hammond connection ultimately became of dominant importance through the sustained generosity of "Uncle" Ben Hammond.

Like old John Preston, "Uncle" Ben Hammond was a dealer, but in cattle rather than wool. Like old John Preston, he amassed a huge fortune, and, like old John Preston, he had no children. Unlike old John Preston, however, he did not hang on to his money to die intestate. He discovered a truth first expounded with great clarity by the French writer François Rabelais. Hang on to your money, said Rabelais, and all your family and acquaintances will gather round like vultures waiting for you to die. But amass huge debts and your creditors will be most solicitous of your health, sending you the best possible doctors in the hope of your surviving to square accounts. "Uncle" Ben Hammond didn't run himself into debt—quite the reverse—but he gave away most of his fortune to his thirty-odd nephews and nieces several years before his death. On one occasion, at a dinner held in 1882 in a Bradford hotel, he gave away some £30,000 in a single evening, which was a lot of money in those distant, preinflation days.

None of "Uncle" Ben Hammond's money ever came my way, however. Because of Ben Preston's thirteen children, his share of the spoils was inevitably thinned down, and what remained mostly went to Ben's younger daughters, who looked after him in his old age. This was seen as fair enough by the family, but, when the youngest daughter then married and willed the money partially into her husband's family, harsh things were said that delighted my young ears.

In the longer run, young John Preston's family did better because young John Preston had only one child—a son who married a Cornish woman, thus breaking out of the local pattern of inbreeding. The result was a considerable family in the second generation with vigorous, good-looking, and long-lived children. The men lived to their eighties and nineties and two of the women to more than a hundred. Young John Preston would have applauded the friend of Franz Schubert's family who, when asked the name of Franz's teacher, replied: "What need of a teacher has a boy who learned his music from God?" The notion that natural ability is best left to assert itself without technical training continued in John Preston's line until my own generation, which was unfortunate, because his granddaughter Nannie had a genuine gift for draftsmanship. Nannie was a cuddly-looking woman who would come to our house and draw for me anything I cared to name—horses, rabbits, and so on—all done in a flash. Nannie did the first color brochure for the original Marks, the founder of the Marks and Spencer chain of shops. True to young John Preston's mystic philosophy, one of his descendants recently gave Nannie's originals to the memorabilia of the Marks and Spencer empire, for nothing. Old John

Preston—"Sell-and-Repent Preston," as he was generally known—would not have approved of such munificence.

How it came about that our side of the family had no truck with the "genius will out" philosophy, I don't know. My mother had already made her views clear when, against all the odds, she made her way to London in order to receive a technical training at the Royal Academy of Music. There was never any thought in our household but that I must receive as severe a technical training as possible—in something, although exactly in what remained to be decided.

But all this is really rather petty stuff, as is every family tree I ever saw. Mankind has not changed much since the days of Cro-Magnon man, days separated from our own times by a thousand generations. Every one of us has an immense line filled with stories of struggle, heroism, self-sacrifice, that if we did but know it would put the parochial affairs of the past few generations to shame. Every one of us has an immense line that would sweep even the greatest dynasties of recorded history into affairs of minor consequence. And before Cro-Magnon there were upwards of ten thousand generations, so many that the imagination is too staggered to conceive of what happened to them all on a generation-to-generation basis. What we actually see are nothing but bits of foam at the surface of the vast ocean of prehistory.

Chapter 4

Creeping Like Snail
More Slowly Than Shakespeare

I COULD WRITE out the multiplication tables up to $12 \times 12 = 144$ at the age of four, but I couldn't really read until I was seven. I can remember precisely when I first learned to read properly. It was in the "bug hole" of the local cinema. Those were the days of silent films with subtitles. To my surprise, I found I was suddenly reading the subtitles without difficulty. Within a week or two, I was reading generally. Rather obviously, I had suffered from an eye-focus problem.

In my late teens, I was destined to suffer from severe headaches. After a remission of forty years, the headaches returned—fortunately, only briefly. On the recent occasion, I made an effort to discover what the trouble was. A highly competent general practitioner, an eye specialist, and a neurologist made the same diagnosis—migraine: "Ah ha! You are an absolutely classic example of migraine," they all said with satisfaction. Since every one of them then went on to say that nobody really understands the cause of migraine, this was actually no diagnosis at all. My own suggestion (applicable to my case, at least) of an eye-focus problem did not win much support, but this remains my opinion. Migraine is a clinical condition that can have many sources, and eye-focus problems may be one of them. It was mainly important to me at the age of nineteen because I had a literally blinding migraine during the last paper of one of the most important examinations of my life.

Between the ages of five and nine, I was almost perpetually at war with the educational system. My father always deferred to my mother's judgment in the several crises of my early educational career, because she had been a schoolteacher herself, whereas he had been obliged, for financial reasons, to leave school for the mill at the age of eleven. The bare recital of events would suggest that my mother was unreasonably tolerant of my obduracy. But, precisely because she had been a teacher herself, my mother could see that I made the best steps when I was left alone. I had learned to tell the time and I had fixed the multiplication tables imperishably in my mind without help; in addition, the circumstances in which I learned to read all in a flash would have been sufficient to bemuse any sensitive parent.

Yet my mother knew, with the certainty that night follows day, that, without formal schooling, I would never hoist myself onto the scholarship ladder, which could take me first to the local grammar school and then to university. Her problem, therefore, was to persuade me to submit to the necessary degree of formal teaching without allowing the streak of originality to be knocked out of me by the use of force. It was due to her forbearance, and to her realization at the biggest crisis point that I was factually correct while the school system was wrong, that things eventually worked out as she and my father had hoped.

The elementary educational system began normally at five, and from five to six you were an infant. In each succeeding year, you passed from one "standard" to another, ending at the age of fourteen in Standard VIII. The infant year was not obligatory by law, so that children were not actually forced into school until the age of six. As soon as I learned from my mother that there was a place called school that I must attend willy nilly—a place where you were obliged to think about matters prescribed by a "teacher," not about matters decided by yourself—I was appalled. Since I made it abundantly clear that I would not accept incarceration in a mental prison house, my mother began by permitting me to ditch the first infant year, in which nothing of substance was taught anyway. But— when my sixth birthday arrived—on 24 June 1921—the nettle had at last to be grasped. The solution that appeared to present itself was the following.

In 1920, there was a temporary trade boom, with my father's business in Bradford doing well for a while, which made it possible for me to begin school in July 1921 at a small private establishment located not far from the mill where Damart clothing is manufactured today. This particular school was chosen because the children of my mother's contemporaries on young John Preston's side of the family had been sent there. The tactic worked in the short term—which is to say, through the month of July 1921. After that, freakish factors intervened.

Following the trade boom of 1920, the cloth trade fell on hard times in 1921, making it more likely that my father would lose money rather than gain it if he continued in business. So, with my mother ill following the birth of my sister, it was decided, at the end of July 1921, that we would all go as paying guests to the house of a man in Essex known to my father because he'd been secretary to Keir Hardie, the first Labour Member of Parliament. The Essex countryside to which we retreated, in the environs of the town of Rayleigh, was quite rural in those days. During August

1921, I had a companion of about my own age, Freddie Clamp, who knew a thing or two about devilment that had never occurred to the village boys of Gilstead. The time came in September when I was required, together with Freddie Clamp, to cross a common (which I recall, in distant memory, as being bright with gorse bushes) in order to attend school in the nearby village of Thundersley. It was during September and October 1921 that Freddie Clamp and I worked out a system of truancy that supplied the essential basis of the technique I was to use in succeeding years. Freddie Clamp's father was at sea, an officer in the navy, and it was then that I acquired a mystical respect for the Royal Navy that I have never since been able to shake off.

It had been my father's intention to stay in Essex until the business climate improved, but, by October, there were disturbing intelligences from my maternal grandmother, the redoubtable Polly, which brought us all hotfoot back to Gilstead in November. To my father's chagrin, we found that the man called Brady to whom the house had been rented had done a flit. The word "flit" was in widespread use in those days. Nobody thought of it as being connected with the verb "to fly." It meant running up debts and then getting out fast, before creditors got clued in to the situation, as Americans say.

Our house must have been below the standard to which Brady and his family were accustomed. But it had a grand-sounding name—4, Milnerfield Villas—which was presumably what really counted in their scheme of things. Brady made the mistake of running up huge grocery bills in my father's name; otherwise, he could have got away with the deception for several months more. Like all successful entrepreneurs, Brady had flair. He entered his younger children at the local grammar school and then had their school uniforms made specially to measure, instead of accepting them off the peg as you or I would have done. After all, if you have no intention of paying, why not have the best? He hired a Rolls-Royce car and chauffeur. His older daughter, whom I remember as being exceedingly good-looking, drove in the Rolls to Bradford, where she cleaned out the best dress shops of their most expensive creations. All this uproar was milk and honey to me, of course, and it occupied the time very handily until the question of school again raised its head at the beginning of January 1922. I returned to the same private school as before, but I returned no longer an innocent child prepared to have irrelevant knowledge poured into my head by the old beldame who ran the place. Thanks to the Royal Navy—at one stage removed—it was an experienced young blackguard who entered the same school one morning in early January 1922.

The situation as it now presented itself to my mind was that you spent the first bit of the morning, from nine to ten, getting interested in something. Then, just as you were nicely into your stride, there was a jump to something else. Once again you cooperated with the teacher by becoming interested in the new topic. But all to no avail. Like somebody with St. Vitus's dance, the teacher was off again into a new subject that bore not the slightest resemblance to anything that had gone before. The thing that eventually finished that first school for me was connected, as you might expect, with numbers. Because I found the sums I was given rather easy, I was told to learn the Roman numeral system, whereby I found to my amazement that VIII stood for simple old 8. How could anybody be so daft as to write VIII for 8, I wondered. Yet I made no instant complaint, for the task was not an onerous one. Besides, I hit a problem with some puzzlement in it—how did you multiply these strange new numbers? When the question proved intractable, I asked the teacher, only to be told that you didn't multiply Roman numbers. When I persisted by asking what were they good for then, the answer was that Roman numbers were very old and that they were sometimes used in books.

This was more than I could reasonably stomach, and the day this outrage to the intelligence was perpetrated became my last at that particular school. The date must have been early March. Using the avoidance tactics Freddie Clamp and I had worked out, I contrived to persuade my parents that I was in attendance at school, and, through the agency of a friendly boy in the village, I conveyed to the school the sad news that I was confined at home with a ghastly illness. Amazingly, the bubble lasted until late April or early May, and it only burst because a contemporary of my mother's with children at the school, hearing I was at death's door, expressed concern about it. As fate would have it, I found myself with the trump cards in the row that followed, because it was during this spasm of truancy that I learned to read, while patronizing the bug hole in the Hippodrome cinema. To this point, my parents had awarded the old lady at the school the credit for teaching me to read. Now they were nonplused to see that this was not so. Ever one to press a solid argument, I emphasized that, if I'd continued at school, I would *never* have learned to read. The bug hole at the Hippodrome was evidently a superior educational establishment, I persisted, and, at 1*d* per admission, a good deal cheaper than school. Whatever way you looked at it, the case was unanswerable.

My parents had no recourse but to explain the law governing attendance at school, which at least set an interesting problem worth thinking about. How was it, I wondered, that the law could pursue so relentlessly a harmless boy like me while permitting Brady and his family to do a flit with all those debts unpaid? After worrying at this problem like a dog with a bone, I concluded that, unhappily, I'd been born into a world dominated by a rampaging monster called "law," that was both all-powerful and all-stupid, a view that has resurfaced from time to time ever since.

The outcome was a compromise. To avoid the family being punitively fined and all of us ending up in the workhouse, I would yield myself up to the enemy, but not until the beginning of the next school year, in early September. Since I would pass the age of seven on 24 June, I had therefore defeated the educational system at this point by essentially two whole years. My parents permitted me to choose the next school to try. I opted for a state elementary school in Bingley, largely because several of my mates in the village went there. It was a good joke to start with that the headmaster's name was Woodcock; he was a walrus-moustached man who always wore a stiff collar. Nobody in the Aire Valley—except, I suppose, a refined one percent—used any word for penis other than "cock." So here was a ha-ha situation from the outset. In addition to this, Woodcock, in his daily morning ramble to the school, persistently referred to the local Bradford newspaper, which everybody knew to be the *Observer* in the morning and the *Daily Telegraph* in the evening, as the *Argus*, which everybody also knew meant "arse." But, although the situation was daft enough, it was at least an improvement on Roman numerals.

State schools were all mixed. There were two spells each of fifteen minutes, one in the morning, the other in the afternoon, when the children were put out into so-called "playgrounds." Playgrounds were rigorously separated between boys and girls by a high wall. At no time do I recall any effort being made by the boys to invade the girls' playground. Since no teacher was ever there to monitor the state of play, it is clear that a strict taboo must have become established early in the school system, presumably by frequent canings and the like. Growing up later with the taboo, we simply accepted it.

At Mornington Road School, the boys went in through big iron gates on a street close by a large Methodist church. The gates led into the boys' playground. There were specially grand stone flags leading immediately to the school, but, over the rest of the playground, the surface was in need of repair. At all events, the playground was hard-surfaced, so that heavy boots running on it were very audible. There was a big slow-witted lad who was approaching the school-leaving age of fourteen. Because of his slow wits, he was the unfortunate butt of his class. The standard practice at playtime was for him to chase after his contemporaries in studded boots, aiming kicks wildly in all directions, to immense roars of laughter from the others. The day came when this lad acquired a horsewhip with which he was able to make immense cracking noises in the chase. My memory is that the thing went on for weeks, with every play period shrill with shouts of laughter. I swear that at no time did any teacher intervene.

For small fry like me at the bottom of the school, it was actually easy to keep well out of the way, and again I swear that I found the situation a great deal more entertaining than Roman numerals.

Although I had as yet received no formal education to speak of, because I could read now and knew about numbers, I was put into Standard II, the normal standard for my age. The written tests that even very young children were frequently given in those days were new to me, and, in the first test or two, I performed indifferently. But, once I had the hang of what the tests were about, I moved upwards in the class and was rewarded by a desk among the top echelon. On one occasion, "Argus" Woodcock even commended my example to the whole school, so that, for a while, it seemed as if at last I was set fair, particularly as the teacher, Sally Pearson, was both competent and pleasant.

I cannot explain why, in looking through some old photographs recently, I found a picture of my class taken towards the end of my first year at Mornington Road School the most poignant. Quite unlike the phantom recollections of my earliest years, the faces of my fellow pupils are etched in memory with total clarity. Locked in a time capsule in my brain, every one of the children in the picture is a Peter Pan who can never grow old, the rough diamond at the upper right of the picture and the gentlest boy in the class crouching at the lower left. There was not a pampered one among them. If you look carefully, you can see the strains of malnutrition on some of the faces. Sadly, I can scarcely remember any of their names.

The real lives of my schoolfellows must now be spent, or nearly spent. I can only hope fate dealt kindly with them, although I know that, in at least one case, it did not. Ours was but one class in one school in one small town, but standing there is a boy who became a member of Winston Churchill's select "few," a fighter pilot who died in 1942 after receiving the Distinguished Flying Medal. Jim Hopewell was not the biggest or strongest boy in the class, but he was always the most active and best coordinated.

My days at Mornington Road School were already numbered by the time this photograph was taken, and I had already begun to suspect this would be so. The teacher in the next class upwards into which I would be passing was of the strong-armed variety. The woman's class was notorious as a regular charnel house of canings and beatings that would seem unbelievable today.

It needed little in the way of common sense to see, once I had moved into Standard III in September 1923 at the age of eight and had observed the first canings, that I would need to spend the coming year not in learning but in self-preservation, and, indeed, I managed to get through to the spring of 1924 with only a few weeks' attendance at school. By now I had learned that illness was the key to absence. Doubtless, I experienced some genuine illnesses during the winter of 1923–1924, but, additionally, I spun out every small sniffle into a week or ten days. The tactic led to frequent examinations by Dr. Crocker, who found it difficult to diagnose malingering since I had never malingered before. Eventually, Dr. Crocker diagnosed tonsils and adenoids and recommended their removal, which minor operation and subsequent convalescence I stretched to two months. Eventually, the sniffles disappeared with the coming of warmer weather in the spring of 1924, so, unfortunately, I ran out of further excuses and had perforce to return to school. By now, however, I could see light at the end of the tunnel. A few months more and I would be gone from the universally dreaded Standard III into Standard IV, which was reputed to be much better. With my eyes ever wide open for trouble, I thought I might avoid it. Then, just as I was picking up a little confidence, disaster struck in a way I had failed to anticipate.

The spring flowers were in bloom now, and we were asked to collect a specified list of about twenty kinds. Because I lived out of the town and had roamed the countryside since the age of five, I knew exactly where the flowers could be found. It was no great trouble, therefore, for me to collect the whole list, and thinking to ingratiate myself with the dreaded teacher, I did so. Therein was my error—for, when the teacher gave a lesson on the flowers, I was able to compare what she said with the specimens in my hand. One particular flower was said to have five petals. Mine had six. Here was a bit of a problem, I thought. If my flower had been a petal short, I could have understood that it might somehow have lost one. But how could my flower have a petal extra? Or could it be that this abominable teacher didn't know the difference between five and six? I had reached this point in my thinking when I received a stinging blow on the side of the head and a strident voice ordered me to pay attention.

The blow was delivered flat-handed across the ear. Since the teacher was certainly right-handed, the blow must therefore have been across my left ear, the one in which I was to become deaf in later life. Since, moreover, I wasn't expecting it at all, I had no opportunity to flinch by the half inch or so that would have reduced the impulsive pressure on my drum and middle ear. Joseph Conrad wrote a novel concerned with an underground political movement in Eastern Europe before the Russian Revolution of 1917. A man suspected of being a police informer is tried and found guilty by a secret court. He is then held by two men while a third hits him a flat-handed blow first across one ear and then across the other, after which he is released. As he emerges from the cellar in which he has been tried into the street above, he realizes, to his horror, that he is totally deaf. The story has the chilling quality that comes, I would suppose, from being based on a real event. Although it may seem fanciful, these several factors have caused me to wonder if the blow I thus received might not have had something to do with the later deafness. Perhaps, perhaps not.

As I recovered from the immediate pain of the blow, it became apparent that a total crisis in my relations with the educational system had at last been reached. Cost what it might in the way of the law, of fines, of the workhouse for my parents, I was done with it. First, I pocketed the flower with its six petals. Then I put up my hand and asked to go "out," which meant going to the bogs—"toilets" or "lavatories" would have been quite misleading words because they were neither. They were bogs, in the good old-fashioned sense. Such a request was never refused, for obvious reasons. When I emerged from the school building into the boys' playground, instead of heading up the playground and slightly to the right, which was the way to the bogs, I kept straight ahead along the grand flagged path that led to the main iron gates, and thence into the sidestreet beside the big Methodist church. Half an hour later, I was home, with the news for my mother that, come hell or high water, I was finished with school.

My mother must by now have suspected something was going badly wrong at the school, and, as she heard my story and examined the flower herself, she could see at last what it was. Like most adults, however, her thoughts were on some form of honorable compromise. I was never one for compromise myself, and, since politics is said to be the art of compromise, I was never one for politics either. Pretty soon I was back at Mornington Road School, not in Standard III, however, but in "Argus" Woodcock's office with my mother, who, to my delight, was in one of her rare angry moods. But, not to my delight at all, "Argus" and my mother were evidently moving to patch up the situation, a promise being offered that there would be no further assaults, provided I returned forthwith to the school. No deal, I told my mother afterwards, no way. To the question of what I would do if I didn't go to school, I replied that there were plenty of things to do. After all, I pointed out, I'd given the school system a try-out over three years, and, if you didn't know something was no good after three years, what did you know?

Two or three weeks elapsed before my parents received a letter requesting that they present themselves at the offices of the local education committee. My father, being occupied with his business, left it to my mother to respond to the letter. She took me in tow, just as she had done on our visit to "Argus" Woodcock. We were shown to an office where there were two men—one, I think, an education officer, the other probably a lawyer. The lawyer wore a big-winged collar, which expanded and contracted like a spring as he moved about the office. After the legal position had been stated—it was an offense for parents not to *deliver* their children at school like sacks of merchandise—my mother tried to tell my story. At this, I produced a tin box, opened it, and took out the now much faded flower. I asked the two gentlemen if they could count the number of petals. There are six, you see? And the teacher said there were five, because she is unable to count up to six. I have no wish to go to schools where teachers can't count up to six, you see? And so on and on I went, grinding away on the theme of the teacher being unable to count up to six.

I suppose neither of the men knew that flowers which normally have five petals may have sports with six. They counted six for themselves, there were six plainly before their eyes, and, with thirty children from my class as witnesses to the teacher saying five, the case was going to appear ludicrous if it ever came to court. My mother sensed that the day had been miraculously won. She pointed out her own qualifications as a school-teacher, and we were permitted to leave with the understanding that nothing would be done, at least for the time being. As I retrieved the tin box with its flower, I felt confident that the millstone of education had at last been lifted from my shoulders.

Each morning, I ate breakfast and started off from home, just as if I were going to school. But it was to the factories and workshops of Bingley that I went. There were mills with clacking and thundering looms. There were blacksmiths and carpenters, but as yet no garages. The miraculous thing about it was that nobody told me to get out. Nobody boxed my ears or shouted that I must pay attention. Indeed, it was just the other way around. Everybody seemed amused to answer my questions.

The Leeds and Liverpool canal passes through Bingley. In those times, there were many horse-drawn barges on the canal. A little northwest of the main town, there are locks where the canal level changes. I spent many an hour watching in astonishment the simple ingenuity of the opening and closing of lock gates and sluices, and how marvelously the process effected the passage of the barges. Nobody had ever told me that these "five-rise locks" were historically famous. In choosing them as a place at which it was worth spending many hours, I like to think I was instinctively displaying a sound sense of engineering, and that the experience was of far more consequence than anything I might have been learning at school.

Indeed, the Bingley five-rise locks are still widely known, and recently they were awarded first prize as the best-kept locks of the still surviving canal system. According to a nineteenth-century historian:

> The Leeds and Liverpool canal was opened 21st March 1774, when great rejoicing took place in Bingley. The church bells were rung, guns were fired by the local militia, and there was a general holiday.

The first boat went down the five-rise locks in 28 minutes amid loud huzzas from the spectators. The locks here, as is well known, are unique on the Leeds and Liverpool system, and when they were completed were considered one of the grandest engineering achievements in the world.

My studies of the five-rise locks were always made in the mornings. Through the afternoons I made my way in the opposite direction, towards the moors, to explore in the woods and fields. It was then that I found the kingfisher's nest I mentioned in an earlier chapter. At the end of the afternoon I would return home for my tea, just as if I had been at school. Although I had no watch, there never seemed to be a problem in judging the time to within fifteen minutes or so.

Unfortunately, I was not old enough yet to have a ticket for the Bingley Public Library. I suppose I could have gone there and asked the librarian for permission to look at the books, but I had an instinctive feeling that the library people would be in league with the school people. Instead, I foraged among my parents' books. The haul was thin. Only two of them interested me. There was a book called *Greek Myths and Legends*, a précis of Homer's *Iliad*. I found it interesting not just for the stories themselves. For the first time, I realized there was a land where the sun shone brightly, instead of perpetually through a veil of soot from the burning of soft coal. I also became aware that human deformity was not a necessary concomitant of life. A major difference from the way things are today and the way things were in the mid-1920s is the many instances one saw then of deformities associated with early industrial conditions.

The other book was of a very different kind. It was an introductory textbook on chemistry, about 250 pages long, which I read from cover to cover as best I could. Compared with the trivia of school classes, this book started a train of events that first secured my scholarship to Bingley Grammar School, which played a considerable role in taking me to Cambridge University, and that, in the 1940s, led me into fruitful research work on the origin of the chemical elements themselves.

The chemistry book was my father's. There were also in the house, tucked away in a remote cupboard, simple bits of chemical equipment that my father had purchased in earlier times. There were flasks and retorts, corks and a cork borer, glass tubing and a Bunsen burner, together with a dozen or so bottles of reagents. From that moment on, until I became established at Bingley Grammar School, I undertook a sequence of steadily more complex experiments, advancing bit by bit through the chemistry text. Indeed, if I had continued my home experiments, granted adequate financial resources and space, I would either have become a tolerable experimenter or alternatively have blown myself to kingdom come.

Although I started modestly enough, experiments of a more violent type soon dominated my attention, with much employment of such oxygen-rich substances as potassium chlorate. Making gunpowder was a natural, especially as older boys also made gunpowder. It was, in principle, easy to grind together charcoal, sulfur, and potassium nitrate. I was canny enough to do the grinding in small quantities—which, I discovered, were not dangerous, as long as the mixture could go off freely without being closely confined. The trick was to pack gunpowder inside the hollow barrel of some large key, and, at this stage, you really had to be mighty careful. Then you went outside and whanged the key as hard as you could from a distance into a stone wall. If a good mixing job had been done, you got a satisfactory sharp explosion, which served to win for you a good deal of prestige among bystanders whom you had thoughtfully invited to watch the proceedings.

The experiments were carried out in the small kitchen of the house, which we called the scullery. My parents did not realize how hairy the experiments were becoming until one day I allowed a girl into the house on an occasion when my parents were out. I showed her one or two bits of alchemy, such as pouring sulfuric acid on sugar, a trick I used, from time to time, to light the Bunsen burner when I ran out of matches. I then made the mistake of taking my eyes off the girl, only for a short time, I swear. She must have started mixing things at random with as sweeping a style as if she were baking. In next to no time, there was a crack and a huge flare-up. Luckily, neither of us was hurt physically, but the girl's dress was a wreck, and, since the kitchen itself bore evidence of the flare-up, there was no hiding what had happened. I suppose it was because at that time there was a fad for what the shops called "chemistry sets," but which were really piffling stuff, that I was permitted—amazingly, in retrospect—to persist, but only on the solemn condition that I admitted no one else to my seances.

The experiment I pursued with dogged persistence for more than a year was illustrated towards the end of the chemistry text. For me, it represented a compulsive dividing line between kid's stuff and adult stuff. It was the preparation of phosphine. The diagram in the book did it for me. There was a flask with two entry ports at the top, out of one of which a glass tube led under water. Emerging from the water were bubbles of phosphine, which turned into splendid smoke rings as they rose up in the air.

My first problem was that I didn't have adequate equipment, so I had to begin by saving up for several things. A day eventually came when I took the tram that ran along the valley to Bradford. I made my way from the tram terminal to Sunbridge Road, where the biggest chemist's shop in Bradford was located—a wholesale chemist's shop, I was informed, not that I could see anything wholesale in my intended purchases. Imagine yourself to be the assistant in that chemist's shop in Sunbridge Road in the late months of 1925. A boy of ten and a half comes in and fixes you with a gaze like that of the Ancient Mariner and says: "Does 'ta 'ave a Woolf bottle?" Rather to my surprise, he did. It cost 3s 6d. "Does 'ta 'ave any glass tubing?" was my next question. The assistant went away and came back with a veritable armful of glass tubing. My cup ran over with happiness, for, to this point, I'd been managing with only a few bits of glass tubing, bending the same bits into varying shapes in the flame of the Bunsen burner according to the changing demands of each experiment. Here were riches beyond the dreams of avarice. "Does 'ta 'ave any concentrated sulfuric acid?" was my next request. At this, the assistant at last looked me over with an element of misgiving, but I repeated compulsively: "Does 'ta 'ave any concentrated sulfuric acid? H_2SO_4, tha knaws."

The chemical formula must have done it, for the assistant now went away and eventually returned with a half-pint bottle of concentrated sulfuric acid. Can you imagine a small unknown boy walking into a chemist's shop today and thus securing a half-pint of concentrated H_2SO_4? To this point, the chemist in Bingley, a kindly old fellow called Cranshaw, had always fobbed me off with dilute stuff offered in a small, green-ribbed poison bottle, saying with a gentle smile: "I think this should do." Yet even concentrated H_2SO_4 was simple stuff to handle compared with my projected preparation of phosphine. The delicious trouble with phosphine was that it would explode on contact with air, so that, if you went ahead and prepared it without first flushing out your equipment with an inert gas, the whole shooting match was likely to go up in small pieces.

The book recommended persistent flushing out with ordinary "town gas," but I didn't like the idea of filling my mother's small kitchen floor-to-ceiling with town gas and then cooking the phosphorus and potassium hydroxide with a Bunsen burner. So, after using just a little of the town gas, I relied on the form of explosion that became known in scientific circles in later times as the "fizzle." I decided the amount of oxygen left in the equipment would be small, so that, if I made all the corks loose, the explosive combination of phosphine and any remaining internal oxygen should dissipate itself in blowing out the corks. The day came at last when I was ready to go. Taking care that my parents were out, and that the corks were nicely loose, I slipped in the Bunsen and let things cook, retreating to a safe distance myself. Things went as planned. There was a soft explosion, which freed the corks, whereupon I rushed out of my hiding place and pressed them in more securely than before, arguing that the oxygen had gone now and therefore my equipment was satisfactorily flushed. Pretty soon the phosphine started to bubble up through the water, and there were my smoke rings, just as the book had said would happen.

I repeated the experiment two or three times more. My mother really did protest strenuously about it, because the decay of the phosphine stank out her kitchen terribly and took surely a week to clear itself. But my landmark experiment was still more than a year into the future when, coming up to the age of nine, I left Mornington Road School for the discovery of kingfishers' nests and for the study of the five-rise locks at Bingley. To this point, I had three blank years to my credit in the war with the educational system. More and more, I felt that I was winning.

Chapter 5

A Scholarship Won from the Grip of Fate

THE NEARBY VILLAGE of Eldwick had a small school with about ninety pupils between the ages of five and fourteen. The school building had just two rooms—a small one for five-year-olds and a larger, barnlike room about thirty-five feet long, which housed all the standards from I to VIII, with I, II, and III separated from IV to VIII by a curtain. When in full complement, the Eldwick village school had three teachers— a woman for the infants, a woman looking after standards I to III, and a male head teacher for standards IV to VIII. The school was often not in full complement, however, because one or other of the two women was almost always off sick. The record of their exits and entrances was complex enough to put Shakespeare to shame.

The head teacher from around 1905 to the 1930s referred to himself as Tom Murgatroyd, although everybody else called him Tommy Murgatroyd. In my time, he was in his late forties, and he invariably wore plus fours. As it happened, he had once been briefly on the same school staff as my mother, and so, in the problem of coping with my determined exit from Mornington Road School, he and Eldwick School were my mother's final backstop. To this point, she had hesitated even to contemplate sending me to Eldwick, because it would be hard from Eldwick to climb onto the scholarship ladder. But now, in the summer of 1924, without any other move left on the board, the position was essentially forced on her. So my mother approached Tommy Murgatroyd with whatever part of my rebellious life history she saw fit to tell him. Sometime before Eldwick School closed for the summer holiday at the end of July 1924, I was taken there. To my surprise, Tommy Murgatroyd's establishment had a better look about it than I had seen before. As I quickly calculated the situation, this chap in the plus fours, with five standards on his hands, would have little time or motive to pester me. So, with the cautious agreement of all parties concerned, it was decided I would enter Eldwick School when it reopened in September.

My mother would soon have lost the limited measure of enthusiasm she'd managed to summon up, had the report on Eldwick School, tendered at almost exactly this time by His Majesty's Inspector T. J. M. More, been available for her inspection. Dated 9 July 1924, it read:

> The teaching conditions indicated in the last report still exist. Until they are remedied it is impossible to expect that the children can be properly taught. The main room is awkward in several respects, and is indeed suitable only for a single class. At present an attempt is made by two teachers to teach eight standards in it.
>
> For the teachers, some of whom stay to dinner, there is no separate room, cloakroom or lavatory. The playground space is encroached upon by ladders, broken iron pipes, and coke and coal dumps. There seems to be no storeroom. The caretaker's cleaning gear is stored partly in the girls' cloakroom, which, even as a cloakroom, is inadequate.
>
> Many of the desks are too small for the children using them. In view of these discouraging conditions it is gratifying to be able to record an advance in certain directions. . . .

Plainly, winning a scholarship from Eldwick, with its broken iron pipes and coal dumps, was going to be no easy assignment, especially as the educational authorities of the West Riding of Yorkshire gave only about a dozen to the whole Bingley area. It was never apparent to me until recently why my parents made such a big thing of my winning a scholarship to the grammar school. The reason, I discovered eventually, had an emotional component. At the age of eleven, my father had won such a scholarship, but, with his own father recently deceased, he had been obliged to go to work to help support his mother and younger brother. There seems to have been an almost compulsive wish to compensate for things in my own generation.

Eldwick was the school I would brave the biting north wind in winter to reach; you turned sharp right at Brick Kiln Row to get there. The fact that I often accepted real physical hardship without complaint to reach Eldwick School shows that I wasn't really difficult. It was rather that, even at an early age, I couldn't stomach stupidity, however much the stupidity might be cloaked in adult authority.

If you could see Eldwick School today, you would like it. It still stands in relation to the wild moorland, just as it did in my time, but with the number of pupils expanded greatly to about 300. A few years back, I opened the school's centennial jubilee. I took the opportunity of my visit to the district to tour around, looking at other schools over the whole Bingley region. Eldwick seemed to me the best of them all, although, unfortunately, it is confined nowadays to pupils up to the age of nine or ten. In my time, there was just the main building as you turned in from the main road—through iron gates, it is true, but lower and less prisonlike than those at Mornington Road. If you turned right on entering, you arrived immediately at the boys' playground, sloping from right to left, and, if you turned left, you arrived at the girls' playground, the two playgrounds being separated by the obligatory high wall, which was standard and regarded as essential by the educational authorities. The strange aspect of the morality of the time was that a teacher could burst your eardrums practically without comment, but, if a girl had been made pregnant at school, the uproar would have continued from the moment the pregnancy became visible to the delivery of the child, and the notoriety and disgrace of it would have vaulted out of the Aire Valley into half the county of Yorkshire.

At the end of the wall, as far from the school as possible, were the bogs, one on each side of the wall. As soon as the nights became frosty with the approach of winter, the older boys swilled a section of their playground with buckets of water so as to produce an icy surface by morning. Then they would launch themselves, preferably with a yell or roar, from the highest part of the playground at its roadside end in the direction of the bogs. Owing to the slope of the yard, the sliding boy picked up speed as he went down the slippery black ice. Because of the curvature of the ground, the boy would be turned gently to the left, and would thus avoid crashing into the bogs, but would collide instead with the high wall that separated the two playgrounds. I watched the sturdy sons of local farmers doing it scores of times—roar, pick up speed, crash.

Since I had missed essentially the whole of Standard III year at Mornington Road School, I was started at Eldwick School again in Standard III, which meant that I began on the junior side of the curtain in the main schoolroom. But the head teacher's logbook records that, on 5 January 1925, I was promoted to Standard IV on the senior side of the curtain, so rejoining my rightful age group. As the succeeding months unfolded, I moved steadily up the desks by which the scholastic pecking order was defined. There was actually a good reason for the positioning of the sharper pupils in special desks, because the special desks were the ones most likely to catch the eye of a visiting inspector. I always enjoyed visiting inspectors. For one thing, you could see your own teacher was edgy, and that was good. For another thing, your teacher, instead of wanting to keep you pegged down as on a normal day, seemed positively to welcome it if you were bouncy when the inspector asked questions. Gradually, I came to see the reason why. The very worst thing, the most boring thing, for an inspector was to be faced by a sea of unresponsive children.

If you even started a row, a row about anything at all, it made the inspector's day, and he gave the school a good report. Gradually, I came to see that this was a real trump card in my hand, as long as it was played with a bit of tact.

I was more popular with the girls at Eldwick School than I've ever been before or since, girls not only of my own age but right up to school-leaving age. Because of the open system, it was easy, when Tommy Murgatroyd was occupied elsewhere, for girls to slip me their exercise books. I would quickly do their sums and then slip the exercise books back. In return, the girls would dry my clothes for me. We often arrived at school with sopping outer garments. In a smaller room adjacent to the main schoolroom, there was a big smoky boiler near which we were permitted to hang our clothes. The girls did much better at this—they were much better at pushing in than I was.

A disadvantage of the open classroom with all ages present is that an older boy can more easily come to have a "down" on (that is to say, a compulsive spite for) a younger boy. Rather remarkably, during my two years at Eldwick School, only one older boy tried to bully me. Over a period of about two months, he set himself to waylay me in the afternoon on my way home, and it became a tense battle of wits to avoid him. This period, in which I learned every wrinkle for varying my route between Eldwick and Gilstead, turned out to stand me in good stead three or four years later, by which time I had begun attendance at the grammar school. In the evening, there would be homework to do, but, since I could never see much point in laboring unduly on homework, I was usually finished by 8:00 P.M., which left an hour before my mother would put on a bit of late supper, as she always did around 9:00. Often enough, I would occupy the intervening hour by taking a walk, and, since the journey to Eldwick and back took just an hour, I would walk there by one route and return by another.

One evening after dark, I was coming into Eldwick by an unpaved road and had arrived at the first gas lamps. As I came under the light, I heard an ominous voice say, "*That's* 'im." Perhaps forty yards away to my right down a lane, I saw two hulking figures, both evidently lads much older than I was. My former experience with the bully told me this was no moment for elegant delay. Instantly, I took to my heels, so quickly indeed that I must have opened up another thirty to forty yards before the two figures reacted. The lights of Eldwick lay ahead. If I could reach a certain point where there were several shops, the bus stop, and the local policeman's house, I felt I would be safe. Because I really knew the ground, and because of my lead, I made it. Then I stopped in the bright light of one of the shops and waited for the two figures to reach me. They were youths of about seventeen whom I hadn't seen before. "Does 'ta know ———?" one of them asked with extreme truculence. (I have unfortunately forgotten the girl's name.) "Naw, ah doan't knaw ———," I replied, speaking as much in the dialect as possible. "Well, if tha does, we'll beat the hell out of tha," offered the second youth unhelpfully. Then the absurdity of a fourteen-year-old boy depriving them of a girl who was good enough for both must have struck home, and, as if a switch had been pressed, they became friendly, telling me they'd been robbed of the girl, but they didn't know by whom: "If tha 'ears tell, just let us knaw," was the last shot.

The injuries that boys of this age group inflicted on each other in their pursuit of girls were something to behold. One evening after dark several years earlier, the call had gone out among us much younger village boys that a real "feight" was on. A guide escorted us at a determined run to the spot—to my surprise, taking the turning down Primrose Lane, past my own house, to where the big gates into the Milnerfield Estate were located. The road was unpaved at that time, and two hefty-looking figures were sprawled there in the dirt under a gas lamp, seeking to choke the lives out of each other. The youth who appeared to be getting the worst of it had his face streaked with blood, and one of his eyes was black and had swollen to the size of a golf ball.

"It's Stalker," breathed a ghoul next to me in hushed fascination. Stalker was a handsome fellow, and it was easy to guess he'd "robbed" his incensed opponent. How the triangular affair eventually turned out I never learned, essentially because, to us, the affair itself seemed unimportant. It was the "feight" that really mattered. The spot where the two figures had sprawled in the dirt lay in the very middle of the ancient traders' route between the Aire and Wharfe valleys, the route dating back to Domesday.

Tommy Murgatroyd lived for his garden. With considerable ingenuity, he had persuaded the local Education Committee to allow him to use the garden for instructional purposes, I suppose because it was thought sensible that the sons and daughters of farmers, who were themselves likely to become farmers, should be taught how to grow vegetables. On every occasion when the weather was remotely tolerable, he had most of the boys out in the garden, set to some job or other. Since the older boys were robust, as their crashing into the playground wall demonstrated, the school made a fine and handy labor force. Tommy Murgatroyd's garden was always in the pink of condition.

In my judgment, some people are born with skeletons suited to gardening and others are not. I was one who was not. So here I played my school-inspector card. In lieu of gardening, I did assigned school tasks—sums and so on. Because Tommy Murgatroyd could make the tasks sound almost like punishments, he lost no face with the other children over the arrangement, and it suited both of us, especially since I was well-nigh useless as a gardener.

Possibly it was usual for there to be lessons in gardening at village schools generally, but my experience at other schools had not led me to expect it. The thing struck even my child's mind as curious, to the extent that clear memories of it have persisted for sixty years. It was beyond my perception, however, to realize just how far gone and acute the situation really was, as Tommy Murgatroyd's logbook makes clear—to the point of painful hilarity. It was already in full cry ten years before my time:

> An extra gardening lesson was taken today to make up for the one lost on 21 November, through fog.
> Miss Carrodus had to leave school today through illness.
> Miss Dawson left school today with very severe toothache.
> Gardening could not be taken today owing to the absence of Miss Dawson.
> An extra gardening lesson was taken today to make up for the one omitted Feb 8.
> An extra gardening lesson was taken today to make up for the one omitted March 2.
> Miss Dawson was absent, visiting dentist.
> Gardening is impossible this p.m. owing to snow and sleet.

Since the gardening fetish must have been known to my mother, it becomes clear why she was skeptical of Eldwick School serving as a red-hot establishment for the winning of scholarships. Precisely the same two issues, the illness of the women teachers and the health of his garden, were still Tommy Murgatroyd's main concern in my own time, as the following contemporary extracts from his logbook demonstrate:

> Gardening is impossible owing to fog and frost.
> Miss Smales absent today, ill.
>   "   "   is absent with influenza.
> An extra gardening lesson was taken this morning to make up for the one lost Tuesday.
> Miss Hey left sick at 10 a.m.
> Gardening could not be taken today owing to the absence of Miss Hey.
> Gardening is impossible today. It is V. Wet.
> Owing to a rainstorm only 53 children were present out of 90.
> Gardening outside is impossible.

Miss Smales has been absent today with bad cold.
Gardening again impossible owing to rain.

The time eventually came, on a cold Saturday morning in mid-February
1926, for me to take the West Riding County Minor Scholarship Examination. The examination was the same for every child in the West Riding
of Yorkshire. If you won a scholarship, you were sent to the nearest grammar school. This was an advantage over winning one of the half dozen
scholarships the Governors of Bingley Grammar School gave each year
on their own account, because, if your parents moved from one district
to another, you simply transferred from one grammar school to another.
If my parents had moved into Bradford, for example, I might well have
been transferred to Bradford Grammar, assuming I won one of the dozen
or so scholarships awarded each year to the Bingley district, with its population, including outlying villages, of about 20,000. I was of minimum
age, ten, whereas you were permitted to take the examination up to the
age of twelve. So, if I failed, I was assured that there would be two more
chances.

The pupils at schools in the town itself, like Mornington Road, took
the examination in their own classrooms, which gave them a considerable
advantage over us children from the rural areas, who were herded together into whatever space could be made available. I was told to go to a
school close by Holy Trinity Church, then a high-spired, black-grimed
building at the hub of the collection of streets of back-to-back houses that
constituted the poorer, eastern part of Bingley. I remember it seemed exceptionally cold, as I made my way—I think alone—down from Gilstead
into the outskirts of Bingley. I crossed the canal bridge and turned right
into a street with the unpromising name of Dub Lane. A hundred yards
along Dub Lane, I turned sharp left, and then up a flight of steps into the
streets with the back-to-back houses, arriving a few moments later at
Holy Trinity Church, and at a room with a high ceiling, a room that had
been cleared to accommodate what seemed to me a multitude of desks
where an apparently monstrous horde of children was assembled. The
scene was about as different as was possible from the small Eldwick
School.

The examination room was cold that winter day. Perhaps because it
was a Saturday morning, the school boiler was not working properly. You
could say a fair amount of heat was put out by the children themselves,
and this had to be true—say, sixty watts per child—but the high roof of
the room, allowing the warmer air to rise way above us, largely canceled
this self-sustaining effect. I know I felt terribly leaden-footed in everything I was called on to do: slow to find my place in the throng, slow to

answer the questions that were asked. In recent years, a German chess grand master walked out in despair from a candidates' match for the World Championship because he was called on to play in poor surroundings, with district trains rattling past most of the time. His opponent was a Soviet grand master who conveniently happened to be deaf, so it was widely believed that the venue had been arranged by the KGB. My situation was not dissimilar, although I could make no such spectacular claim as to its cause.

The arithmetic paper had two parts. Part 1 said, "Answer only 4 questions," and Part 2 said, "Answer No. 6 and 3 others." The time allowed was one hour and twenty minutes. For a trained child, it would not have caused much trouble, but, for a child sporting my formidable truancy record from a country school whose head teacher had a gardening fetish that was quite out of control, the paper was something of a corker. The easy opener read:

1) (a) $\frac{5}{12} + \frac{7}{15}$, (b) ($\frac{3}{5}$ of $3\frac{1}{3}$) $- 1\frac{11}{15}$ Divide (a) by (b).

The last shot was:

10) A boy was told to multiply the sum of 0.028 and a second number by 0.035. Instead of this he multiplied the second number only by 0.035 and then added 0.028 to the product. His result was 0.05012. What result should he have obtained if he had done as he was told?

Nobody had ever told me the trick for answering questions like this last one. It is simple: Name the quantity you require—which is to say, give the unknown number a name (x, say) and work it out from there. It was only some months after the examination that I discovered this trick, and then I felt really angry at having been tricked myself.

Following the arithmetic paper, we were required to answer questions in English grammar and to write an essay. Throughout my school days, I was always unenthusiastic about English grammar. It was what I called a daft subject. So I did it because I was required to do it and for no other reason. My point of view was that, either you knew how to write and talk or you didn't, a view confirmed in later years by a friend in the English Faculty at Cambridge, who said: "Every child of eight has an almost perfect command of syntax in speech but not of word order or vocabulary." It would be interesting to know how it first came about that grammar became such a feature of so-called elementary education. The thing must have started on somebody's arbitrary say-so, and, having become established, it simply went on and on. One might rather say the distinguishing feature of the English language is the lack of grammar, with word order and the subtle use of vocabulary taking its place. One can also say that pedagogues do not invent language. New uses in language spring from the instinctive perceptions of people needing to express an idea or an emotion that cannot be expressed adequately by previous usages, as in the famous example when Christopher Marlowe broke suddenly through to the discovery of blank verse in his play *The Tragical History of Dr. Faustus.*

The subjects offered as topics for an essay were distinctly peculiar. As I remarked before, the examination took place in early 1926, shortly after the signing of the Locarno Pact. There was no offer of this grandiose subject as a topic for an essay. Instead, there was the following absurdity, to which nobody I ever talked with afterwards was able to offer much of an answer: "Write a short story which brings in the following: As Freddy stepped out of the train, his sister ran to greet him. 'How you have grown,' she said, and then, 'What! Have you left your bag behind on this day of all days?'" Months, even years after the examination, I puzzled away to myself over what Freddy had been up to. But I never managed to fit the growing and the bag and the day of all days together into a plausible scenario.

During the arithmetic paper, I was conscious of being unaccountably slow. I did only five questions out of the required eight, which was disappointing because, in arithmetic, I usually worked rather quickly. During the English paper, I was vaguely conscious of writing a lot of nonsense, which I tried to pass off to myself with the assurance that the paper itself really was a lot of nonsense. The very next morning, I woke with my face feeling a mile wide. When my mother brought a mirror, I laughed unrestrainably, even though laughing was quite painful. I was down with an attack of mumps, an attack that had been an unsuspected joker in the pack throughout the morning in the chilling environment of the examination hall.

Because I had done only five arithmetic questions, and because I suspected my English paper was harum-scarum, I insisted to my parents that I'd done badly. But my father was not so sure. I'd brought the arithmetic paper home with all my answers jotted in the margin. It proved they were uniformly correct, so that, although I'd done only a fraction of what I'd been asked to do, everything was properly done, which an examiner might be expected to judge a hopeful sign from a child who was as young as it was possible to be for the scholarship examination. It never occurred to my parents to obtain a doctor's certificate and to ask Tommy Murgatroyd to forward it to the educational authority. I doubt it would have done much good if they had. Viruses were still unknown as the cause of many childhood illnesses, which were then seen as mysteries one simply came to accept. The general idea was that you were "well" up to the outbreak of clinical symptoms. So the idea would be that I was "well"—lucky for me, people would have said—during the examination itself and had only become "ill" the following day.

Results appeared some two months later. If you won a scholarship, the school received a colored letter—blue, I think. Otherwise, the letter was white. There was no blue letter at Eldwick School. I realized it the moment Tommy Murgatroyd came in one morning with the relevant letters. So I'd failed and that was that. The situation had to be accepted both at home and at school. Yet my father had plenty to talk about in connection with the examination, because, as soon as the results became known, a considerable scandal broke loose. Results everywhere in the middle part of the Aire Valley were poor, with no more than half the usual number of scholarships awarded to the Bingley area—six instead of the usual ten or more. Other districts in the West Riding had extremely large numbers, however, which raised the possibility—nay, the certainty—that there had been serious cheating. The scandal escalated rapidly, with members of the Bingley town council threatening action against the county. In these circumstances, it became inevitable that the county education authority took a second look at children from the Bingley area. I know from later personal experience that there is all the difference in the world between the routine marking of examination papers and the hunting through of examination papers on the look-out for exceptional cases. The papers of a young candidate who wrote quite deliriously on the English paper but was almost totally precise in the arithmetic paper must surely have caught the eye. About three weeks after the first announcement, there came a request, through Tommy Murgatroyd, that I present myself to the grammar school for an interview with the headmaster there.

The headmaster, Alan Smailes, had come through the mathematical tripos at Cambridge, the university examinations then being known as tripos, with the math one being the most famous. Yet I do not recall him asking anything at all about arithmetic. He asked me to read a passage from a book, which I did easily enough. Then he asked what books I had read myself. In retrospect, it is clear that Smailes had been requested by the county authority to verify that I really was in possession of my faculties. I still never thought to tell him of the mumps, but I did tell him about the chemistry book I'd read, and about books on stars I'd borrowed from the public library. Pretty soon he packed me off for a word with the chemistry master. Herbert Haigh was in his thirties, a dark-haired man with a pleasant face who had received terrible wounds in the 1914–1918 war. In later years, he told me an incompetent nurse had poured caustic on the wounds, thinking it a disinfectant. At the height of the depression, he bought chemistry books out of his own pocket to help those of us who were then seeking entry to universities. He was the only teacher I ever came across who kept perfect discipline in class without any physical demonstration or even a harsh word. On this occasion, he asked me what experiments I'd been doing, and it was natural for me to tell him about the phosphine preparation I described in the previous chapter. He was incredulous and demanded to know how I'd done it. So I launched into a description of the loose corks and the fizzle of an explosion I'd provoked. At this, a smile flashed across his face, and with twinkling eyes he said: "Well, you'll not be doing that here."

I pondered this remark on my way home, and I thought it might mean I'd won the scholarship after all. In a curiously roundabout way, I had. A blue-colored letter appeared at Eldwick School shortly afterwards, and my name eventually went up in gilt letters on a board in the schoolroom that held all the scholarship winners there had been in the history of the school—at that time, about twenty.

Chapter 6

Education at Last, and a Scholarship Lost to the Grip of Fate

I BEGAN ATTENDANCE at Bingley Grammar School in September 1926, almost four centuries after the school was founded in 1529. The school must have been active in around 1645, when Oliver Cromwell had his considerable nest of supporters in Bingley. No such profound historic thoughts entered my head, however. The most immediate effect for me was that I now had to walk nearly eight miles each day instead of five. I continued to return home for lunch, since there were no organized school meals. The annual rainfall for the district being about thirty-five inches, I must have got wet—sometimes soaked, sometimes merely damp—on hundreds of occasions over the seven years from eleven to eighteen, by which time I had walked a total of some 10,000 miles. The odd thing is that I have no memory of sitting damp and miserable at school, although I do remember exceptionally rainy periods when I set out from home wishing it would be fine on at least one day in the week.

My father maintained a never-ending struggle with the West Riding educational authorities. In contrast to their generosity in every other respect, the authorities were very rigid about bus fares. The rule was that a scholarship holder who lived more than two miles from school could claim transport costs. The distance from our house to the grammar school along the shortest route by which you could take a vehicle was within a cat's whisker of two miles. It is actually so close that it would need a commission of inquiry to decide the matter, and the outcome might even turn on which classroom and building at the school you were heading for. The shortest route for vehicles from Gilstead went downwards at a gentle angle past the Cottage Hospital. It turned a corner, and there on the left was a largish house with a walled garden where lived a girl who, at about that time, married Wally Hammond, the great English batsman. The streets of Bingley were lined for the occasion by thousands of people, because a wild and absurd rumor was abroad that Wally Hammond was going to ditch his career with Gloucestershire and come to play cricket for the town of

Bingley. Then you went gently upwards and around a long bend to Park Road, where a horse trough stood on an open bit of ground—my relative Fred Jackson used to pad around its iron rim in triumph until, one day, he slipped and broke a front tooth clean across—then left down Park Road into Bingley until you came to the T-junction at the Bradford-to-Keighley road just by Leach's, the high-grade sporting-goods shop, which used to stock Meccano sets and Hornby trains, but is now given over to the sale of baby wool, and there you turned right and continued for two-thirds of a mile until you reached the grammar school.

This was my father's route. But the educational authorities probably argued that there was a shorter walking route. Actually, there was—branching sharply left at the Cottage Hospital and down past the Catholic school, whose pupils were stoned periodically. This brought you to a cliff-like drop of fifty feet, which, at that time, you had to scramble down. Then you continued along and around streets with high-sounding names, like Belgrave Road, which were actually lined with back-to-back houses, until you reached Mornington Road School. Here you circumnavigated the high wall enclosing first the girls' and then the boys' playground until you reached the iron gates, where you stopped for a moment to deliver a big ha-ha. Reaching Mornington Road proper, you then turned right and continued across Park Road, past a large mill, and down to the Leeds and Liverpool canal. After crossing the canal, pausing to gaze in awe deep down into three locks, some way below the famous five-rise locks of my earlier truancy, you turned right and followed a narrow lane that took you under the railway through a dark passageway smelling not unlike a urinal. Eventually, the passageway debouched onto the Bingley-to-Keighley road, more or less opposite the parish church. Turning right, you finally reached the grammar school in about a third of a mile.

This, apparently, was the route the educational authorities wanted me to take. It had the advantage that, once a week or so, my walk home in the afternoon chanced to coincide with the passage of a high-speed boat train. Many were the occasions in winter when I saw brightly lit carriages sweep past and wondered if the day would come when I would take the train myself to join a boat at Southampton bound for the United States. This cherished dream never became reality, since, by the time I eventually visited the United States, airplanes had largely replaced boats for trans-atlantic travel. Nobody, as far as I can recall, ever predicted the use of airplanes for mass transportation—not, at any rate, over large distances. All the talk in my youth was of airships. This notable failure to foresee the march of technology has made me skeptical of technological prediction, which is fraught with uncertainty in every field because of unexpected discoveries and unexpected snags. It is not so hard to foretell that

such-and-such will be possible in the future, but the best technical way to
achieve it is very difficult to judge in advance of events.

The school uniform was not obligatory, and in those times of depression nobody wore it. But we were required to wear the school cap. For a
year or two, I did so with reluctance, until going bareheaded became common generally. The cap often got me into "feights" I would have preferred
to avoid, a point in the bus-fare controversy the education authorities had
presumably failed to consider. If they wanted to toughen up their scholarship boys, this was certainly the way to do it, for a grammar-school cap
was like a red rag to a bull to the boys from every intervening school on
my route. I was never actually stoned like the Catholic children, but the
"feights" would have been a daily occurrence if I hadn't displayed a little
of the subtlety in varying my route that I'd developed in earlier years.

Actually, I agreed with the brusque lads from the intervening schools.
The cap was a mark of class distinction that I disliked as much as they
did. I was still two or three years away from reading the works of George
Bernard Shaw, but I was already struggling towards a concept that Shaw
expresses admirably towards the end of his play *Arms and the Man*. The
Swiss mercenary Captain Bluntschli is in competition with Sergius the
Balkan aristocrat as to which of them has the greatest possessions and the
highest social standing. Bluntschli draws himself up and says: "My rank
is the highest known." When this immediately rivets the attention of
everybody on stage, Bluntschi continues: "I am a free citizen of Switzerland." Throughout my life I have never ceased to regret that such a point
of view is not ingrained deeply within the British people. Had it been, I
doubt whether the position of Britain in the world would ever have retreated from the high point it once occupied.

When I began attendance at Bingley Grammar, I tried to persuade myself that changing schools would have no effect on my relationships with
the village boys. Yet it did. Not through money, or through a cap, or an
old school tie, but through ideas developing in the head, a process that
had begun already by the time I was coming up to ten. A story from those
earlier years may be worth telling, since it had some effect on my life,
perhaps subconsciously a dominating effect.

Once we returned to school in the autumn—at the time in question, it
must have been Eldwick School for me—there were set activities we engaged in from year to year as standard procedures. One was to begin collecting wood for the village bonfire on Guy Fawkes night. Another was a
game we played every fine night over a period of three or four weeks,
pretty well through the month of October, once the evenings became dark
enough. A band of us, perhaps twenty strong, divided into two halves.
We had a center point by a particular gas lamp with a fine, large piece of

sandstone below it, which is there still today. One half of us went off into the darkness and was given a few minutes' start. Then the other half went after them, but leaving behind a guard stationed perhaps fifty yards from the sandstone block. If any member of the search party came near enough to recognize you and call out your name correctly, that counted one up to the search party, but if you managed to get back to the sandstone block without being recognized, that was one up to the field party. We made the game self-adjusting, in the sense that we learned to judge the distance of the guards from the sandstone block so as to ensure that the tallies came out fairly evenly. The roles of the two parties were alternated from night to night. We called the game by the unlikely name of "Bed Socks," obviously a corruption of something or other. I just grew up with the name and accepted it, without ever thinking it a problem, as I would have done if something physical had been involved. Already at the age of nine, I felt that the world of people and their ways was too complex and arbitrary to be sorted out, unlike the world of things, which seemed to have an attainable rationality to it.

I happened to be in the field party one perfect starlit night. I had a special friend I went about with most of the time. Even to this day, I'm pretty sure of the route we took from the sandstone block. We ran off about twenty yards down Primrose Lane and then ducked left into the darkness of the "Sparable," along which we continued to its second sharp-angled turn. Then we climbed a wall, low on our side but with a biggish drop for us on the opposite side, then down into the valley where the stream ran that passed through the Milnerfield Estate and on the banks of which I'd found the ill-fated kingfisher's nest. There was first a wood and then a wall over into the "bull field" of Robinson's farm; then up the bull field and over into the roadway close to where Herbert Haigh, the chemistry master at Bingley Grammar School, would come to live a year or two later. Then we doubled back to the sandstone block in the shelter of more of the fields of Robinson's farm.

When on top of a wall that perfect starlit night, I seemed to be in contact with the sky instead of the earth, a sky powdered from horizon to horizon with thousands of points of light, which, on that particular dry, frosty night, were unusually bright. We were out for perhaps an hour and a half, and, as time went on, I became more and more aware—awed, I suppose—of the heavens. By the time I arrived back at the sandstone block, I had made a resolve. I remember standing on the block and looking upwards and deciding that I would find out what those things up there were.

This resolve became more apposite when I reached the age of thirty than it was immediately. Even so, we had an old encyclopaedia in the

home with simple articles on stars and planets. And in 1927, the book
Stars and Atoms, by Sir Arthur Eddington, became available in the town
library, which to me was another big turning point. At the age of ten or
eleven, I tried to tell my friends in the village something of all this, but
either they weren't interested or they were so incredulous as to be derisive,
which indicated that a barrier was inevitably opening up between us,
quite apart from any difference in the schools we attended.

The way people enter our lives with apparently the tightest of relationships, which are then gradually but inexorably dissolved away by age, by
a divergence of interests, or by external events over which we have no
control, is for me one of the saddest aspects of things. You have no wish
for it to be so, but you are powerless to stop it from happening. Even the
relationship with parents, which seems so overwhelmingly strong in our
earliest years, is first weakened bit by bit as we "go out into the world,"
as one says, and is finally terminated by death. Likewise, the seemingly
unbreakable relationships with our own children are steadily weakened
as they themselves go out into the world. Parents may see their grown
children every few days, if they all chance to live in the same district, or
every few months, if in the same country, and the feeling is always that
nothing has changed. But it has, just as it had for my relationship with
my friends in the village.

Like bits of jetsam in a fast-flowing river, we come together and are
then parted by the flow of events. There is just one relationship that can—
and that does, in many instances—stand rock solid: the bond between
husband and wife. It is a curious thought—if we happen to be among the
lucky ones for whom this relationship becomes a reality—that we go
through our youth with relationships to parents, friends, teachers, all
seemingly of dominant importance, and yet the person who is to mean
most in our lives has not yet appeared, as it would be in a play if the most
important character did not come on stage until the second act.

All school classes I had attended, to this point, had been mixed. At the
grammar school, however, the sexes were rigorously separated in class as
well as on the playground. Although the girls' school was immediately
adjacent to the boys' school, over seven years I did not have a single conversation with any girl, as I had done so often when doing sums for girls
at Eldwick. Somewhere around 1930, a new building for the girls was put
up and was even joined physically to the boys' school. But, throughout
my time, there were always locked doors shrouded by a big, black, impenetrable curtain across the joining corridor. In later years, when I described this situation in an address to a later generation of pupils at Bingley Grammar, then happily mixed, there was merriment from the boys
and girls sitting interspersed in the audience. What really brought the

house down on that later occasion was the case of my relative Biddy Jackson, who was severely reprimanded by her headmistress for walking to school on the public highway beside Bill Jackson, her cousin.

Early in the first term at the grammar school, the members of my class were given a kind of pecking-order test, and I was sixteenth out of thirty-two. The stronger scholarship holders were ahead of me, as were a few scholarship holders who had come into the Bingley area from outside, and some of the fee-paying pupils who had already been at the school for a year or more. Most of the fee-payers were from better-off homes than the scholarship boys. Indeed, the historic raison d'être for the school was to provide education for those who could afford it, with scholarship holders introduced in ones and twos at first, and in my time in roughly a half-and-half distribution. As one might have expected, the fee-payers were less uniform in ability than the scholarship boys, with the best being very good and the worst not so good. My position in the middle of the class, called "forms" now instead of "standards," was thus about right. The annual intake of schoolchildren for the whole Bingley area must have been around 400. So I was sixteenth in about 400, which was a fair reflection on the irregularity of my education to this point. Or, in terms of population, I was sixteenth in the annual output of children in a town of about 20,000.

Let me interpolate what a thorough investigation of standards would have revealed. There were, at that time, about 500 fellows of the Royal Society of London, elected over the years on grounds of scientific merit from the Commonwealth as well as from the United Kingdom. The election rate was about twenty-five new fellows each year out of an educated group of about 100 million people, for a production rate for the whole Commonwealth of about one fellow per four million people per year. The production rate for scientists of international repute was of a still lower order of magnitude—say, one per 100 million people per year. It will therefore be clear how important it was to have no idea at all of the length of the road ahead if ever I were to become a scientist, which, by now, was beginning to suggest itself to my mind. The way to perceive the road ahead is not in terms of a far distant goal, but as a sequence of nearer steps by which you aim to improve yourself by a single order of magnitude at each stage. This is just the way mountaineers scale a high peak. The trick is to divide the route into a sequence of smaller objectives; otherwise, the whole would seem overpoweringly vast.

By the end of the first term, I had improved my form position to fifth, by the end of the second term to second, and by the end of the year to first, thus pulling up one of the many orders of magnitude I had to go. In the following year, my positions were first, first, and second, but in the third year they became first, second, and third, which led to some disagreement with my parents who probably felt I wasn't trying hard enough, which in some degree was true.

There were several reasons why I appeared to be in decline. A simple reason was that I always had to overcome a fairish handicap in not being able to draw. In the term when I was third, my percentage in drawing was forty-seven, while the top marks were in the seventies. This cost me two or three points when the average percentage for all subjects was taken. But, more importantly, I had ceased by the time I was thirteen to bother much with homework, which carried half the weight in our results. I preferred to score seven or eight out of ten for homework done quickly rather than spend a further hour or more in scoring an extra mark. There was a boy who came in from outside the Bingley area during this third year who simply slaved at his homework, and good luck to you, I thought, because it won't do you the slightest good when the real examinations of the so-called matriculation year begin.

By the age of thirteen, I had begun to read widely—borrowing the books from the town library—not only science books, such as Eddington's *Stars and Atoms*, but even such far-ranging and unlikely works as T. E. Lawrence's *Seven Pillars of Wisdom*. I made a considerable effort to trace Lawrence's activities in detail, finding some of his episodes cloudy. When claims were made many years later that certain of the episodes might not have happened, I wasn't altogether surprised. My parents felt this extracurricular reading to be unwise, since they would have liked me to remain a bright and shining first, an example to all beholders. I remained deaf to their protestations, however, preferring obstinately to trust my own judgment.

Another possible reason for my easing off was that I didn't particularly like the form master in the third year. While he had good qualities, there was an irrational irritability in him that showed itself by a perpetual nagging of particular boys. A day came when a big lad higher up the school than I was, provoked beyond bearing, felled him good and proper. Although this was represented to us as a crime Agatha Christie might be expected to write about, I felt it was long overdue.

Looking back over my time in the educational system, I would say that the attitude of pupils towards teachers who became angry and violent was almost always fairminded and psychologically accurate. Charlie Hulme, our English teacher, was a big, intrinsically gentle chap whom we teased unmercifully. Whenever we did readings of a Shakespeare play with the parts allotted to the class, there were invariably cries of, "Please, sir, I want to be a lewd strumpet." The cries were repeated ad nauseam until Hulme let out a roar, followed by a frenzied tour up and down the aisles between the desks, aiming blows randomly as he went. Then there was a smallish boy, Tich Taylor, who carried a red diary in the breast pocket of his coat. Whenever Hulme told us some book or other was important, Taylor would slip out his diary and say: "Please, sir, I have a little book . . . ," which was usually good enough to start an enraged tour, beginning in Taylor's direction but spreading like a tornado through the classroom. Nobody felt the slightest ill will for any random blow they might receive in these periodic uproars, and Hulme was one of the best-liked teachers in the school.

The anger of Eddie Dodd, a fiery little Welshman, the senior master after Smailes, was something to behold. Dodd was our history and Latin teacher. One or possibly two of his younger brothers became university professors, and my suspicion is that our Dodd might well have done the same, had there not been economic reasons why the eldest brother in the family had to secure a job as soon as possible. He was the most erudite person I came across in my school days, and, as befitted his intellectual qualities, his anger had nothing down-to-earth about it. Anger was not, in his view, something to be dispersed lightly. It was reserved for ludicrous mistakes, such as claiming Caesar to be a Gaul, or the Diet of Worms a new medical treatment. This sort of anger from a teacher never troubled me unduly, even though, on occasions, I found myself at the receiving end of it, especially as Dodd was my form master through my last three years at school. Almost all my early perceptions outside science I owed to him.

I didn't worry too much about boys who sought the top position in the class through meticulously prepared homework, but there was one boy in the class whom I did watch closely, and, if he had begun to beat me consistently, all such extracurricular activities as the *Seven Pillars of Wisdom* would instantly have gone out of the window. His name was Chester. I can't remember his first name—largely, I think, because it was never used. He was dark-haired with a rather round head, as I recall, and more robust than I was. He was always quiet and friendly, but neither I nor anybody else I can remember ever penetrated a sort of withdrawn reserve. Mostly one knew the home background of the boys, certainly of any who were outstanding either in sport or in their work. Yet I never learned anything of Chester's background. It was consistent with his having been given a brief opportunity by a mother who was a war widow, such as might have been my own case, if my father had not survived the Ludendorff attack of 21 March 1918. Be this as it may, Chester left school, despite great promise, immediately after the matriculation examination. He took a job in an industrial chemical company. Years later, I learned he had risen to become its manager, which was no surprise at all. He treated the educational system more ruthlessly than ever I did, by paring down the subjects he would need for the matriculation examination to their bare bones, and these subjects he went after with unremitting determination. For instance, only one foreign language was required, which, at our school, meant either French or Latin. Quite why Chester chose Latin I never learned, but, once he had chosen it, I could never touch him. In my second year, my class position in Latin was first in each term, but, in my third year, Chester toppled me, and I couldn't cope with the challenge, coming second in each term.

Chemistry was another of Chester's special choices, but here I considered any degree of effort to stay top was not only desirable but essential. Over the same six terms, the honors were equal between us. It was a dog-fight down almost to the last mark, with neither of us giving or receiving any quarter.

Physics during my first three years was a noninspirational subject, and so it must have been for Chester, otherwise I would surely have made more effort at it. The experiments we were asked to do required neither planning nor dexterity. You did the experiment, and then you wrote it up. Worse, you drew diagrams to show what you had done, and my drawings were pretty poor, as usual. My class position in physics over the same six terms was between third and sixth. Actually, there was not much to separate the sheep from the goats, since I had eighty percent even in sixth place. Since the work was straightforward, the marks were all crammed together between eighty and ninety percent, with those who took most pains over drawing and homework securing the highest places.

At the end of the third year, three or four of us skipped the normal fourth year, so arriving in the form destined for the matriculation examination a year earlier than usual. I thus began my fourth year at the grammar school eighteen months younger than the average age for the class, instead of six months younger, as had formerly been the situation. No more reading of library books, for now I had the opportunity to see if I could lift myself by another order of magnitude. My class positions in my fourth year for the two terms before the matriculation examination were:

> All subjects: fourth; second.
> Chemistry: fifth; first.
> Physics: first; first.

Although the standards were as yet nothing to speak of, these placings were beginning to become significant, really more significant than the public examinations themselves. They demonstrated an ability to increase pace, with the results in physics perhaps of most interest. The hope was that, if I could spurt once, I could spurt twice, or perhaps three times, as levels of difficulty continued to rise.

I was just on my fifteenth birthday when the time eventually came for the matriculation examination. We were not specially drilled in the kind of questions to expect. Indeed, I do not recall our being shown the papers that had been set in previous years. So it was all rather like shooting at an unseen target, when high scores are not to be expected. This relaxed policy was, I suppose, consistent with the history of the school as an establishment for the education of the children of the well-to-do, who must not be put through the hoop too severely. It was not a policy likely to secure optimum results; the outcome for a class of thirty-odd would typically be about five matriculations, fifteen awards of what was known as the School Leaving Certificate, and, for the rest, nothing. Such an easygoing attitude would surely be impossible today, but, in times when matriculation was sought by pupils as a complete entrance qualification to universities, five out of thirty would be about right.

A paper-backed book of the results of the examination could be ordered through the school, with delivery at your home promised for the official announcement date in late August. I didn't order a book. Nor at any time to the end of the public examinations I sat in my youth did I ever make special arrangements for results to be sent to me. It was a sort of superstition, like touching wood, or like not counting your chickens before they were hatched, or like the old saying that no news is good news. There was another boy in the village who was involved, a boy whom I'd caught up when I jumped the fourth year. I was passing his house when the mother, a small ginger-haired woman, shouted: "You've matricked. But Stanley has only certificated." She invited me inside the house, and I tried not to appear overpleased. I looked through the results book and saw that Chester had also "matricked." Unfortunately, I was never to see him again.

Just over fifteen years old, I returned to Bingley Grammar in September 1930, to find myself now in a different regime. Over the past four years, I'd been taught systematically, subject by subject, according to a specified timetable. But this discipline could continue no longer, because the school had no special teachers for its sixth form. There were only eight regular teachers to about 200 pupils, so instruction in the sixth form had necessarily to be sporadic—extremely so in mathematics, because, in addition to teaching the younger pupils, Alan Smailes had his administrative duties as headmaster to attend to as a first priority. In effect, the sixth form was an addition tacked onto a school whose pupils traditionally left immediately after the matriculation examination. Laboratory space was also a problem, with the main laboratories occupied by junior classes for most of the time.

This was hardly a favorable situation for winning a university scholarship, but, as a training ground for research, it was surely excellent, in that it threw us heavily on our own resources. The system was for us to be given tasks by the masters, which we set about solving in the spirit of research projects. Formal lessons were comparatively few, and our number was so small that a lesson in chemistry, for example, would be attended by everybody aiming at the Higher Certificate (A level), irrespective of which particular year you were in. The other two pupils following a path similar to my own happened to be my relative Fred Jackson, the one who broke a tooth padding around the horse trough in Park Road, and a big strong lad with whom I would eventually do quite a bit of hill-walking, Edward Foster. Fred Jackson was in the same year I was, but Edward Foster was a year ahead, and it was he who now set the pace for us.

Situated off both the main chemistry and physics laboratories were smaller rooms, reminiscent of my mother's kitchen but somewhat larger. We were given these rooms in which to set up our own laboratories; this procedure had the advantage that the time we could spend in our little private laboratories was almost unlimited. Money for equipment and chemicals was always a problem, just as it had been when I bought my Woolf bottle for 3s 6d. It was impossible for the school to buy all the chemicals we might need, so, in organic chemistry especially, we often had to synthesize even the reagents before we could begin assigned experiments—not the quickest way to prepare for the Higher Certificate, but mighty good for our eventual self-reliance. Edward Foster was to become a Reader in Physics at Imperial College, London, while Fred Jackson became the Head of a strong cardiac department at Newcastle General Hospital.

Major Scholarships, as they were called, were awarded by the West Riding educational authorities on the results of the Higher Certificate examination, and there were also national State Scholarships. Few counties were as generous as Yorkshire in education matters, so the competition nationwide for State Scholarships was very keen, forcing a higher standard than in the county. I was told that an average of 67 percent or thereabouts in mathematics, physics, and chemistry would give me a good chance of winning a county award, but that an average of 70 percent or more was needed for a State Scholarship. Consequently, we all opted for the county system and geared ourselves to it.

How about personal finances through all this ("the penny in your pocket," to paraphrase Harold Wilson, who, I suppose, was working his way towards Oxford at that time, not too far from Bingley, in Huddersfield)? We were now mired in the worst years of the Great Depression. My father had retreated from his business in Bradford—luckily for the family, since people who persisted in private business through those years mostly lost their capital. Bankruptcies occurred frequently throughout the whole Aire Valley. Money was therefore tighter for my parents in 1930 than it had been in 1920. Up to the time when I "matricked," which is to say up to the age of fifteen, I was supported entirely by my parents, with a shilling a week given as pocket money. After matriculation, the West Riding authorities more than made up for any reluctance they'd previously shown over the bus-fare issue by awarding me £15 annually as a contribution to the expenses of further education. It was given as an encouragement to parents to permit their children to remain at school. My own parents reacted with a like generosity by leaving the £15 under my control. I used it mostly to buy clothes, which perhaps explained why I wasn't knocking my shoes out so much these days by climbing walls. The difference between £15 and the cost of my clothes therefore became the penny in my pocket.

The money had to be watched somewhat carefully, but I don't remember making a big production out of it by fussily keeping detailed accounts. At all times, I knew to within a shilling or two what I had, which was entirely sufficient. One extravagance I could not resist was a season ticket to six concerts given in Bradford by the Hallé Orchestra. The bargain offered to pupils in schools throughout the whole Bradford area was simply too good to be missed—six shillings the lot. Fred Jackson and I stumped up our money in good heart for what was to prove one of the best investments of our lives. The return tram fare to Bradford was sixpence, so it was really one shilling and sixpence per concert. At such a price, you might have expected the concert hall to be overflowing with bargain hunters, but I suppose it was known from previous years that this would not be so. Pupils from schools occupied the first three or four rows of the audience, which, being almost on top of the orchestra, gave us an excellent view of whatever might be going on. Pupils from Bradford Grammar had their seats just in front of ours, and it was immediately noticeable to the eye that the girls were not rigorously segregated from the boys, without any suggestion of mayhem breaking loose, so far as I could see. Perhaps because it was the first time I'd heard a full orchestra, I still remember the program on that occasion. It had been billed as a symphony concert, and a symphony concert it certainly was: Schubert's *Unfinished*, Sibelius's Second, and Beethoven's Fifth. And the conductor? None other than Sir Thomas Beecham himself. It surely was a good shilling's worth.

Forty years on, I attended another performance of Beethoven's Fifth, given at Cornell University by the Baltimore Symphony. As the first movement got under way, I gradually realized, from the responses of students around me, that many of them had never heard a real live performance of the Fifth before, just as it had been for me so many years earlier. By a mysterious process of communication that undoubtedly exists but that is hard to explain, the sense of excitement in the audience became apparent to the orchestra itself. The players began to work much harder, and, it being warmish in the hall, I remember the sweat on the musicians' faces as they reached the triumphant coda of the symphony. Beethoven lived in a brief slice of time between the court orchestras of the eighteenth century and orchestras organized by municipal societies in the nineteenth. He gave his own concerts, hiring the musicians and paying them out of his own pocket, which explains why he always made them work hard. He would have enjoyed the sweat on the faces of the Baltimore Symphony, just as he would have enjoyed the regal gestures of Sir Thomas Beecham on the occasion of my own first symphony concert.

The time came at last for the Higher Certificate Examination. My performance was almost exactly as I'd aimed it. By now, I'd taken so many tests that it had become second nature to judge how well or how badly I'd done with quite a measure of accuracy. On one occasion, on my way home in the evening, I'd fallen in with E. A. Kaye, the physics master, just after Kaye had finished marking a term examination. Throughout my later years, Kaye and I never ceased trying to get a rise out of each other.

"How d'you think you've done, position and mark?" he asked.

"Top, with 75 percent," I replied decisively.

"You *are* a conceited blighter, aren't you, Hoyle?" Kaye then said in his not so popular southern accent, adding, with a somewhat twisted grin, "Top, with 76 percent." So I knew I was about right, just OK, so far as a Major Scholarship was concerned, but not with any great margin to spare. And I knew, moreover, what I intended to do the coming September. I was entering Leeds University to study chemistry. I would be just past my seventeenth birthday by then—a bit young perhaps, but still consistent with my policy of getting on with the job. I almost had my bags packed, so confident was I that it would work out that way.

And, so far as my total score of marks in the examination was concerned, it did work out that way. I was over the average I had aimed for, which, in previous years, had been considered sufficient for a senior county scholarship. What I hadn't reckoned with was the return of the "Geddes Axe." In response to the mysterious economics of the depression, education budgets were suddenly cut throughout the country. In effect,

Yorkshire was asked to close up the gap between its standard and the technically somewhat higher standard of the National State Scholarships, and I had fallen headlong through the gap.

This was one of the episodes that have convinced me that blows in life, even hard blows—so long as they are not crushing—can be turned to advantage by suitably reorganizing oneself. Momentary episodes of embarrassment in personal relationships have something of the same quality, when just the right words can turn a social disaster into a triumph. The classic example is that of Samuel Johnson at a dinner with many notables around him, male and female. Deep in conversation, Johnson unwittingly took a very hot potato into his mouth. Immediately and instinctively, he spat it openly and noisily back onto his plate. To the startled aristocracy around him, he said: "A fool would have swallowed that."

I was too angry to notice an almost pleasing symmetry about my career at Bingley Grammar. I had won my scholarship there in curious circumstances, and I had now been excluded from a leaving scholarship in something like equally unusual circumstances. My overdraft from the Bank of Good Luck, which we all hope to find around the next corner in life, had been repaid.

Chapter 7

First Journeys to Cambridge

O F T H E T H R E E of us who had worked together, only Edward
Foster escaped the Geddes Axe, so it was he who traveled to London
in September 1932 to study physics at Imperial College, leaving Fred
Jackson and me to return disconsolately to school. I told Alan Smailes,
the headmaster, that I was unhappy with the situation. I felt that simply
grinding again over the previous year's work might lead to boredom, in
which case there would be no guarantee of an improved performance.
Smailes took my point, and within a short time he handed us a packet
containing the question papers in chemistry, physics, and mathematics for
the Cambridge scholarship examination that the St. John's College group
held each year in December. A glance showed them to be in a different
league from anything I'd seen before. So far from being able to answer
the questions, I didn't even know what most of them meant.

Fred Jackson decided to join in an assault on the same objective, although from September to December 1932 was a pitifully short time to
make the upward leap to Cambridge standards. But it is the great advantage of youth that such things do not seem impossible. Fred Jackson's position was fortunately different from mine in the important respect that
the financial position of his parents was better. Fred's father was manager
of a branch of Barclays Bank, and he was indeed soon to become manager
of the main local headquarters in Bradford. At that time, the cost of a
Cambridge education was around £200 per annum, and this was not a
sum beyond the bounds of possibility for Mr. Jackson—especially, I am
sure, as Alan Smailes was always seeking to persuade him that Fred
should go to Cambridge. Yorkshiremen are canny folk, however, and Mr.
Jackson wanted to be assured that the extra cost would be worthwhile,
and the proof of this was to be in the Cambridge scholarship. Fred's task
was therefore to satisfy his father by giving a satisfactory account of himself in the December examination.

It was now at the height of the depression that our chemistry master,

Herbert Haigh, bought us books out of his own pocket—and what tremendous books they were. For physical chemistry, we had the classic text of G. N. Lewis. Not that I understood Lewis in any strict sense, but I soon had a general qualitative picture of the various types of chemical bonds. For organic chemistry, our project was little short of fantastic. Lacking substances of appreciable complexity, we began with just two organic compounds, methylated spirits and (I think) benzene. Starting from them, the aim was to synthesize chains of substances, building one on another, rather as one proceeds from proposition to proposition in Euclidean geometry. It was a valid glimpse of the method of operation of the modern chemical industry.

We dug around separately for clues to the answers to the Cambridge questions, pooling our knowledge as we acquired it. Frequently, the questions would run us into a brick wall. I still remember struggling vainly to understand the operation of a diffraction grating. At length, I asked our physics master, E. A. Kaye, if he could explain. The answer was short: "No, I can't, but I can tell you it's difficult." So, far from avoiding the question, Kaye's reply was really a good one. He had taken a first-class (an A grade) himself at Cambridge, in part I of the natural science tripos. To understand anything of the correct operation of a diffraction grating demands a standard higher even than that. Indeed, the question I had been trying to answer had no business at all to appear in the scholarship examination.

Mathematics was the worst problem, however. Unless one is gifted with exceptional talent, mathematics is difficult to learn alone. Nevertheless, the school had an excellent mathematics teacher. Alan Smailes had been a student at Emmanuel College, where he had taken the more advanced second part of the mathematics tripos, including the still higher level known as "Schedule B." Given time, Smailes could have guided us with ease through the Cambridge examination. The trouble was that there was no time. Our own time was short enough, but Smailes himself had little or none. The depths of the Great Depression was not a period when the education authorities supplied headmasters with secretaries and typists. Smailes had to do every scrap of administration work himself. His days were crammed with minutiae, and the real ability of the best-trained teacher in our school went a-begging.

In this situation, twice a week after dinner, Smailes would have us round to his own house. Although his instruction was entirely unprepared, he would invariably write out the solutions to problems in an unblemished hand. He always went at a slow, methodical pace—excellent for us—digging into his memory as he went along. But it was scarcely enough for us to cope with the meticulously trained products of Britain's specialist schools. Many years later, I would myself be an examiner of mathematics papers in the same scholarship group of Cambridge colleges. I found that sets of candidates would tackle questions in precisely the same way, even using identical abstract symbols in their work. For a while, I thought there must have been wholesale copying between them. Then I noticed they all belonged to the same school, and had obviously been drilled together like recruits on a parade ground. Against this kind of competition, our position in 1932 was extremely weak, but it was a weakness of a somewhat unusual kind. Some things we knew well. But against the things we knew well there were great gaping holes in our knowledge. If we were lucky, we might find two or three questions out of ten on a Cambridge mathematics paper that could be summarily dispatched, but the rest would be quite beyond our comprehension.

So the day came, in early December 1932, when Fred Jackson and I began our first journey to Cambridge from Bingley Railway Station. In those days, you left the local shuttle train at Shipley, where you picked up a connection to Doncaster. There was another change for Peterborough, another for March and Ely, and then a final unhurried link into Cambridge. Fred Jackson and I crowded to the window of the train for our first sight of Cambridge—high spires on the distant horizon. The buildings must have been King's Chapel, John's Chapel, and the Catholic church. A few minutes later, we discovered that Cambridge Railway Station is a mile and a half from the town center, a fact made singular by the unique peculiarity that the "center" of Cambridge lies on its circumference. The bus fare changed at Emmanuel College, and, because I was going to Emmanuel, I paid a penny, whereas Fred, who was going to John's, paid a penny ha'penny, much to his chagrin.

The Cambridge colleges were mostly empty of their undergraduates during the first week in December, and Emmanuel seemed deserted and forbidding in the fading light of a winter afternoon. It was a curious quirk of psychology that later, when I spent two of my three undergraduate years living in rooms in "Emma," I could never bring back to memory exactly which rooms I was given on that first occasion. I remember being thoroughly homesick, for this was the first occasion, in my seventeen and a half years, that I had ever been alone and away from home. I also remember eating dinner in hall for the first time. There was much to amaze me, not least the manner in which some of the regular students who were still in residence made their way in and out of their seats. The tables were wooden, broad and of great length, some arranged parallel and close to the walls. Those who were on the wall side of such tables appeared to be thoroughly wedged in. But not so. At the end of the meal, they stood up on their seats, hopped up on to the table, and simply marched its length until they could jump down. It seemed that the better-mannered did their best to avoid kicking the plates of those who were still eating.

Such, then, was my background in early December 1932, as I ate dinner that first night at Emmanuel College. I knew nothing then of the parade-ground drilling provided by schools that prided themselves on the achievements of their pupils, schools competing fiercely for the reputation of winning most scholarships at Oxford and Cambridge. Yet I sensed it even on that first night. I listened to the conversation around me, and the gulf between my homespun knowledge and the best from great schools of long experience and tradition was far too obvious to be missed.

It might therefore be supposed that I went along to the first paper—in the hall at John's, I think—in great apprehension. Yet I was entirely relaxed. This Cambridge expedition had not been my idea in the first place. My idea was to become a chemist at Leeds University, and the real problem was to raise my standard to make certain of winning a Yorkshire West Riding scholarship in the summer of 1933. If a miracle happened and I won something in Cambridge, well and good. I would be glad to accept it, but my real aim, as I set about answering the first of a number of question papers, was to prove to myself that the efforts of the past three months had really improved my standards.

In this spirit, I had a considerable advantage over well-drilled candidates, especially over those who were under pressure to do well. And Cambridge examiners were no easy pushovers for those who only repeated what they had been taught. Every effort was made, as I discovered in later years, to award scholarships to pupils, not to schoolmasters. This was the reason why Cambridge entrance papers contained so many unexpected (and sometimes hopelessly awkward) questions. The aim was to get us onto questions that we hadn't seen before so that the examiners could separate spontaneous ability from repetitive efficiency. But, of course, schoolmasters sometimes became experts themselves at guessing what unbeaten tracks the examiners would choose to follow. Then the unwary examiner could come to believe that he had a genius on his hands. The distinguished mathematician A. S. Besicovitch always believed himself to be discovering genius, and, if you happened to be one of Bessi's fellow examiners, you risked summary decapitation if you so much as dared suggest that it might not be so.

The chemistry paper fell extremely well for me. The questions in physical chemistry were such that I could throw G. N. Lewis again and again at the examiners, without acknowledgments, I must confess. The physics paper was not quite so good, but tolerable, and the mathematics was just as I had expected. Two or three questions out of ten I could do quickly and well, and then, for the rest of the time, I tried vainly to guess what the others meant.

There were sessions in practical chemistry and practical physics, the latter held in the old Cavendish Laboratory, and the chemistry in the old Pembroke Street Laboratory. Coming from the tiny cubbyholes at Bingley School, these places seemed vast beyond belief. The boot was on the other foot, now, and with a vengeance. Those who had been carefully trained seemed to know their way around those big laboratories. I knew nothing of laboratory conventions, nothing of where I might expect to find reagents or the simplest equipment. Whenever anything had to be queued for, I always seemed to be the last in line for it.

There was, moreover, a fundamental difficulty. In the theoretical papers, the examiners had purposely left their questions a little vague, to give some scope for originality. Practical examinations cannot be left vague, yet the precise specification of what one is expected to do involves technical jargon, and jargon involves conventions. Because I had not been carefully trained in the conventions, there were inevitable uncertainties in my mind, semantic uncertainties about exactly what I was expected to do. This difficulty was compounded in the physics experimental tests by certain of the instructions being given orally, so that I had nothing written down that I could read through several times to clarify the meanings of words. Although I didn't realize it then, the same lack of training in laboratory procedures, and especially a lack of training in keeping laboratory notebooks, had almost certainly cost me a West Riding scholarship the previous summer. We had performed many quite difficult experiments in our cubbyholes, but our style would have been more appropriate to the old alchemists than to these well-scrubbed Cambridge laboratories.

It was while I had something or other "cooking" in the chemistry practical that a hand tugged at my sleeve and a voice said something in my ear about a chemistry scholarship. I knew nothing about a *chemistry* scholarship. All I knew was that I was seeking what Cambridge called a natural science scholarship. I realized that the hand and voice belonged to a messenger, not to one of the university scientists—you could tell from the smarter suit and the voice. So it would be pointless to ask *what* chemistry scholarship. I simply listened, to learn that I was required to present myself for an oral examination at such and such a place and time.

I noticed the messenger going up to one or two others. Among them was another candidate from Emmanuel, a man called Leben. He was taller than I was, heavier in the shoulders, and he walked with his head thrust a little forward. I had noticed him because, unlike most of the others, he had been quite silent at dinner in the evenings. Of course there were candidates in a wide variety of subjects. Those of us who were "up" for chemistry, physics, and mathematics were, I suppose, about a quarter of the total. There was much crosstalk over the dinner table—not on the first night, but as the examinations proceeded. If all the talk was to be believed, Emmanuel would soon be awarding a powerful lot of scholarships. But, of course, this was just a way for the candidates to keep up their morale. If they announced to the world how well they had been doing, perhaps some fairy godmother would hear them and it would come true after all.

When the chemistry practical was thankfully over, I asked Leben if he understood this business about a chemistry scholarship. He told me there was to be an award specifically for chemistry, and it must be that the two of us were short-listed for it. We walked back to Emmanuel together. Later we were to have neighboring rooms on a staircase in the Emmanuel hostel overlooking Drummer Street. I turned up for the chemistry oral at the appointed time. It was in Emmanuel too. I went up the appropriate staircase and knocked at the appropriate numbered room. There was a shout and I went in.

I was not to get that chemistry scholarship, but I can scarcely complain at the competence of the man who now sat facing me as I stood awkwardly before him, since thirty years or so later the man would receive a Nobel Prize in chemistry, something that even he could not guess, because the work for which the prize was to be awarded still lay far in the future. Nor could I guess that the day would come when I would be Ron Norrish's guest when he was president of one of the most famous of London clubs. The Savage Club does not provide its president with a toy of a gavel. It furnishes him with a massive knobkerrie that could smash your skull as easily as a hen's egg. My overwhelming memory of Norrish is of the smile on his face as he pounded furiously with that lethal instrument in prelude to the speech I was to make. The thunder of it still rings in my ears, quite excluding all efforts to recall just what happened during the oral examination back in 1932. I could never get Norrish himself to remember the occasion, and I myself remember only a single odd detail. It had never occurred to me that the word "halogen" could be pronounced with anything but a short "a." Norrish pronounced it hale-ogen, astonishing me mightily, although I would not wish to claim that it was my bemusement on this matter that cost me the chemistry scholarship.

Cambridge did not keep either its own undergraduates or its scholarship candidates on tenterhooks for very long. Within not much more than a week of my return home to Bingley, I had a letter from Emmanuel giving highly detailed marks of my performances. In total, I was a mere handful below what was called "exhibition standard," which was the lowest level

at which Cambridge awarded money. My theoretical paper in chemistry
had been of scholarship standard, but I had, of course, been pulled down
by the practical test; but, in total, I was still above exhibition standard
for chemistry. Likewise in physics, I was just above exhibition standard.
It was mathematics that pulled the whole thing down below this apparently mythical standard. The few questions in mathematics I had understood and answered had given me solid marks, but they were just too few
to maintain the standard.

Thinking I had lost only £40 a year, which was the extent of an exhibition award and which would not have taken me to Cambridge anyway,
I went to see Alan Smailes and showed him the letter. As he read it
through, he came as near swearing in real anger as I had ever observed
him. Only then did he explain what that standard would have meant, and
what those few missed marks had cost me. The point was not the £40.
The point was that *if* I had obtained the standard, and *if* I should obtain
a West Riding scholarship in the coming summer, then Yorkshire would
have met all my Cambridge expenses. Without the mythical standard,
they would not do so. Like the axing of scholarships the previous summer,
it was another of fate's minor digs in the ribs.

I made my way to the public library in the center of Bingley and found
there the copy of *The Times* with the "John's group" scholarship results
in it. As my eye traveled down the list of awards, I saw that Leben had
won the special chemistry scholarship, and in that moment I found that
I envied him. It was with this momentary surge of emotion that, for the
first time, I realized that I wanted to go to Cambridge.

Left to myself, I would have been beaten, but Alan Smailes was not.
He soon discovered that Pembroke College held its scholarship examinations in March. The situation could be recovered, he told me, if I were
to attain the standard there. So the pressure was on. I knew the stake now,
and I had three months more to force the position.

Oddly, I remember very little of that second attempt. Nothing of the
train journey, which I made alone, since Fred Jackson's father had been
happy to accept his entry to St. John's. I remember nothing of the dinners,
only of the room I was allotted—not of its interior, but of a climb up a
long flight of stone stairs to reach it. I knew nothing then of the great
ghosts who walk the cloisters of Pembroke—of John Couch Adams from
Laneast, Cornwall, or George Gabriel Stokes from county Sligo. I do not
even remember the details of the written examination itself, only that my
performance was somewhat less brittle than it had been in December.

I have just one clear memory—of another oral examination to which
I was called, this time in physics. I do not know if I was called because I
was a contender for some award. If so, I didn't get it. The examiner was

Philip Dee. Later I discovered that Dee had something of a reputation for giving students a hard time, and he quite certainly gave me a hard time. When I was seeking to answer a question on absolute temperatures, he got me in something of a tangle and then exclaimed: "Is there *any* physics you *do* know?"

This was the sort of remark that gets me fighting mad, and instantly I dumped my deferential attitude and waded into Dee just as hard as I could. As with Norrish, I was never able later to get him to remember the occasion. He had no recall of the broad-spoken candidate from Yorkshire who had sparked back at him. I had been but a momentary irritation, an unwelcome interruption from some more important activity: "Glad to have finished with these damned scholarship candidates," was a remark I was to hear often at college dinners in later years.

In spite of this encounter with Philip Dee, I traveled back to Yorkshire with cautious optimism. There was no letter from Pembroke setting out my results in detail as Emmanuel had done. There was simply a short communication, regretting that I had not been given an award. Smailes wrote off in great anxiety about the elusive standard. Soon there was a further reply, saying that it gave the tutor at Pembroke pleasure to report that I had indeed attained the exhibition standard. This can happen even though there is no actual award, because there are often more candidates above the standard than there are awards available. Yet, since I never saw the actual marks, I do not know to this day what the situation really was. A kind-hearted tutor might quite well have given Smailes the answer he so obviously wanted, since it cost the college no money to do so. If so, the kind-hearted tutor made a profound difference to my life.

But all this would be a pointless charade unless I should win the West Riding scholarship, which, by now, Smailes and my parents seemed to be taking for granted. The Higher Certificate examinations started in late June, when circumstances fell out that my parents and young sister had a chance of a holiday in Cornwall. But the period of the holiday would overlap with the two-week duration of the examinations, and my mother would not hear of being away at such a critical time. I made strenuous protestations that the opportunity should not be missed—holidays that could be afforded, as this one could, were scarce in our family. And, if the truth were told, I was not unhappy at the thought of our small house being quiet when I needed it to be quiet. Still more important, there would be no one to offer consolation and advice if I should happen to hit a bad paper. Too much solicitude leads inevitably to a crack-up, with worse to follow.

I had no bad papers that summer, and I obtained the Yorkshire scholarship by a safe margin. Yet fate still could not quite let me off its hook.

With six months' further experience after reaching scholarship standard in chemistry at Cambridge, I might have been expected to run away with the chemistry paper, and this is exactly what I thought I had done. I did not seem to find a difficulty anywhere, and the questions with definite answers had all been verified later—the answers I had written on the margin of the printed question papers were correct. Yet my marks were completely the reverse of the Cambridge results. In mathematics, my percentage was up in the eighties; in physics, in the upper seventies; and in chemistry, down just below the seventy level. From 1930 to 1933, I sat four chemistry examinations set by the Northern Matriculation Board, and each time my result was about fifteen percentage points lower than I expected it to be. Once or perhaps twice I might have misjudged the situation, but not, I think, four times. There had to be something wrong for the Northern Board examiners about my style, and something right about it for the Cambridge examiners. The Northern examiners apparently did not want G. N. Lewis—evidently, they had little use for differences between chemical bondings. The year being 1933, only eight years after the beginnings of modern quantum theory, and hence of modern chemistry, such matters were probably too new. What the devil the Northern Board really did want I never discovered, for nobody could tell me. Perhaps it was carefully drawn, ruled diagrams, instead of my freely and rather badly rendered sketches. The situation was interesting because of what it revealed about the vicissitudes to which youngsters are exposed, even in an apparently strict science. The vicissitudes of examiners must be greater in history, economics, and English literature, and they could even be important in mathematics—if A. S. Besicovitch happened to be around, although Bessi's vicissitudes were sharply in the upward direction, never downwards.

There is no doubt at all that it was much harder for students dependent on scholarships to gain access to universities in my time than it is today, and that the situation in this respect was particularly severe in the early thirties. But there was now a big compensation. It would be unheard of today for a student, completing his financing as late as September, to enter a Cambridge college less than a month later. Nowadays, entry would inevitably be delayed a further year. But the pressure on the colleges was less in 1933, and now my marks at the December examination stood strongly to my advantage. Emmanuel wanted me because, for a "commoner" rather than a "scholar," my standard was quite good. Besides which, I am sure the letters written by Alan Smailes to his old college had maintained a stubborn vagueness about my prospective finances, to such an extent that the college had been persuaded to keep a place open for me.

So it came about that, at the beginning of October 1933, I journeyed again to Cambridge, again by train, changing at Shipley, Doncaster, etc., and again in the company of Fred Jackson. We parted company at Cambridge Railway Station. In the months ahead, we were to see each other only occasionally, for the logistic reason that our "digs" were widely separated on opposite sides of town. His were situated in a maze of streets to the north of Chesterton Road, and mine were far south, close to where the railway bridge crosses Mill Road. There was no thought at all of our hiring taxis at the station, as I was to do so often in later years. Fred took the bus, paying twopence, I suppose, to Chesterton Road, while I humped my heavy suitcase the half mile into Mill Road. The route was unfamiliar to me, and I remember twice asking the way.

One of the first things to be done was to have a personal interview with one's college tutor. This does not necessarily mean one's teacher or "supervisor." A tutor's job is to look after an undergraduate's worldly well-being, like bailing him out of jail. It is perfectly possible for a science student to have a historian or a classics don as tutor, although, by and large, an attempt is made to minimize such big differences. Thus, although I was a science student, apparently destined to work mostly in physics and chemistry, my tutor was a mathematician.

P. W. Wood was a striking and tragic Cambridge figure. He was tallish, quite thin, with a slight stoop. He mostly walked around the room as he talked, likely enough jangling keys in his pocket. He had a big beak of a nose through which he sniffed frequently and very audibly. He wore casual "bags" and a tweed jacket liberally studded with leather patches. These leather patches were a kind of badge of the Cambridge don. You could always distinguish them from the servants, who were more neatly dressed in dark suits.

At the risk of incurring the scorn of mathematicians more profound than I am, I have no hesitation in saying that P. W. was the finest Cambridge geometer of his generation. The trouble lay in his style of geometry. Nowadays, mathematicians use an algebraic style, just as Wood did, but, in Cambridge up to 1939, the preferred style was different: almost everything was argued in words, which was really a negation of mathematics, for one of the most powerful weapons of mathematics lies in its notations, and the preferred style of Cambridge geometry (whose protagonists referred to themselves as "pure" geometers) made little or no use of notations.

P. W. Wood, on the other hand, was an outstanding master of notation, a breathtaking wizard at it. He would begin by writing a seemingly straightforward equation on the blackboard. Now came the tour of the lecture room, talking the whole while with head down and keys jangling.

When the tour reached the blackboard again, he would add some apparently simple superscript or subscript to the symbols in his equation, which he did casually as he passed by. The tour would be repeated, with a further casual addition to the equation at the second passage by the board. And now, at the third tour, the real sniffing would begin, so that you knew the dénouement to be fast approaching. On reaching the blackboard yet again, he would delicately adjust his subscripts and superscripts, the notation, and—lo and behold—the proof of some startling proposition, which you might have puzzled about all morning without being able to prove, was suddenly before your eyes. It was characteristic of P. W.'s marvelous sleight of hand that you were always convinced, at the moment of revelation, that you could perform the same rabbit-out-of-a-hat trick for yourself, but, when you tried a couple of hours later, you couldn't. Nor, I think, could P. W.'s fellow geometers, and this was exactly why they condemned his methods.

The Cambridge fad for "pure" geometry—or "projective" geometry, as it is more properly called—had been given impetus earlier in the century by H. F. Baker, whose several unreadable volumes on the subject I still keep on my shelves to remind me of how far a fad can go. There is a damaging modern fad abroad in physics and engineering that goes by the name of the S. I. System. A book entitled *S. I. Units, a System for Fools*, which I wrote some years ago, still lacks a publisher. Indeed, whenever I mention it, publishers always show me the door. This too was poor P. W.'s position. Nobody would take him seriously, except a yearly sprinkling of discerning students. Wood had really only himself to blame, I suppose. He was at the height of his powers at just the time Einstein published his general theory of relativity. Instead of continuing to fight a worthless battle with H. F. Baker, he should have transferred his interests to Riemannian geometry and general relativity. If he had done so, his impact would quite likely have been sensational. It was similar unfortunate judgments, allied to a sharp tongue, that caused him to be passed over several times for the mastership of Emmanuel.

My last memories of Wood are from the autumn of 1945. I had just been made a junior lecturer at Cambridge, and my first task was to give a course of lectures in geometry to science students. The number of students was so large that it was necessary for the course to be given in parallel by two lecturers (in different rooms, of course). The other lecturer was P. W. I had been away from Cambridge from 1940 until the early summer of 1945 and so had not seen him for some time. He was on the verge of retirement now, looking older than he really was, more stooping and thinner than before. I walked into the lecturers' common room to find him already there, walking around in the same characteristic pose, head down, keys jangling. As I came in he looked up with the wry twist to the corners of his lips that was the nearest I ever saw P. W. come to a smile. Eyeing me ironically, he said: "Well, Hoyle, we meet again under changed circumstances."

Indeed, it was then a far cry back to the day, almost exactly twelve years before, when I first called on P. W. Wood. After turning in at the front gate of Emmanuel, you took the staircase immediately on the left. P. W.'s rooms were one flight up. I knocked, waited for the shout, and then went in. P. W. didn't waste any words. From a file, he took a sheet that showed my marks from the December scholarship exam. He studied it with a loud, disconcerting sniff. I waited in silence until the process was complete. At length he said: "Not good in mathematics. Not good enough for a real scientist." I was tempted to blurt out that my recent Higher Certificate performance was better, but then I reflected that Alan Smailes must certainly have written about the Higher Certificate results. So I asked what would he, P. W., do about it? "You might consider Mathematics, Part I," he replied.

In a few tersely worded sentences, he told me that the first part of the mathematical tripos was a one-year course, and this would leave me favorably placed with two years to devote to natural sciences part II, instead of the usual one year, which was much too short for effective specialization in either physics or chemistry. He told me that, in any case, I must already have covered a fair portion of the first-year chemistry and physics, and that, if I opted immediately for science, I would spend most of my time studying subjects I didn't really wish to do, such as botany or geology. "Go away and think about it," he concluded.

Two days later, I returned and told P. W. I would accept his advice, whereon he sniffed and said that it wasn't advice he offered, only the facts of Cambridge life. Neither of us thought it a particularly important decision. For me, however, it was to be a watershed. It was to lead, not to a leisurely two years of chemistry or physics, as we both thought then, but to the full rigors of the later parts of the mathematical tripos. I was either too late or too incapable ever to become a creative mathematician, but the training I would receive in the next three years would have one inestimable value. When I eventually returned to science, it would be with a thoroughly sound mathematical technique. It was the real beginning.

Chapter 8

Then the Undergraduate Seeking a Bubble Reputation in the Cannon's Mouth

MATHEMATICS HAD always been the jewel in the Cambridge academic crown. In the good old days, you had to acquit yourself first in mathematics before you were permitted to proceed to other subjects of study. In 1933, students in mathematics constituted about 10 percent of the whole university, with an intake of about 150 new faces each year, divided into two roughly equal streams—a fast stream and a slow stream, taking different examinations and attending different lectures. In effect, the slow stream was one year behind the fast stream, with the fast stream made up, about half-and-half, of exceptionally gifted students of about my age, for whom Cambridge was their first university, and a considerably older group who had already completed a first-class degree at some other university, British and Commonwealth, and who then came to Cambridge for a final topping-up. In particular, there had always been a strong intake from Scottish universities. Because they had already received good training—not quite up to the Cambridge level, but not too far short of the first two years' work—these advanced students from Scotland always scored very well in the mathematical tripos.

Even this powerful contingent of older students from other universities did not represent the ultimate pinnacle of the competition, however. At the very summit were dazzlingly able students who showed their ability immediately, without needing to pass through another university. These were the ones who appeared at the top of the Cambridge scholarships. Mostly the bigger colleges, Trinity especially, grabbed these nuggets of gold straight from school. Such a one was a friend of later years, who was starting his last year as an undergraduate at the time, in 1933, when I was just beginning. With respect to the problems that constituted a mathematics examination, it was said of Maurice Pryce that, every time he looked at the question paper, you knew he'd done another one, which was surely depressing for the other candidates who happened to be sitting near him.

The most awesome piece of mathematical virtuosity I ever saw was done by Pryce. It happened around 1941, in the following circumstances. As you go further and further away from a radio transmitter, the direct signal from the transmitter weakens. The calculation of how the signal weakens with distance is quite easy, so long as the transmitter remains in view, but the calculation becomes increasingly awkward as you pass more and more over the horizon. Indeed, with the method formerly used, long and tedious calculations were needed to determine the answer for even a single over-the-horizon case. So what, you might ask? Well, in 1940–1941, the Germans had a scheme for guiding their night bombers using intersecting radio beams from two such over-the-horizon transmitters. They destroyed the center of Coventry with it, and fear instantly erupted that the same would happen in every English city. R. V. Jones describes, in his book *Most Secret War*, the ensuing political panic at the highest level of government, with nobody able to say just how far the Germans would be able to go. Jones tells how it became a critical question of whether the signals from the German transmitters would extend as far as Wolverhampton. He blames the British Broadcasting Corporation's engineers who were consulted for giving equivocal and sometimes misleading answers, but this, I think, may not be quite fair, because, before Pryce's solution of the problem, nobody could give reliable answers without long and intensive calculations.

What Pryce set himself to do was to rework the mathematics so that answers to such questions as those that tied the government in knots during the "Battle of the Beams" could be given in only a few minutes. He did so not from a don's pleasant rooms overlooking Great Square in Trinity but from a small, harshly furnished office, shared with two others and with telephones ringing perpetually, in a building in war-torn Portsmouth, where bombing raids occurred pretty well every night. I was working at a country outstation at the time. One day, I received a telephone call from Pryce, telling me what he had done and asking if I would be willing to check the details for him. We arranged to meet, and Maurice handed me about a dozen sheets of white, unlined paper, not quite as large as the student pads we bought from Heffers' shop in Cambridge. Everything was there in remorselessly logical sequence, set out in neat handwriting.

Why is it, one wonders, that the ablest people always seem to be the neatest? A friend who has studied Mozart's manuscripts told me Mozart always wrote the notes down beautifully, even as the music came out of his head for the first time. Except on one occasion, my friend said, when Mozart crossed the waltz with the minuet in the ballroom scene of *Don Giovanni* and things didn't work out on paper quite as he'd expected. So, my friend added: "There were crossings-out, just like the rest of us." Luckily for Maurice Pryce, he never tried to cross a waltz and a minuet, at least not to my knowledge.

I think it is correct to say the staff of the Mathematics Faculty in Cambridge was overwhelmingly the most powerful of any British university. In 1933, there were five professors in the faculty, together with something approaching thirty lecturers, many of the lecturers, in addition to the professors, being fellows of the Royal Society. Cambridge mathematics was a lodestone for the majority of mathematically able students throughout the United Kingdom, and indeed for some far beyond that. If one sets the catchment population yielding the annual intake of 150 students at thirty million, the estimate would, I think, be conservative. It would give an average production rate of one student per 200,000 population per year. For the best ten students, on the other hand, the production rate would be one per three million population per year, while for the outstandingly best, such as Maurice Pryce, it would be one per thirty million population per year. Remembering that the production rate in those days for fellows of the Royal Society was one per four million per year, one could easily have predicted that, except for those who deliberately left the mathematical field—for example, to enter the Civil Service—the best few each year would inevitably have been elected to the Royal Society, a prediction that, generally speaking, was verified in later years.

I could not claim myself to be one of the average mathematics students in the one-in-200,000 class, since I came to mathematics because I was weak in it. In chemistry or physics, I might reasonably be said to have demonstrated this kind of an average standard, but my mathematics in October 1933 was evidently at a statistically lower level. My entry point into Cambridge mathematics was, therefore, at the bottom of the slow stream, and the question in my first year as an undergraduate was whether I could manage to come up to speed with the slow stream. This was the challenge my tutor, P. W. Wood, had set for me.

In later years, I became an examiner myself in the various parts of the mathematical tripos, so I had eventual access to the examiners' record books. Such information is, I suppose, confidential, strictly speaking, but I see no profound objection to saying what my performances were, provided I say nothing about anyone else's. In the examination held at the end of the first year in May 1934, I was three-quarters of the way up the list from the bottom, or a quarter of the way down from the top, according to how you prefer to look at it. In effect, I had come up to speed with the slow stream, and it was time for me now to return to physics and chemistry, as I had planned with my tutor at the beginning of the year.

Inevitably, however, the entire style of my educational history prompted this question: If I were to stay in mathematics, how far, in the next two years, could I catch up with the leaders? To have any chance of doing so, I would evidently need to leap immediately across from the slow stream to the fast stream, which meant jumping a whole year all at once. At the level of Cambridge mathematics, that would be a far greater feat than the jump I'd made at school in my matriculation year. Besides, wasn't physics my real line, not mathematics? Yet could I really return to physics without feeling I'd ducked a challenge? These were the questions that filled my mind over the summer of 1934.

I remarked in an earlier chapter how from Rombald's Moor, which could be seen, it will be recalled, from the windows of my home, it was possible to walk almost the whole way to Edinburgh without passing anything more populated than remote farmhouses and hamlets. In my later years at school, I had begun to explore the vastness of the territory that had lain so long on my doorstep. Now, in the summer of 1934, I penetrated much further into the wild country. Together with Fred Jackson and Edward Foster, I made an extensive walking tour of the North Yorkshire Dales. Then came two tours of the Lake District with Edward Foster. These hill-walking trips provided ideal circumstances in which to mull over the choices before me.

Although we lived economically in youth hostels during our walking trips, the total cost ran to perhaps £10, and the question of where a sum of that magnitude came from naturally arises, especially as there was also excess wear and tear on clothes to be considered. My grant of £225 per year from the West Riding of Yorkshire was nicely tailored to include lecture fees (which were very modest), college fees (which were not quite so modest), food and lodgings, clothes, and subscriptions to those university and college clubs to which an undergraduate might normally expect to belong. The latter, such as the Cambridge Union, were not obligatory, however. Since the Yorkshire authorities simply made a flat grant without calling for detailed accounting, I had a free choice of whether I would plump for the optional extras, so running my finances essentially to zero, or whether I would forgo the extras in the interest of establishing an annual contingency fund of about £25. In October 1933, I chose the latter course, and very glad I eventually became that I had done so. The money for the walking tours came from my contingency fund, while in the university itself there proved to be worthwhile activities I could not have noticed at the beginning, which I would have been precluded from joining had I consumed all my resources at the outset. Shortly after my first visit to the University Chess Club, I was asked to become a member of the university team. Over the winter of 1933–1934, the cost of travel to away matches was about £5. One of the away matches happened to be with the

House of Commons, so permitting me an intimate sight of the Mother of Parliaments I couldn't easily have obtained otherwise; and, at a later date, I joined a fairly vigorous walking club, which was to establish lifelong friendships of considerable significance. Although the expenditures required by these activities were but a few pounds, even such modest sums are decisively prohibitive if you are entirely out of funds. So, although I had to forgo some of the standard attractions of undergraduate life, my fiscal policy proved wise.

As I walked the Yorkshire Dales and the mountains of the Lake District in July and August 1934, the decision I had to make came to seem less difficult than it had seemed in Cambridge. A decision in October 1934 to join the fast stream in mathematics would not be irreversible, because, up to Easter 1935, I could always go back to the slow stream. What was irreversible was a decision between mathematics and physics. If by mathematics one had meant mathematics in a European or American sense, my choice would undoubtedly have been physics. But mathematics in Cambridge included what elsewhere was called theoretical physics. In essence, therefore, the choice lay between theoretical and experimental physics. If I stayed with mathematics, I could evolve eventually into theoretical physics, but, if I wished to become an experimental physicist, I must return to the laboratories without delay. Which way did I wish to go? My choice, I came to feel, was decided by history, by the fact that probably the major fraction of outstanding Cambridge physicists had evolved through mathematics—Newton, of course, Maxwell, Kelvin, Thompson, Cockroft, Eddington, and Dirac. This seemed decisive, so mathematics it was.

With the principle of the matter thus settled, I argued, on my return to Cambridge for my second undergraduate year in October 1934, that the situation was not too much different from the previous year. I was now at the bottom of the fast stream instead of the bottom of the slow stream. If I could work my position, by the time of the next examination in May 1935, three-quarters of the way up from the bottom, or a quarter of the way down from the top, according to the way you care to look at it, I could reckon I had come satisfactorily up to speed. This, then, was my objective for the second year.

In the event, I did not succeed. In the examination of May 1935 I came three-fifths of the way up the fast stream. This, however, was the examination in which I had the bad migraine headache I mentioned earlier. It occurred on the very last paper, just as I was beginning to feel I'd attained my objective. Just as with the attack of mumps on the occasion of my junior scholarship, I felt there was no point in making a public issue of the matter. Surely but for the migraine, I told myself, I would have scored better on that last paper, and so, psychologically if not objectively, I counted my real position a little higher, perhaps even approaching the objective I'd set. At all events, I could plainly see now the leaders ahead of me on the running track. For the first time I had a really clear aim—to spurt in my final year into the top ten, because, of course, it was the top ten who would receive further grants to continue in research, if they so wished.

Through June and July 1935, I gave attention to the kind of work I would do in the final year. It was on a bigger scale than before, more like playing a whole sonata instead of concentrating on student exercises. I became generally hopeful that the bigger scale would suit me, and that the individualistic aspects of my early education might now at last become an advantage.

Together with Edward Foster, I took off in August on another walking tour. Edward had just finished his normal three undergraduate years, and he was mighty pleased to have secured a grant that permitted him to continue further at Imperial College in, as I recall, vacuum spectroscopy. I hoped the next year I would be able to say the same. Instead of confining ourselves to week-long tours, we were six weeks in the Scottish Highlands. At the end, I was considerably fitter and stronger than I'd ever been before, and it was in this condition that I journeyed to Cambridge in early October 1935.

Lectures were attended now by research students as well as undergraduates. Imperceptibly, as time went on, I came more into contact with research students than with my fellow undergraduates. My joining the walking club led to several firm and lasting friendships, particularly with several fellow members who were research students. Through them, especially through George Carson, a biologist, and Charles (E. T.) Goodwin, a theoretical physicist, I came to know still more of the research students. Charles Goodwin had come through the mathematical tripos himself, just as I was doing, and so there was much to talk about.

The rambling club went far outside Cambridge each Sunday, usually some thirty miles in a hired bus from Drummer Street, immediately behind my rooms in the Emmanuel College hostel. We brought our own sandwiches and would eat them in a village pub somewhere in the wilds. The main business of the day was then to walk for perhaps three hours along small paths and often across ploughed fields thick with soft clay. To obviate arguments over the route, a "leader" chosen in advance was supposed to have been over the route previously to ensure that we didn't end up in the wrong place on a short winter afternoon, the "right place" being the spot where our hired bus had been instructed to meet us. The right place was also some pub, inn, or café, where an excellent tea was to be had by one and all. After returning to Cambridge, we took a bath and ate dinner individually, meeting afterwards for coffee in the rooms of the club's founder, Harry Marshall. Harry was the only one of my Cambridge friends to die in the Second World War. He entered the Foreign Service and was sent to Malaya in 1937. I was told he was upstate at the time of the Japanese invasion from the north in early 1942. Although not trained as a soldier, he volunteered for a holding operation and was not seen again.

As the year 1935–1936 progressed, I never lost the fitness I'd acquired in the Scottish Highlands. Indeed, from Easter onwards, instead of retreating into my books, as in former years, the amount of outdoor activity increased still more. As well as the Sunday walks, I would often go canoeing on weekdays with Charles Goodwin and George Carson. We were emphatic in preferring a heavy wooden affair, which we hired on a long-term basis from a boathouse just below Magdalene Bridge, to the light-weight jobs that rode higher and higher as they picked up speed, making it impossible to get what we considered a satisfactory bite from the paddles. The paddles were one-sided, with handles. Eventually, there became only one possible arrangement—Charles Goodwin in the front on the left; I in the middle, paddling on the right; and George Carson in the guiding spot at the rear, paddling on either side, according to the expediencies of the moment. You grasped the crosspiece at the end of the paddle with one hand and delivered the stroke with that hand and corresponding shoulder, the other hand beginning each stroke firmly, holding the paddle shaft immediately above the blade. At the last moment, you had to whip your lower thumb away from the shaft to prevent its being crushed against the side of the boat. The problem was to get our timing coordinated, with the strongest strokes we could manage, and without ruining the lower thumb and fingers. From that time in 1936 through two consecutive summers, until Charles Goodwin left Cambridge, we prided ourselves on being the fastest craft afloat on the river Cam. I can't be certain we were, but I can say that nobody ever beat us in a challenge. Our time from Byron's Pool to the Garden House Hotel became incredibly short.

This might seem a strange way to prepare for the most important examination of one's life. But, as the year had progressed, I had become steadily more confident. The curious aspect of it was that, in my very last examination of all, I had begun to appreciate how it would have felt to have been a notably gifted child, if only it had fallen out that way earlier in my career. What had actually happened by May 1936 was that I'd once again exercised my penchant for jumping. My acquaintances and friends were all now research students, so that, so long as pride did not go before a fall in the examination itself, I had jumped entirely out of my own year.

Lectures for the final part of the mathematical tripos had finished now, but there were still special courses organized for research students. Together with Charles Goodwin, I attended two of them, one given by Rudolf Peierls and the other by Max Born, both on quantum field theory. Born lectured on Monday, the day after the walks of the Sunday rambling club. One Sunday, a dozen or so of us had decided, on a whim, to walk forty miles. We started early from Cambridge. Instead of taking a bus in the usual way, we walked from the city center out along the Roman Road as far as Haverhill, where we had lunch. In the afternoon, my rucksack seemed unduly heavy, until I discovered it to contain stones, put there, as I recall, by a friend who was a printer's son from York, Bill Sessions. We joined the rest of the club for tea with the best part of thirty miles behind us. After tea we did the last ten miles back to Cambridge, with Charles Goodwin and me accelerating over the last three miles because we didn't want to miss a performance on the radio that night of Beethoven's *Choral Symphony*. We just had time for a bath at Charles's digs in Station Road before the music came on. Then, after a meal with Charles, I stumbled in a thoroughly stiff condition from Station Road back to Emmanuel College.

In these days of widespread media coverage, we hear of lots of people walking fantastically long distances. Such people must be exceptional. There are very few indeed who can walk long distances without ill effects, and these few are mostly small or thin people with comparatively light impact forces on the feet. We were quite fit young people at the prime of life, but I think only one of us, a young woman, suffered no noticeable ill effects from our forty-mile walk. Bill Sessions was a big man, very springy on his feet, and in winter he was the goalkeeper for a well-known amateur soccer club. Bill suffered badly in the final stages of that walk, and it was some revenge for the stones in my rucksack when, in the final miles, I could hear him muttering, "What does a ghost care about its feet?"

The following day, I had quite seized up, and, because of the slowness of my walk, I was late at Max Born's lecture. He had not started yet, but he was just on the point of doing so when I appeared. Let me draw the picture: My academic gown I had bought secondhand for thirty pence at the beginning of my undergraduate life in 1933, and, although it did not have the savagely ripped appearance of the gowns of the young bloods about town, it had very definitely seen better days. In the matter of shirts, this was about the time when several of us found that Selfridge's in Oxford Street had a line of them going at twopence ha'penny each. We bought a pile with the idea of giving each one a thoroughly good wear, and then simply chucking them out to minimize laundry expenses. It was the idea of the disposable razor, but, in 1936, it was an idea before its time.

Shoes were an impossibility for me that day—I was reduced to wearing a hastily borrowed ancient pair of gym shoes, one of them without a lace. In addition to all this, I couldn't walk properly. The best I could do was to sort of drag myself across the small lecture room, immediately past the lectern where Max Born was standing. Afterwards, Maurice Blackman, later a professor of physics at Imperial College, said with a big grin: "It wasn't like that for Born in Germany. He was as near as dammit to hoofing you out." This was as near as I ever came to exchanging a word with Max Born.

The two weeks before the examination, I spent the mornings working and the afternoons on the river. The week of the mathematical tripos was exceedingly hot in 1936. The heat was an advantage, however. Being exceptionally fit now, I felt no oppression or heaviness in the head, as I would have done the previous year, and the warmth allowed me to write quickly for long periods. After the two papers on the first day, I joined George Carson for tea in one of the town cafés. To his inquiry, I committed myself as far as to say that I thought I'd done reasonably good papers.

The final undergraduate year was different from the previous two in the important respect that one expected to receive the B.A. at the Degree Day ceremony held in the fourth week of June. This left a three-week period between the end of examinations and the ceremony itself. Not wanting to kick our heels around Cambridge over these three weeks, a party of us decided on a hitchhiking trip to Cornwall. Hitchhiking was a new idea at that time, and it had been imported into Cambridge from the United States by Ivan Inksetter, a very tall, thin fellow.

Inksetter's first name was not Ivan. He came by his pseudonym through a misguided attempt to row in the low echelons of the Lady Margaret Boat Club. The story was that one day the Captain of Boats caught sight of Inksetter at work. Stopping dead in his tracks, the Captain exclaimed, "God, who's that, Ivan the Terrible?" So Inksetter became Ivan throughout Cambridge. I never knew his real first name. He was a student in English, and so he had natural entry to the Marlowe Society, the undergraduate theatrical group. Probably he is the only member of the Marlowe Society ever to have caused a sensation from the stage without speaking a word—as the corpse in an Aristophanes comedy.

Before setting out for Cornwall, I went to see Rudolf Peierls in his office in the Mond Laboratory to ask if he would agree to be my supervisor, should I be able to obtain a research grant. Peierls said he would consider the matter. I learned afterwards that he had subsequently inquired around Cambridge to find out who I was, but nobody knew. Particularly, I had not applied for a research studentship from the Department of Scientific and Industrial Research (DSIR). The trouble with those studentships was that you had to apply for an award at Easter time, two months or so before the examination, which seemed an obvious invitation to Providence to hit me with another migraine or with a refurbished Geddes Axe. It was now too late to obtain a DSIR studentship, no matter what the result of the examination might be, since, in those days, there was no late appeals system, such as has been set up in more recent years. If you didn't apply in April, you couldn't get a studentship. My ace in the hole, however, was that the West Riding of Yorkshire had provision for the continuation of their scholarships over a further year, in "exceptional circumstances." A good result in the examination would, I knew, be construed as an exceptional circumstance.

Even this final contretemps, involving the apparently considerable risk that, after my first year of research, I would find myself without support, turned out an advantage. My contemporaries who were awarded DSIR studentships were required, as you might expect from anything connected with bureaucracy or government, to write frequent progress reports. Because I became favored with a more aristocratic style of support, I did not have to follow this procedure. I was able to go as I liked, and, because it seemed that the best way to go was to be on my own, I was able to cut myself free at last. The bureaucracy would never have permitted me thus to stand on my own feet, which, being unusual, was against the rules. To satisfy bureaucracy, you had to be everlastingly subjected to instruction, which is why science today, being almost entirely controlled by bureaucracy, is mostly an unproductive activity.

I had not hitchhiked before. There were many fewer cars on the roads in those days, often almost none on the small roads. But a far higher proportion of motorists would give you a lift than will do so today. Ivan Inksetter and I had finished our examinations ahead of the others, and we set out for Cornwall a week earlier. Inksetter was the original master hitchhiker. He dressed for the part. Very tall and thin, he was wearing shorts, a Tyrolean hat with feathers in it, and a jacket of the same vintage with bells and big silver-looking buttons on its front. Few cars, when we really wanted one, ever passed us by on that trip.

Inksetter was highly skilled at choosing his cars. In poor districts, like the mining districts of Cornwall, he was happy enough to tuck his long legs into a small Ford. But, in wealthy areas, he had a quick eye for the Bentleys and Rolls-Royces, being especially fond of the latter. It was, as I recall, in the region of Ascot on the very first day out of Cambridge that Inksetter landed an exceptional ride. The lady owner of the Rolls-Royce invited us to tea at some large estate, where we were served tiny sandwiches on a silver salver by a butler in tails. Although hungry, since we had forgone lunch, I remember thinking that, as the sandwiches were scarcely worth eating in a nutritive sense, I could afford to be conspicuously polite by eating them very slowly.

The situation was distinctly horsy. The horse is a topic on which I have never at any time excelled, since I have always agreed with the man who said: "I know before it starts that a horse race is going to be won by a horse, and I don't much care which." This view not being popular on that occasion, I had to leave the burden of conversation to Ivan Inksetter. He had a big, resonant bass voice, and I think he really enjoyed the situation, blarneying his way through it. We ended our first day out of Cambridge at a youth hostel in Winchester. Being exceedingly hungry by then, we consumed a mixed grill for 1s 3d in a café, not so politely, with thick slices of bread and butter.

It was my first visit to Devon and Cornwall. The second day, we reached a youth hostel at Dunsford, a country village about halfway between Exeter and Moretonhampstead. The third day, we were in the region of St. Austell. Thereafter, we moved around the Cornish coast. We had no preconceptions about what to expect, and it was with profound surprise and delight that we came to the coastline between Newquay and Padstow, hitting it at just about its finest stretch, at Bedruthan Steps.

Inksetter took great pleasure, until his stomach turned a bit acid, in the genuine rough cider that was easily available at country pubs in those days. The weather was excellent and the days soon slipped by, until it was time for us to forgather with the others at the Boswinger youth hostel, a few miles to the south of the picturesque fishing village of Mevagissey. My memories are of days spent swimming and sunbathing on the local beach, and of several of us following the agile Bill Sessions on low, waterline traverses of mile after mile of rocky coastline. The youth hostel had a notable peculiarity. Each morning, the farmer's wife would ask us whether we wanted ham or beef for supper. Day after day we said beef, and day after day it turned out to be ham. It was like Henry Ford's remark: "You can have it any color you like, so long as it's black."

It was while we were at Boswinger youth hostel that the results for the mathematical tripos were published. George Carson sent me a telegram. I had what was known as "honors with distinction," which meant that I had attained my objective of being in the top ten. What I hadn't quite expected was to receive the Mayhew Prize, given for the best performance on the "applied" mathematics side—which is to say, in theoretical physics. More precisely, I had a half share of the prize, the other half going to Stanley Rushbrooke, later professor of theoretical physics at Newcastle. Only a year or two back, I was out walking the hills on a winter day with Stanley. In later years, we have often laughed at how we damned each other in the summer of 1936, for the prize was worth £25, and losing a half of it to somebody else was no light matter. The examination marks were not published in detail, but the share of the prize had to mean that I had managed to lift myself into the first three or four places on the list.

Not all members of the party at Boswinger were in their third year, needing to return to Cambridge for the Degree Day ceremony. The time came for those of us who were so involved to leave. Ivan Inksetter and I once more paired up for the trip. Although returning to a small personal triumph in Cambridge, I left Cornwall for the heavily populated eastern region of England with regret, an indication that big city life was not for me. The return journey was an anticlimax after the exuberance of our outward journey three weeks before. My one vivid memory is of a very long ride that ended at a transport café near Stamford, where we got a bed for the night. The ride was memorable for the tremendous thunderstorm through which we passed, and for a windscreen wiper that ceased to function.

The following morning was brilliantly fine, and we were in Cambridge by 9:00 A.M. Inksetter made his way to St. John's and I to Emmanuel. In the following years, I had a number of letters from Ivan, but I never saw him again in person after we parted that morning. In retrospect, I realize we have a built-in wrong attitude towards human relationships. While they exist, we think they are forever. If one travels abroad with friends or with one's family, the tendency is to take pictures of unusual inanimate objects—ancient temples or fine mountain scenes. The feeling is that the people around us are eternal and can be ignored, whereas a record must be made of the remarkable inanimate things we see in our lives. But just the opposite is true. The ancient temples are likely to be around for centuries to come, and the mountains for millions of years. It is the people who will soon be gone.

I had a day to spare before Degree Day on the morrow. I checked with Rudolf Peierls that he would accept me as a research student, and I checked with P. W. Wood that a letter would be sent on my behalf to the West Riding of Yorkshire. P. W. told me the master of the college wanted to see me urgently. It was to swear me in officially as a Scholar. After my first year, I had become an Exhibitioner, the old exhibition standard that I had sought to achieve from school. And now, with just one day to spare before graduating, I had at last achieved the scholarship standard. It seemed a long time since my oral interview with Ron Norrish on the occasion of that ill-fated attempt at the chemistry scholarship.

It was even a long way back to the day, in October 1933, when I had arrived in Cambridge a frail young fellow weighing about 115 pounds, knowing little except the few things I had managed mostly to teach myself over the years. As I prepared for the Degree Day ceremony on the following day, I stripped off and bathed. I was now about 150 pounds, quite hard physically after all the walking of the past years and the canoeing of the preceding weeks, and strongly browned by many days just spent in the Cornish sun. I had started on the mathematical tripos somewhere near the bottom of the slow stream, and I had ended my undergraduate career somewhere near the top of the fast stream, almost infinitely higher than I had ever dared to hope.

With this said, I should admit there have been times when I have looked back to my school and university years with the niggling regret that, until possibly my very last undergraduate term, I never displayed anything of the competence of the wonder child. My regret was partially allayed one morning in 1944. There were about a dozen of us in an office within a large, rambling school building that then housed the radar department of the British Admiralty. The purpose of our meeting was a review of the work of all the hundred or so persons employed in producing radar equipment for the Royal Navy. The review was made with a view to recommending upgradings and promotions in categories where they were deemed to have been well earned. There was one particular chap in his early twenties who had shown considerable experimental skill and who, I thought, should be promoted to a category normally reserved by the Civil Service for university graduates, which the young chap wasn't—not, at any rate, at that time. I couldn't see why, with a war on and with the lad's work, in my opinion, of an appropriate standard, he shouldn't receive the promotion. The only other person who supported this view was Maurice Pryce. After the meeting, as we walked together along a corridor, Maurice said to me with his wry smile: "You know, Fred, the only two people who thought a degree wasn't important were the two with the best degrees." My heart gave a bit of kick, for I thought to myself: If Pryce thinks my degree good, it surely has to be.

In later life, Pryce and I got around one day to admitting that each of us had both envied the other. Why I had envied him was obvious. Why he had envied me was that I had somehow managed to reach much the same position as he had, but seemingly out of nowhere. He told me of the psychological problems the exceptionally gifted child has to contend with. Once such a child becomes known outside family and school circles, the pressure never to slip becomes unremitting. If the widening circle of people who watch the wonder child were truly sympathetic, the situation would be fine, like a home football team urged on by its supporters. But there are too many actively hoping for a slip, hoping they will be able to say to themselves: "Ha ha! He's no better than I am."

Ray Lyttleton, who will appear in Chapter 10, once told me that, in the year Bobby Jones won all the major golf championships, Jones found himself obliged to hire bodyguards to prevent people from invading the course, ostensibly to advise him on how he should play his shots. I myself remember a letter in the national press from a reader who wished to inform the world that, while he would travel a hundred miles to watch Don Bradman field, he wouldn't walk ten yards to watch the man bat. The world is well stocked with people like that who feel compelled to degrade excellence of any kind in all manner of subtle and unsubtle ways. Tommy Gold refers to such people as those who "seek to make a mark on history by making a mark on those who have made a mark on history." The more I have seen of the world, the more I have come to think I was very lucky to have been permitted to develop in my own way, setting my own objectives instead of having them decided for me by the roar of the crowd.

Chapter 9

Dark Clouds across the Sun

I RETURNED to Cambridge from Cornwall with a sense of anti-climax. My emotion was real—I know that for sure. The reason I gave, a preference for rural rather than urban areas, seems the most innocent explanation, since to say anything different would be to claim a measure of prescience that appears unlikely. Science in Cambridge in the summer of 1936 was like a stock market riding high before a crash. By 1938–1939, there were more visible causes for disquiet, but it seems rather far-fetched to suppose I could have seen anything untoward in 1936, at the very threshold of my research career.

Let me take the biggest matter first. From 1926 to 1936, physics had enjoyed a golden age. By 1938–1939, however, Paul Dirac was to say: "In 1926, people who were not very good could do important work. Today, people who are very good cannot find important problems to solve." This was the first clear warning I had that the euphoria that was still widespread in the summer of 1936 was misplaced.

Everybody had the idea that the triumphs of the previous decade in atomic physics were going to be repeated in nuclear physics, an illusion that continued to bedevil science worldwide for another three decades. People found it easy to be deluded because of the apparently great success of empirical nuclear physics in the forties and fifties. An empirical approach to a subject is simply a look-see method. Suppose you want to know if salt dissolves in water. An obvious way to decide would be to try it and see. This would be the empirical approach. But, if you understood, in sufficient depth, the basic nature of the atoms that constitute the molecules of salt and water, you could settle the matter by calculation, without bothering with look-see at all. This would be what scientists call a fundamental approach. The success of nuclear physics in producing energy from reactors, and in producing devastating weapons—if you call that a success—came from the look-see method. To many, these successes appeared so striking that the illusion persisted.

The example of salt dissolving in water suggests wrongly that an empirical approach is simpler than a fundamental one. At the outset, it is; but, as time goes on, the empirical approach becomes more and more complex, with epicycles piling on epicycles. The fundamental approach, on the other hand—when at last it succeeds—resolves into an amazing simplicity. The route towards the fundamental resolution of the nuclear problem has not turned out to lie in studying atomic nuclei at all, as scores of laboratories and hundreds of physicists through the forties and fifties believed would be the case. The route began in the study of cosmic rays, and then, from the fifties onwards, in the building of particle accelerators. From experiments with accelerators, it became clear that the particles that make up atomic nuclei, protons and neutrons, are themselves composite. The components came to be called quarks, and, from the mid-sixties, theories of the properties of quarks were developed. Many forms of theory appear at first sight to be possible, although, in the end, one particular theory will almost surely come to seem "deeper" and more elegant than the others, with a future generation thinking the outcome should have been obvious, just as we tend to think the outcome of struggles in physics in the past should have been obvious. Physicists in the nineties have come to feel this eventual outcome is now within sight, and, from their work, a fundamental understanding of atomic nuclei is at last emerging, half a century later than we thought in 1936.

The coping stone of the golden age was not yet in place in 1936, the coping stone known as quantum electrodynamics. We all thought this eventual achievement to be just around the corner. It seemed to be something that might appear in the physics journals at any moment. The lectures I'd attended by Rudolf Peierls in May 1936 gave an up-to-the-moment discussion of where quantum electrodynamics then stood, while the course by Max Born gave his own attempted solution of the ultimate problem. In effect, there seemed to all of us to be a big nugget of gold lying around somewhere, if only we could find it. This, too, was an illusion, but not as distant an illusion as the nuclear problem. Quantum electrodynamics was to resist the assaults of the world's foremost physicists of the time—Dirac, Heisenberg, Bohr, Pauli. When it did eventually fall, in about 1950, it would be to a postwar generation of theoretical physicists younger than myself. It should be said, however, that Dirac, who had done most in the early stages of addressing the problem, was never to accept the validity of the later solution, and that there are considerable grounds for thinking that he may have been correct.

A generation of theoretical physicists is much shorter than a social generation—ten or fifteen years at most. I do not think physicists today would deny that my generation was the unluckiest of the past half century. We were too late to receive anything but crumbs from the rich table of the years around 1926, we were too early for quantum electrodynamics, and very much too early for quarks. Additionally, from 1939 to 1945, we lost six years to wartime, half the life of our generation. A perception of the way things were likely to go came gradually to me over the period from 1937 to 1939—mainly, I think, because, in 1937, I made a decision to stand on my own feet. Others of my generation who continued regardless, sometimes because they were under the close direction of a "supervisor" or research director, were, I think, less fortunate in not making a change of direction soon enough, or not making it at all.

At the time I returned to Cambridge from Cornwall in June 1936, the Cavendish Laboratory was one of the most famous, perhaps *the* most famous, center of experimental physics in the world. There was nothing to tell us that the great Rutherford would die in 1937 or that, so far as fundamental physics was concerned, the Cavendish would soon opt out thereafter. It was to be like a highly prosperous company going into sudden liquidation. All of Rutherford's associates of any appreciable seniority were to leave, to be replaced from outside Cambridge by physicists who were not concerned with the more fundamental aspects of their subject. For form's sake, it has often been claimed that the exodus took place because the individuals in question saw opportunities elsewhere. But a group of men, mostly already well known to the world, would not lightly abandon *the* most famous laboratory. The exodus occurred cooperatively because Rutherford's associates were dismayed at decisions taken over the direction of the laboratory.

The demise of fundamental experimental physics in Cambridge inevitably had repercussions in theoretical physics. After the 1939–1945 war, Cambridge was to produce a number of outstanding theoretical physicists, but every one of them left, never to return. A major reason for this second exodus was a gross efflorescence, in the Faculty of Mathematics, of a subject known as "fluid mechanics" or "continuum mechanics." This subject had been given impetus by the need, during the war, for an understanding of aeronautics, and it continued growing thereafter due to the illusion that Britain would become a major exporter of commercial and military aircraft. It was a case of Gresham's law, the bad driving out the good; or, as John Gerard wrote in the sixteenth century about the spreading of ground-elder or goutweed: "It is so fruitful in its increase, that where it has once taken root, it will hardly be gotten out again, spoiling and getting every yeere more ground, to the annoyance of better herbes."

Although, in 1936, all this was but a distant cloud on the horizon, other, darker clouds were nearing the zenith. The remilitarization of the

Rhineland by Hitler in March 1936 came as a dividing line for many people, myself included, between feeling incredulously that "Surely war can't happen again" and feeling that, inexorably, "It *is* indeed going to happen again." From 1936 to 1939, we oscillated between a despairing hope that war might be avoided and the equally despairing conviction that, the sooner it came, the better. To many people, again including myself, it seemed that the reoccupation of the Rhineland provided Britain and France with a last opportunity for cutting the Nazis down to size without involving the world in full-scale war. Disgust at the weakness of the British government was felt in many quarters. Ray Bell, later a Treasury mandarin and vice-president of the European Investment Bank, was a fellow student at Cambridge. His parents lived in Bradford, and, while on vacation in April 1936, he and I did a longish walk in the environs of Ilkley, just before returning to Cambridge for my last undergraduate term. I was invited to dinner that night at the Bell household, and I still remember the father saying, with intense regret, but prophetically: "If only Winston Churchill were prime minister now."

Philip Dee, the same Dee who, early in 1933, had given me a hard time in the Pembroke scholarship examination, was the most belligerent of the senior Cavendish staff. It must have been a year or so after the Rhineland fiasco that Dee said, at teatime in the Cavendish, "I think it's time for a crack at them," and I remember thinking to myself, "He really means it!" When the war actually came, Dee was the head of a group that produced a novel form of airborne radar, not for detecting a moving target, such as a plane, ship, or submarine, but for obtaining a detailed picture of the ground. It was the father of the kind of radar used in modern times for determining the surface topography of Venus.

At this point, with wider issues now raised, I should perhaps say something of student life in general. The periods 1930–1933, 1933–1936, and 1936–1939 were, I think, quite distinct. Those who passed their undergraduate years in the 1930–1933 period were dominated in their thinking by the social misery of the depression years. This was the period that produced intellectual socialists of such great conviction that they maintained their views unswervingly for a lifetime. By 1933–1936, the worst of the depression had passed, however. There was economic expansion with improving conditions and with lower unemployment. Interest in socialist economics waned. I remember being fiercely concerned with such matters in my later years at school (which is to say, during the depression years themselves), but by 1933–1936, the efforts of undergraduate representatives of the main political parties to canvass my interest were always met with a polite but firm *no*. The situation in Germany had begun to appear more relevant, but not yet grossly threatening. Consequently, my own period, 1933–1936, was one of exuberance, of economic improvement, of expansion—a cause for rejoicing generally.

Students today can hardly imagine what a difference the trappings of Empire made. For one thing, we had a wide spread of incoming students from all over the earth. For another, there were opportunities in the Foreign Service for any competent student who wished to move out into a world much larger than Britain itself. The face of Cambridge itself was subtly different from today, but since all this is excellently documented by easily obtained photographic collections, it would be rather pointless to launch into descriptions of old Guildhall, or the old Festival Theatre on Newmarket Road, which we patronized before the advent of the Arts Theatre. Rather, let me dwell a moment on other issues—clothes, for one. Ours was a woolly, baggy generation—huge baggy trousers, baggy caps, baggy jackets, with lots of air pockets and wool in them because of the cold buildings. We did not attend lectures in centrally heated university buildings but in large, unheated rooms within colleges, which have now been subdivided and have consequently disappeared as lecture rooms. Almost every college had one such lecture room used by the general university population, and some colleges had several.

The tight-fitting clothes made of artificial fibers that are such a visually evident feature of undergraduate life today were impossible for us. Instead of simply mooching from one lecture room to another in the same university building as at present, at the end of each lecture we were thrown in all weathers literally into the street. Lectures were supposed to start five minutes after the hour and to end five minutes before the hour. We used the intervening ten minutes to walk, run, or cycle from one college to another—the longest stretch for me being from St. John's to Peterhouse. Because lecturers often infringed the five-minute rule, our ten-minute intervals could be much reduced, so that I would frequently be obliged either to run or to cadge a lift on somebody's bike step. A bike step was a solid steel extension to the axle of the rear wheel about three inches long on which you stood behind the rider. The trick was for the rider to get started slowly, and for you to run behind and lift your foot up onto the metal projection more or less in midstride, steadying yourself by putting a hand on the rider's shoulder. Under favorable conditions, it wasn't a hard maneuver, but, in wet weather in a crowded Cambridge street, it needed a bit of practice. There were few motor vehicles in those days, apart from the buses, which ran about as frequently as they do now. So the streets were dominated by bicycles, particularly on the hour, as undergraduates sped from lecture to lecture. Nobody took off his gown as he rode through the streets, not even if he was standing on somebody else's bike step.

Most undergraduates lived in so-called "licensed lodgings," which were "digs" in which your landlady or landlord kept a book on you. It was essential to be indoors and locked up for the night before midnight, and, if you did not get in before 10:00 P.M., a note was made of what was considered a slight misdemeanor. The same system continued for a while after the 1939–1945 war. There was a Girton girl who ran up a fearsome total of late nights. Since she was not only very good-looking but was also placed exceptionally high in the mathematical tripos, her case provoked another of Tommy Gold's classic remarks: "All human abilities are positively correlated."

The normal arrangement was for a male undergraduate to spend his first two undergraduate years in digs and the third year in college. Because my examination result in 1934 had been considered reasonably good, I profited by being assigned rooms in college already in my second year, thereby saving me quite a bit of bother. The digs in my first year near the railway bridge in Mill Road had involved me in a morning rush to reach a 9:00 A.M. lecture, and often enough an evening rush to avoid a late night. Too high a score of late nights could lead to being hauled in front of your tutor, although such things never amounted to much for students whose examination results were strong, as with the noteworthy girl from Girton. Added to other misdemeanors, however, too many late nights could lead to the phenomenon of "rustication," and even, in extreme cases, of being "sent down." Being sent down amounted to your being summarily booted out, while being rusticated meant being temporarily dismissed from the university for some assigned period—usually the balance of a term, but, in more extreme cases, the balance of a year. The most famous misdemeanor in my time at Cambridge was perpetrated by a St. John's undergraduate, the future England cricket captain Freddie Brown. At a special football dinner, he took it into his head to jump up onto one of the long wooden tables and then dribble everything in sight on the table tops into the laps of fellow undergraduates sitting on either side of the table.

The university and college punishments were rudimentary survivals of a system that had been considerably more severe in earlier centuries, before the coming into existence of the police. The rules had been designed to prevent clashes between students and the indigenous population of Cambridge. They permitted tutors to keep students constantly in sight, to make sure, I suppose, that a tutor could efficiently inform parents whenever one of his charges had his throat cut by the locals. While you were in the system, you were said to be *in statu pupillari*. You didn't get out of being *in statu pupillari* simply by graduating, as I had done in June 1936, when I had received the degree of Bachelor of Arts. To escape from being *in statu pupillari*, you had to receive the Master of Arts degree, for which no examination performance is or was required. To become a Master of Arts, you simply had to wait a further three years—in my case, until June 1939.

To my blunt Yorkshire mind, the system appeared little short of ludicrous, although I accepted it and steered my way around it without difficulty—except on one occasion. This happened late in my years as a research student, just before I was due to receive elevation to Master of Arts. In a rush to attend an evening lecture, I walked unwittingly onto the streets without my academic gown, and was hauled before a Junior Proctor for the oversight. Since the man was not much older than I was, and quite likely by that time did not have as good an academic record, I was quite annoyed about the incident. Yet I was never much of a sea lawyer. So, after swallowing hard once or twice, I accepted the situation, looking forward to the day, not so far off now, when I too would be *ex statu pupillari*.

The Cambridge disciplinary system even survived the postwar intake of students who had served with distinction in the armed forces, although it was severely tested one Guy Fawkes night by ex-army engineers who set off real high explosives in Senate House Passage. What eventually did it in, or greatly attenuated it, was all-powerful economics. Licensed digs were expensive in my time, and eventually, in postwar years, they became prohibitively so, forcing colleges to provide internal accommodation, thereby effectively tripling the number of students housed on the spot. Inevitably, colleges have a generally more populated look today than they had in the thirties. During the mornings, then, there were crocodiles of students making their several ways into and out of colleges as they attended lectures in college classrooms, but, in the afternoons, things were much quieter. The dispersed state of the student body in the old licensed-lodging system made coming together at night for dinner in college halls more important than it is now. We were obliged to pay for five such dinners a week—and, having paid for them, we ate them, of course.

A comparison of college halls today with the way they used to be provides much evidence against this monopoly. Emmanuel Hall in 1933–1936 was dark and forbidding for us captives: the food was not good, and it was thrown at you with lightning speed, so that you simply gobbled the stuff and were out of the place long before the dons, dining at their special raised table, were halfway through their meal. Today, the same hall is pleasantly decorated, students do not walk about freely on the tops of tables as we did, and they eat in a leisurely fashion, taking about as long over it as the dons. Moreover, a fair number of women are to be seen among them, so that you are not obliged to look down from the high table on yet another of those Cambridge all-male dinners.

Dividing the thirties as I did before—1930–1933, 1933–1936, and 1936–1939—my undergraduate years of 1933–1936 were the lucky ones. We were not overshadowed by the gloom of the depression years; nor were we exposed to the overwhelming threat of war, which became more and more repressive of spontaneity and ebullience as the thirties advanced. Yet, even as early as June 1934, we were feeling some threat. I know this, because I still remember the flicker of hope with which I read garish headlines in the press on the weekend of 30 June in that year. This was the weekend when Ernst Röhm, the leader of the Nazi storm troopers, and his lieutenants were executed without trial. This was the "Night of the Long Knives," as the newspapers called it. The thought was that thieves had turned at last against themselves. Over previous years, Röhm had figured largely in the public eye, with details of the brutal behavior of his storm troopers extensively reported in the British press. But, instead of breaking up summarily, as we had hoped, the power of the Nazi party continued its mushroomlike growth unchecked, with the reoccupation of the Rhineland in 1936, the annexation of Austria in March 1938, and the invasion of Czechoslovakia in March 1939. If these issues had been hard fought, as the contemporary battle at cricket between England and Australia was hard fought, that would have been acceptable. What did it in for us students was the intense feeling that we had of being sold down the river by our political leaders. We felt that the Nazi regime could have been suppressed early on, with not much more than unpleasantness. If our political leaders felt legal justification to be necessary, it was amply provided by the reoccupation of the Rhineland.

By 1938–1939, we felt that, through total ineptness, a game that might easily have been won had been converted into a game we were likely to lose; it was as if the defenders in a football match were to stand immobile, permitting the other side to score goals at will. The height of student disgust was reached during the Spanish Civil War (1936–1939), a war that began as a military revolt against a legally elected republican government that should, we students felt, have received whatever international support was going because of its legality. Instead, Italy and Germany were permitted to send overt aid to the rebels, while the British government refused to provide escorts for ships carrying supplies to the republicans, and even, at one point, impeded a ship's captain who tried to carry supplies unescorted—or, at any rate, so it was reported in the press. Anthony Eden was foreign secretary at the time. Eden resigned in February 1938, according to his biography in the *Encyclopaedia Britannica*, to "protest at Prime Minister Neville Chamberlain's appeasement of Nazi Germany and Fascist Italy." But, according to my information, Eden actually resigned because of more personal differences with the prime minister. Since my informant is a modern historian of high reputation, I would suppose his view of the matter is likely to be correct. If it is, we have an example here of how Neville Chamberlain has been cast as the goat for all the disastrous politics of the times. I will say a little more about Chamberlain in a moment.

It was in these developing circumstances that the students' debating union at Oxford passed its since-famous motion that its student members would not fight for King and Country. Much has been written—erroneously, I believe—about this motion by later commentators who evidently do not understand the state of mind of those who supported it. Without actually experiencing the intense frustrations of the times, it is even hard to understand what the students really meant by their vote. The vote was not pacifist in the usual sense. The students really meant that the inept politicians who had wantonly landed themselves in an unnecessary mess had better contrive to get themselves out of it without calling on the blood of a younger generation to make good their own mistakes and deficiencies. When war eventually came, this well-justified point of view was still hard to slough off, and indeed there was a long, inactive introduction to the war, referred to in the press as the "phony war," while psychological adjustments were being made.

The reference to the King in the Oxford motion was perhaps understandable in view of the problems experienced by the monarchy in the thirties. The turnaround in popular esteem for the Royal Family came later, with the war itself, when George VI and the present Queen Mother refused to be budged out of London during the German blitzes.

Emmanuel College holds a commemoration feast annually in late November, and, because of my now improved status in the college, I was invited in December 1936. It was the first formal dinner I had attended. Luckily, I had acquired evening dress for a function connected with the B.A. degree ceremony the previous June, so, together with a newly purchased B.A. gown, I was more resplendent than I'd ever been before. I was placed close to the don's table, towards the left-hand end facing up the hall, and so was in a good position to observe the high table when the time came for the Royal Toast. Deciding this was an occasion to disregard old Mr. Bartle's advice never to let a drop of alcohol pass my lips, I duly drank the toast, my memory of Mr. Bartle being suddenly erased by cries around me of "The King, long may he reign, God bless him." One old boy appeared to be in tears.

Three days later, the news of the abdication of Edward was all over the newspapers. I was dumbfounded at the swiftness with which the notabilities seemed to have switched from their erstwhile cries of loyalty during the toast to outright condemnation. If the truth be told, I was suddenly now more on Edward's side than I'd been at the feast. If these are the sort of friends one acquires higher up the social ladder, then I don't want them, I thought to myself.

But the most distraught man in Cambridge was, remarkably enough, a Turk. His name was Ali Irfan. He had an enormous chest, and he had astonished the University Athletics Club on his first visit there by putting the heavy shot further than anyone in Britain had ever put it before. In short, he looked a professional Turkish strong man, the kind who, under Mustafa Kemal, had hauled the guns that prevented the British and Allied troops from scaling the heights at Gallipoli in 1915. Together with a number of other young Turks, he had been sent as a student to Cambridge by the government in Istanbul, and, as young people from alien cultures sometimes do, he had absorbed himself deeply over the past two or three years into our own culture. On his return to Turkey, Ali Irfan was to become a professor of English—at Ankara, I think. But now he was wrapped in gloom, declaring the situation to be among the most tragic in history. When George Carson and I visited him with the charitable aim of cheering him up, it was to find Ali Irfan playing ceaselessly on the gramophone the gloomiest of his considerable stock of Sibelius recordings.

In retrospect, the only politician in the British governments of the thirties about whom I have second thoughts is Neville Chamberlain. He became prime minister in May 1937—which is to say, after the main mistakes of British foreign policy had already been made. To the extent that Chamberlain had been chancellor of the exchequer under Baldwin, he carried some responsibility for the previous mistakes; but I would suppose that, while a chancellor of the exchequer can make his views on foreign policy known, he cannot regard foreign policy as his primary concern, otherwise he would soon be at loggerheads with the foreign secretary. While the Chamberlain government of 1937–1939 outwardly pursued an appeasement policy, Britain underneath was actively preparing for war. The fighter planes and the radar screen that saved Britain in 1940 did not come out of nowhere. They were planned, designed, and manufactured by the Chamberlain and earlier governments. Britain survived the air battles of 1940 because the dates at which designs were fro-

zen turned out to be optimally chosen. German designs were frozen a little earlier, and their products were therefore not quite so technically
good—in the air, at any rate. All this eventually accrued to the advantage
of Churchill's government, which, in my opinion, did not have the same
positive technical flair as was shown in the Chamberlain era. Churchill
did things politically right but often technically wrong, whereas Chamberlain did things technically right but politically wrong.

I have always been hoping for a well-informed reassessment of Chamberlain, but I suppose would-be biographers find it impossible to move
the millstone Chamberlain had hung around his own neck when he returned from Munich at the end of September 1938. Arriving in London,
he waved a paper at the waiting crowd, saying, "I believe it is peace for
our time." Against such a clangor, given enormous coverage in the newspapers, of course, even biographers must fight in vain.

I have often wondered why the 1939–1945 war broke out when it did.
The explanation usually offered—namely, that Germany invaded Poland
in early September 1939—is a truism, not an explanation. Germany had
already invaded Austria and Czechoslovakia, both in bad faith, without
war breaking out. The answer, I believe, lies in the club atmosphere that
leaders of nations generate for themselves, an atmosphere that obviously
manifests itself nowadays in their widely advertised summit meetings.
The qualification for club membership is, and was, that a leader be legally
elected by his or her country. In this respect, Hitler took great care to be
elected to the German chancellorship through properly constituted elec-
tions. Once elected, a club member is immediately accorded a considerable measure of trust and good will, a measure so great that Hitler did
not dissipate it entirely, in the eyes of Western political leaders, until
1938–1939. The reason why we students took such a very different view
was that we believed that all trust and good will should have been dissipated already as early as 1934, when the German police state was set up.
The fact that we were proved correct should be no surprise because, to
reveal a deep truth, young people do most everything better than old
people.

I come now to a question I have been asked many times. Was there
anything I saw over the years 1933–1939 to suggest that Cambridge was
harboring a nest of traitors owing allegiance to the Soviet Union? The
answer is a plain *no*. There was, of course, plenty of evidence of Soviet
propaganda. There was the *Daily Worker* on sale every weekday, with
yards of Soviet propaganda plastered all over it. The biologist J. B. S. Haldane wrote science articles for the *Daily Worker*—little gems that many
of us enjoyed reading. Haldane was a convinced Marxist who gave lectures to university audiences up and down the country. In 1938, I at-

tended one myself, and I can say, with total conviction, that no prospective traitor could ever have been won over by such stuff. Haldane's lecture was so full of Marxist jargon as to be incomprehensible, unless you were a confirmed Marxist already. I was amazed that a man who could write such beautiful science articles could talk such codswallop. Like George Orwell, Haldane eventually turned away from Marxism, and for the same reason—the Soviet police state. In the fifties, I had a long correspondence with Haldane, in the course of which he remarked: "Every one of my friends among Russian biologists has been removed from his post, and some have disappeared." This caused Haldane to resign his foreign membership in the Academy of Sciences of the USSR. As he said himself, it was his second conversion.

Soviet propaganda made great play with the fact that the strongest opposition to Hitler in the pre-1933 period had come from the German communist party. Soviet propaganda also claimed, I suspect with more cunning than reality, to be sending arms to the legally elected republicans in Spain. Since the first of these protestations was true, we were mostly tricked into believing the second was true. So it came about that the Soviet Union acquired good will from next to nothing it had actually done itself. Indeed, the Soviet Union was able to represent itself for a while, especially in 1938, as the only source of determined opposition to the rise of Hitler. But, in August 1939, Stalin concluded a nonaggression pact with Hitler, which showed all the previous claims to have been a sham. British communists did what they could to save something from the wreckage by strenuously maintaining that the Soviet Union had been forced to this drastic step through the refusals of Western governments, and the British government in particular, to take effective steps to limit German aggression. That this was a flat communist lie, of the kind that turned George Orwell around in Spain, was plain for all but a nincompoop to see when, on 3 September, only a week or two after the Stalin–Hitler pact, Britain declared war on Germany.

In retrospect, the wild lurchings of the communist newspaper the *Daily Worker*, particularly over the month preceding the outbreak of the war, made hilarious reading. Here are a few headlines:

> 3 August: FEARS OF NEW BETRAYAL BY PREMIER
> 7 August: STEP TO SURRENDER DANZIG
> 13 August: DRAMATIC MOVE TO STOP PREMIER'S
> SURRENDER

To this point, Chamberlain was a peacemonger condemned for refusing to join the Soviet Union in its resolute stand against fascist tyranny. Then, most unfortunately for the *Worker*, came the Soviet–German nonaggression pact, which really opened the route to war. The *Worker* instantly went into reverse with a stripping of its gears:

23 August: SOVIET'S DRAMATIC PEACE MOVE A THUN-
DERBOLT FOR THE CHAMBERLAIN CABINET
26 August: RUSSIA SHATTERS AXIS
28 August: CHAMBERLAIN, WRECKER OF PEACE MOVES

So, in only a couple of weeks, the wretched Chamberlain was transmogrified by flagrant witchcraft from peacemonger to warmonger. And, on 1 September, the *Worker* exceeded itself by shuffling a further miracle from its sleeve, a document that "reveals the extent to which the German government has been shaken by the Soviet–German Non-Aggression Pact." What the *Worker* should properly have told its readers, of course, was that the so-called nonaggression pact was actually an aggression pact, whereby Russia and Germany had agreed to carve up Poland and the Baltic States between them. So, far from the German government having been shaken, by 1 September the German war machine had moved to full speed ahead, with consequences that, by 3 September, had become very clear to all of us.

How, in these circumstances, any intelligent person could bring himself actively to spy for the Soviet Union remains hard to understand. Of course, anybody who, in 1938, had become politically and economically convinced that the Soviet Union was destined to be the savior of the world could be expected to make all sorts of excuses for Soviet behavior in 1939–1940. There were such people in fair numbers around Cambridge, but none of them looked then—or even now, with the benefit of hindsight—remotely like a spy. Some character trait quite different from those I encountered in my day-to-day experiences at Cambridge was needed to produce a spy.

There was a bald-headed communist member of parliament called Pritt, D. N. Pritt, who went around during the so-called phony war of 1940 addressing audiences in universities. He published a book called *Must the War Spread?* Its purport was that the war had been started by Western capitalists in order to fuel their profits, for which reason an expansion of the early, inactive period of hostilities into full-scale war should be prevented. I attended one of Pritt's addresses, and I had difficulty, at the time, in understanding his motivation. Presumably, there was already a fear in communist circles that, if the European conflagration really got under way, the Soviet Union would eventually become involved willy-nilly. So British policy, the firm policy we had all wanted to see adopted for so many years, was to be negated, according to Pritt, to suit the interests of the Soviet Union. In Britain, such talk was permitted as fair comment, but, if Pritt had been Russian and had gone around trying to subvert Soviet policy in the British interest, he would surely have found himself on the wrong side of a firing squad. So, even in wartime, British tolerance was still broad, and it certainly took something quite out of the ordinary to produce a person who was overtly a spy by our permissive standards.

Rather the same, up to the summer of 1939, could be said for Bertrand Russell as for Pritt. In the spring of 1939, I attended a crowded evening talk given in Trinity College by Russell. He was a slender man with the thinker's classic forehead, thoughtful eyes, aquiline nose, and a firm mouth strengthened by being frequently clamped on the stem of his pipe. It was a face that, from the perspective of my own pudding-basin features, I gazed at in envy. Russell's speech was incisive, his sentences well formed and completed—unlike Eddington, who never, in my experience, finished a sentence even in the course of a whole hour's lecture. If there was a flaw at all in Russell's manner of presentation, it was that his voice lacked the deeper timbres, but it was certainly not squeaky, like the voice of H. G. Wells.

Russell had perceived what later became known as the strategy of deterrence—the idea that each nation in a political stand-off possesses such terrifying weapons of destruction that both are obliged, in the common interest, to maintain the peace. This was fine as an intellectual concept, and in the later era of nuclear weapons it became apposite. What was not so fine was Russell's identification of the bomber forces of Germany and of the Western allies as weapons of terrifying deterrence. If war came with Germany, every major city on both sides would be totally destroyed, he told us, an opinion that turned out to be untrue. Being at the receiving end of a bombing raid was always highly unpleasant, but in Britain, fortunately, it was not the overwhelming disaster that would have been needed to give strategic substance to Russell's argument. Since I was to spend most of the war years close to the south coast of England, I was to experience many bombing raids, fortunately without personal disaster. Statistically, too, the number of British lives lost in the Second World War was to be much less than the number lost in the 1914–1918 war. So it remains a question how Russell came to make such a critical, morale-destroying mistake. I can hardly believe he was privy to the latest technological aspects of airplane design, nor does it seem likely that he was in receipt of special information from high-ranking officers of the Royal

Air Force. Therefore, he must simply have been guessing, but to what purpose? To urge on us a continuation of appeasement? What else? If Russell had toured German universities expressing the same point of view, that would have evened out the situation, but, had he done so, he would undoubtedly have been given very short shrift. There was the same lack of balance as in the case of D. N. Pritt, but with Germany profiting from the propaganda instead of the Soviet Union. With this said, let me emphasize that, whereas Russell immediately desisted once the war had started, the communist propaganda continued unabated.

The big surprise eventually proved to be that the Soviet spies were not at all at the *Daily Worker* level, with connections to the official communist party, but in comparatively high society, in comfortable armchairs at the Foreign Office and elsewhere. How could this possibly have come about? Attendance at Bertrand Russell's talk in the spring of 1939 gave me my first glimpse of a new kind of person, one I hadn't encountered before, a person with the divine right of class. The divine right of class does not necessarily mean a person overtly throwing his or her weight around. In Russell's case, it was quite the reverse. Russell did not throw the considerable weight of his class around, which was why he was widely and affectionately popular in academic circles. Typical of his popularity was a remark of Harold Jeffreys, the St. John's mathematician, who said, apropos of certain of Russell's earlier somewhat lurid activities, "What Bertie never understood is that in British society you can either advocate adultery or practice it, but not both." The divine right of class implies the possession of an aura, an aura that gives others the feeling that, in any matter of real urgency, you can get things done behind the scenes in ways that are not open to discussion.

The divine right of class was the component in the Cambridge spy story that remained hidden from me during my student years. Nobody I knew had it. There were plenty of people I knew who were disaffected with the political situation, plenty who were in some degree deluded by Soviet propaganda, as I suppose I was myself. But, without the divine right of class, without the notion that you could manage to get things done subtly behind the scenes, the chaps I knew raged impotently.

There is no mystery about why the spies were concentrated in Trinity College, because the divine right of class was itself heavily concentrated there, with Russell and scores of others who were not spies. There was a type who wore jodhpurs and what then seemed very minute flat caps, and who shouted in remarkably penetrating voices at each other across Trinity Street, totally ignoring everybody else in the vicinity. This type would shortly be flying planes and driving tanks, for it was not disaffected or disgruntled. The disgruntled spy type did not shout across Trinity Street for the world to notice. He closeted himself indoors, plotting, out of sight of students like myself. Although we were disheartened by the political situation, we were not disgruntled because we felt, with just cause, that we were lucky to be there at all.

Chapter 10

The Last of the Old World

ALTHOUGH THE STEP from undergraduate to B.A. status in 1936 did not permit me to escape from being *in statu pupillari*, it did bring a release from the discipline of late nights, so I could avoid the expense of living either in college or in licensed digs. The shift to unlicensed lodgings saved me about £50 a year, while the disappearance of college teaching fees saved another £20. Instead of the difference between £225—the amount the West Riding education authorities continued to pay me—and my unavoidable expenses being only £25 as it had been before, my cash in hand soared, in June 1936, to an annual amount approaching £100. That wasn't sufficient to compete with the lads in jodhpurs and small flat caps, but, after the necessary frugality I'd developed over the years, there seemed to be a touch of real wealth about it.

Until 1937, I shared digs with Charles Goodwin in Station Road. I learned from Charles that the next thing to aim for was one of a number of research prizes awarded each year. Usually two first-level awards, called Smith's Prizes, were given, but there were also two or more Rayleigh Prizes, and gaining either a Smith's or a Rayleigh was considered to be almost a guarantee of a post in some university. This, then, was the objective as I began my research career early in July 1936.

The time available for preparing a research essay was only four terms. A research essay had to be finished and submitted by December 1937, so it was essential to be away to a fast start. Partly because Rudolf Peierls was an enthusiastic research supervisor who fed me with two good up-to-date problems, and partly because I was already talking freely with research students during my last undergraduate year, I was able to hit the ground running, as Americans say. This was a considerable initial advantage, because most research students for whom Cambridge was their first and only university found the four terms too short. Why the university had decided that its regulations should be a handicap to its own graduates, who were greatly disadvantaged compared to research students coming from other universities, was a mystery. The slow undergraduate stream in which I had started was indeed entirely excluded by considerations of timing. Had I not made the risky jump in the summer of 1934 from the slow undergraduate stream in mathematics to the fast stream, I would never have been given a chance at these later research prizes, which were to prove pivotal for me. An important immediate advantage was that I could validly represent the prizes as the last performance test of my educational career, an argument that the Yorkshire West Riding authorities accepted with characteristic generosity by continuing my scholarship to the end of my second year of research—which is to say, until the summer of 1938.

The day came when I handed in my shot at the prizes. The day also came, towards the end of the Lent term in 1938, when the results were announced. As I had dared to hope, I had been awarded one of the two Smith's Prizes. As I had not expected at all, the two Smith's Prizes had, atypically, been placed in order of merit, a first and a second, and I had the first. It had been a long climb at Cambridge—three-quarters of the way up the slow stream in the first year, three-fifths up the fast stream in the second year, to among the top three or four in the third year, and now, halfway through the fifth year, to the top position itself.

It was twelve years almost to the day since I had walked, accompanied by the mumps virus, down from my home village to the cold schoolroom close by Holy Trinity Church in the poor eastern section of Bingley, when I had just squeaked through, by a hair's breadth, to my first scholarship award. I had been supported ever since by the West Riding of Yorkshire, with its educational headquarters at Wakefield, a town centrally placed among the "dark Satanic mills" covering much of the region. Now, at the end of it all, I wrote to the Chairman of the Educational Committee in Wakefield with my thanks for the huge measure of support I had received, support that I doubt any other county in England would have given me.

This happy termination of my connection with the West Riding of Yorkshire meant I must look for a new source of support, rather like a floppy dog, with a big tail a-wagging, looking for a new master. My eye lit on the Goldsmiths' Company, a rich master if ever there was one. The Goldsmiths' Company offered a magnificent research grant of £350 a year, but it was open to a wide range of candidates, unfortunately. Applications had to be submitted by Easter with the names of two referees. Peierls agreed to be one, and I had the idea of asking R. H. Fowler to be the other. This was a little unusual because I was not one of Fowler's own students, of which he had many. But I felt I had a small measure of credit with Fowler, for a reason I will now explain.

There was a room in the old Cavendish, built by James Clerk Maxwell in about 1875, where a great deal of blood was spilt in 1937 on a weekly basis. You went up two flights of stone steps to the first floor and then turned right up a further short flight of wooden steps, again turning right shortly thereafter into what was called the colloquium room but was actually a torture room—at any rate, it was for those who gave theoretical physics colloquia in it in those days. Since no less a person than Eddington had spilt a vatful in that room, there was no reason why so minor a person as myself should escape the savage treatment handed out by the fearsome front row—Fowler, his buddy C. G. Darwin, A. H. Wilson, Dirac, and, of course, the sharp-witted Maurice Pryce, who was especially severe on poor Eddington. Eddington was never at his best in verbal battles, and, on the occasion in question, he mistakenly tried to maintain that black was white, with truly awful results. The issue was the correct pressure formula for relativistically degenerate electrons. E. C. Stoner had used the formula physicists believed to be correct to show there was no permanent condition for a large enough quantity of nonrotating matter, a quantity now known as the Chandrasekhar limit, except as a black hole. Eddington thought black holes should be a physical impossibility, so he inferred that the correct formula must be something else—actually, the nonrelativistic formula first used by R. H. Fowler himself.

In my very first term as a research student, somebody had the idea of assigning, to a chosen student each week, a chapter from a big article on nuclear physics that had just then appeared in the *Reviews of Modern Physics*. Eight victims had to be found, mostly second- and third-year students like the three Maurices—Pryce, Goldhaber (later the director of the Brookhaven Laboratory on Long Island, New York), and Blackman. I believe I was the only first-year student to be assigned a chapter—to come late in the term, Peierls assured me. This was a shocker of a situation, partly because the later chapters were pernickety details rather than interesting issues of principle, and partly because the big shots did their worse to bait Pryce, whose turn came early in the term, and this opening fracas set the tone for the no-holds-barred violence of succeeding weeks. When it came to my turn, I was saved from being mangled to death only by the gong. Time ran out, permitting Peierls to suggest that my chapter, being a long one, should be finished the following week. He insisted I should spend the whole week of grace thus vouchsafed to me in further preparation, which I did, viewing the thing more as a research project than as a simple reporting of what somebody else had written. This helped, and, in the event, I escaped without too many more deep wounds. It was this escape out of the bear trap that had given me a little credit with Fowler. Now, as I tapped on the door of his office clutching an application form for the Goldsmiths' exhibition, was the time to use it.

There was a bark from inside the office, which I took to mean "Come in." I went in and asked Fowler if he would act as a referee. He agreed immediately and gave a nod, as if to say, "That's that, out you go." So out I went, with my form duly completed. A few weeks later, I had the Goldsmiths' exhibition, a grant of £350 for the year 1938–1939. Suddenly, I was wealthy, fully able to buy jodhpurs and a small flat cap, if I'd been of a mind to do so.

Getting into Fowler's office or rooms was no mean feat, actually. Pryce always told how, when Fowler was his undergraduate supervisor, he would go along each week to Fowler's rooms in Trinity. Fowler would respond to his knock by opening the door a few inches, just sufficient to admit a hand and arm. Out would come Fowler's hand clutching the exercises—examples, we called them—that Pryce had handed in the previous week. Pryce would take them and push into Fowler's now empty hand his current week's work. The hand would then be withdrawn, and a voice from the far side of the door would bark: "Excellent, Pryce. Work more examples. Work more examples." Then the door would shut firmly.

I have remarked before on how people enter our lives, how they fill the stage brightly for a while, and then fade away as in a dream. So it was now with many friends I had known in my undergraduate years. Each year, I had gone to the mountains in the summer. In August 1936, a party from the rambling club, joined by Edward Foster, my companion of previous years, made a trip to harder mountains than I had seen before—the Cuillin Hills on the Isle of Skye. We toughed it out in a midge-ridden camp in Glenbrittle. Although we now were putting our lives literally in each other's hands, since we were climbing on ropes, even so tight a bond was unable to withstand the swirling currents that dominate the river of life. After two years at school with Edward Foster and three summers of walking, I was never to walk the hills with him again—or indeed with any except one of that party of 1936. The exception was Joe Jennings, who emigrated to Australia in the mid-1950s. Whenever I visited Australia thereafter, I contrived to see Joe and to make a trip with him, and, whenever he visited Britain, we did the same. In August 1937, I was again in Glenbrittle, again with members of the rambling club, but the composition of the party had changed. They were a year junior to those of the previous year—or, more accurately, they were the same with respect to university status, but I had grown a year older. And in August 1938, I was in Glenbrittle once more, alone. My undergraduate friends had vanished into the mists of life, just as surely as had the village lads with whom I had once spent so many vibrant hours. While mountains, buildings, and organizations persist, friends will be gone pitifully soon, like mayflies on a warm spring day.

In 1937, Rudolf Peierls was also gone—to the Professorship of Applied Mathematics at Birmingham University. I now had an important decision to make—to stay in Cambridge and go it alone, to stay in Cambridge with a new research supervisor and research topics, or to follow Peierls to Birmingham. Because I owed much to Peierls for the way he had pressed me quickly forward the previous year, and because he seemed to want me to continue research with him, I felt I should give the possibility of a move to Birmingham first trial. But, when I returned to Cambridge, ostensibly for just a weekend in mid-February, and found early crocuses bright and shining all along the avenue of Trinity Backs, I knew it couldn't be. I have two remaining memories of the several weeks I spent in Birmingham. One is of eating meals in a boardinghouse with other guests, breakfast and dinner together every day. It wasn't that I had anything against any one of them; it was just that we had no interests in common except cricket and the weather. As a substitute for nightly dinner at the Friar House with research students from the Cavendish, it didn't fly, as people say. My other memory is of hearing Felix Weingartner conduct the Birmingham Municipal Orchestra, so establishing a bridge in time between Richard Wagner and my own day. People who nowadays appear to judge the quality of conductors from the extravagance of their gestures would have been amazed at the almost minute signals Weingartner gave to the orchestra. He remarks in his autobiography that anybody who conducted a different opera every night, as he had done as a young man in one of his first posts, would soon become economical of gesture. The program opened with Schubert's *Rosamunde* overture, but the rest of it escapes me now.

The thought of switching to new research topics under a new supervisor didn't fly either, so it almost amounted to going it alone—but not quite, as it happened. A second problem Peierls had started me on went under the name of analyzing the properties of wave equations of higher spin, higher spin than in Dirac's famous equation. It amounted, in modern terminology, to determining the irreducible representations of products of the spinor group with itself. In those days, unfortunately, I knew no group theory and consequently made a fist of the thing. The accidental feature was that Maurice Pryce also had an interest in the problem on his own account. So, too, had Wolfgang Pauli in Switzerland. There was a rumor that Pauli, together with a student, Marcus Fierz, had obtained a certain result. I remember sitting in the Whim Café with Pryce and Peierls. They disagreed over whether the result was true or not, and Peierls said he'd bet a bob on it, but I can't remember which way round the bet went or who won it.

The supervisor problem, once I'd decided I wasn't leaving Cambridge, was simply that the university liked every research student to have a supervisor, although it wasn't absolutely essential, unless you held a government grant from DSIR, or unless you wished to acquire a Ph.D. degree. The Ph.D. was not looked on as a crucial qualification in those days, but, if I wanted to keep my options open on it, I needed a supervisor. In view of Pryce's interest in the wave-equation problem, and Pryce having recently become a Fellow of Trinity, so removing him from being *in statu pupillari*, Peierls suggested to Pryce and to me that he should become my supervisor, which suggestion we both accepted.

The situation proved ironic, for, if it had ever been my intention to seek the Ph.D., it was Pryce who persuaded me out of it. He had a dislike for the degree, which he regarded as a debasement of the academic currency. He showed his opinion by fulfilling all the technical requirements but then omitting ever to go to the Senate House to formalize the situation in an official ceremony. As it turned out, I did the same, but only partly for doctrinaire reasons. I discovered the Inland Revenue distinguished between students and nonstudents by whether or not you had acquired the Ph.D., and, since the distinction affected my tax quite substantially in the period 1939–1941, I had a more earthy motive for avoiding an official ceremony in the Senate House.

It would be a mistake to think this precious. It was an accurate perception of the damage to originality that slavish pursuit of this degree has caused. If a student wishes to study for the degree, particularly in cases where there is a change of university, well and good. But students in general who have gone from the age of five to the age of twenty-one through a never-ending sequence of examinations should, by then, be released from formal processing. It is high time at twenty-one to begin thinking for oneself. More and more, as the years pass by, research students have come to accept close direction from supervisors, so consuming still another three or four years of the precious few in which the originality of childhood can reassert itself. The mere fact that government bureaucracy demands the Ph.D., and has demanded it pretty well from the first moment it was introduced from America, is sufficient to condemn it. The American educational system is different. It puts less pressure on the earlier years, better permitting originality to survive, and so giving the technical aspect of the Ph.D. more of a raison d'être than it has in Britain.

I continued for a while on a problem I had begun with Peierls. It involved beta decay occurring through intermediate states of the daughter nucleus. I am not sure if it was the first investigation of its kind, but it was new to senior physicists in the Cavendish Laboratory. They became interested, and, as a consequence, I was invited to attend their weekly meetings, just as Peierls had done until he left for Birmingham. Since Rutherford had died in 1937, the laboratory was being run in 1938 (before the arrival of Rutherford's successor) by John Cockcroft, together with a consortium including Norman Fowler, Bennett Lewis (who was later to be the designer of the highly successful Canadian nuclear reactor CANDU), and Philip Dee. Seeing the laboratory was eventually to be shifted drastically in a direction away from nuclear physics, Chadwick, the discoverer of the neutron, had left already. My association with this senior group up to the summer of 1939 had a determining effect on my relations with Cambridge physics. I began in research by feeling that the Cavendish Laboratory was my home. When the big changes came, the Cavendish was no longer my home. Quite the reverse. I eventually became one of the few people left in Cambridge with emotional ties to the prewar Cavendish. My feeling of eviction became the source of a tension that could never be resolved and that led eventually, far down the years, to my leaving Cambridge.

This was the opportunity for me to ask Philip Dee if he remembered the Yorkshire boy who had sparked back at him in the Pembroke scholarship examination, when he'd asked: "Is there *any* physics you *do* know?" There was no way I could now dig the incident out of his memory. Dee brushed it aside with a wave of his arm: "Oh, I say that sort of thing all the time," he remarked airily. John Cockcroft was the money man of the group. Whenever an experiment was proposed, Cockcroft would set to work to figure out the cost. The conversation would linger on for a while, but, as John came towards his final addition, a dead silence would fall on the room. The usual thing was for Cockcroft to give his friendly, half-sad smile, and shake his head, saying: "Too much lead." I should explain that lead was the standard material for shielding against nuclear radiations, and that its cost was often critical in relation to the minuscule sums then available, a far cry indeed from the unctuously rich later days at the Science and Engineering Research Council. The small committee was forever jumping on Bennett Lewis, demanding that he produce electronic equipment faster than he could produce it, and Philip Dee was forever exclaiming: "Let's get on with it!"

The news was out that Erwin Schrödinger would be paying a visit to Cambridge, and I was allotted half an hour of his time at the Cavendish to explain beta decay through intermediate states. I had barely begun to talk about the thing when I could see that Schrödinger wasn't really interested, and I had the flash of common sense to wonder why a man who had discovered the famous wave equation should be in the least interested in a question of whether nuclei decayed through intermediate states or not. My next meeting with Schrödinger would not be for some fourteen years, and it would then be in circumstances more different than you could imagine.

To take in this later occasion, my narrative must make a wide sweep. In 1950, the Royal Astronomical Society held a meeting in Dublin, which I attended with my wife to be. At a premeeting gathering, I was introduced to Monsignor Browne, President of University College, Galway. Paddy Browne, as he was known everywhere throughout Eire, had been gifted with talents that were almost too great for any one man to assimilate. He had begun with a Ph.D. in mathematics from Göttingen, which meant that he spoke German fluently, just as he did French, Italian, Gaelic, Latin, ancient Greek—anything at all, in fact, that he happened to take a look at.

It was Paddy Browne who introduced me to the poetry of Erwin Schrödinger, which appears to be all but unknown to the scientific world. Schrödinger was then at the Dublin Institute of Advanced Studies, which fell under the University of Ireland, and so, since Paddy had, a year or two later (around 1950), become Chancellor of the University, the Institute was under Paddy's jurisdiction. One day in 1952, I was with him in Dublin when he said: "We're going out this afternoon to tea with Schrödinger. I'm going to give him a dressing down."

Getting a dressing down from Paddy, if he was in earnest, would have been a formidable business. He was a big man, about six feet four inches, I suppose, with a big voice and a truly awesome command of words. In the car on the way to the Schrödinger home, I asked what the trouble was. "Oh," said Paddy, "there is to be a great reception by the President [of Eire]. We each have an invitation, for ourselves and a lady guest. Schrödinger wants me to take his wife, so that he can take Miss Y."

Now I knew why Schrödinger hadn't been interested in beta decay through intermediate states. "Miss Y?" I asked. "Ah," continued Paddy, "I must tell you that Miss Y is one of the beauties of all Ireland. Why else do you think Schrödinger would write poetry?"

Schrödinger was heartily glad to see me that day, for I was a kind of human tree behind which he could circle to dodge away from Paddy. He did this by talking about science, any old science, by handing round cups of tea and sandwiches, playing for time and, I suppose, for Miss Y. It has always been a matter of regret to me that I never saw a picture of Miss Y, but I still have a clear picture of Schrödinger dodging Paddy's assault with the agility of a matador in front of a bull.

Charles Goodwin had gone now, in 1938, to a post in the mathematics department at Sheffield University. So I moved from Station Road to the Chesterton region at the opposite side of the town, where I shared digs with George Carson. Gone now was our swiftest canoe on the river. George and I were still fairly fast, but it had been Charles's enthusiasm that had really kept us up to concert pitch. My extracurricular conversations these days were mostly in biology, and it was through George that I met one of the outstanding biologists of the time, C. D. Darlington. This meeting would exert a considerable influence on me down the years.

With Charles Goodwin gone, it was natural that I should take advantage of the circumstance that I was in a similar age relationship to Maurice Pryce, and that I could now drop in, from time to time, at his rooms in Trinity. Thirty years on from those days, Pryce was to tell me how much he regretted he hadn't published more. Pryce had a perfectionist attitude, which made it hard for him to publish steps along the road in the manner of progress reports, the way most of us do. With Maurice, it had to be a perfect and complete solution of a problem, usually a big one. In view of the fallow period into which physics was then running, this attitude must have made life tough. Pryce would also circulate his manuscripts freely and generously, and, in at least one case, the work eventually appeared under other authorship. Around 1959, I received several significant and useful pages from Maurice, with an accompanying note saying he'd come on them in clearing out his files, and that "possibly I might find the pages of interest." He also added that he hadn't published the work, which dealt with certain mathematical aspects of the creation of matter in cosmology, because he'd not been successful in finding a connection to the then known theories of particle physics. The connection came eventually in the early 1980s, with the development of what are called "grand unification theories." This was typical of Maurice—always looking for connections that lay too far ahead in time.

He also had what seemed to be a compulsive wanderlust—first Cambridge to Liverpool, then back to Cambridge, then to Oxford, to Bristol, to the University of Southern California, and finally to British Columbia, which, curiously enough, he told me, even in those early years, he thought the most beautiful place on Earth.

I have only one thing to say against Pryce. Because of his wanderlust, it chanced from time to time in later years that I would be giving a lecture in a university where Maurice happened to be at the time. He would then take it on himself to introduce me to the audience with stories of how I had been a hellbroth of a research student, who had been wished onto him because nobody else at Cambridge could cope with me. This was pure calumny. I was always a model research student, and if any tricks were played at all, it was Maurice himself who played them.

There was a research group with a limited membership that met in the evenings—once a fortnight, I believe—called the Delta-Squared V Club, a name derived from nineteenth-century electrical theory. Whenever a place in the membership fell vacant, candidates to fill it would be proposed, and a vote by existing members would then decide the winner.

During 1936–1937, I had twice been proposed but had not made it on the vote. In 1938, Pryce, who was then the club's secretary, again proposed me, and, to my surprise, I won in a near landslide, which seemed to amuse several persons present at the meeting, which I had attended by invitation. The system in the club was that, after serving a specific time, which I think was a term, the president stood down and the secretary took his place. So it was with Pryce at this particular meeting. Once in the presidential chair, Pryce immediately proposed me for the now-vacant post of secretary, and I again won in a landslide, amid much laughter, as it was realized how I'd been limed and snared into a job with a considerable amount of work to it—arranging the venue of meetings (which, by tradition, swung from college to college as the secretary changed), writing up minutes, obtaining speakers and making abstracts of their talks, and generally standing up to the barrage of pointed comments with which meetings always began. On the way out after the meeting, Pryce grinned and said: "Let that be a lesson not to trust anybody, not even your supervisor."

The most awkward part of the secretary's duties was to find speakers. Since Pryce had just been secretary, I asked if he had any suggestions. "You might try Dirac," he replied, "and there's a chap in John's called Lyttleton who has interesting ideas about planets." The next day, I took my courage in both hands and phoned Dirac at his house. When he had understood my request, Dirac made a remark that nobody else, in my experience, would have conceived of: "I will put the telephone down for a minute and think, and then speak again," he said.

The upshot was that Dirac agreed to give a talk to the club, and a famous talk it was, for it was the first occasion on which Dirac explained his idea for introducing advanced potentials to solve the paradox of the self-action of a classically accelerated electron. My attempt to get Ray Lyttleton to give a talk to the Delta-Squared V Club was not so successful. I visited him in his rooms in the New Court of St. John's at about three o'clock one afternoon. In reply to my request, he explained that he was snowed under with work from which he could in no circumstances be dragged. In 1945–1946, I was to have those same rooms myself, above the Bursary (as it was then) on Staircase I. Withdrawing from my failure to entice Lyttleton into appearing before the Delta-Squared V Club, I was crossing the yard or two of a small hallway when a thought occurred to me. Almost certainly, my life would have gone differently in its details if, in that fleeting second, the thought had not been there. I turned impulsively on my heel, moved the yard or two back to the inner door, half opened it without knocking, stuck my head inside, and said: "I hope I can do as much for you another time."

I hadn't realized it, but my remark was exactly the sort of thing that sets Ray Lyttleton off laughing. He called me back immediately and put on the kettle for tea. He was friendly now, but, try as I would, I couldn't persuade him to give that lecture to the Delta-Squared V Club. He talked of being on the verge of great things, of vast depths, in the fashion of Owen Glendower in *Henry IV, Part I*.

The immediate problem for Lyttleton had to do with the rate at which the gravitational influence of a large body causes it to pick up material from a diffuse gas in which the body is immersed—the "accretion problem," as it subsequently became known. Lyttleton had the feeling that a formula in Eddington's book *The Internal Constitution of the Stars* gave considerably too small a rate. The upshot of our conversation was that I agreed to take a look at the thing. It was from this problem that my shift to astronomy from theoretical physics began. It did not take long to see that Eddington's calculation was correct for disconnected lumps of material, but, for a gas in which internal collisions were important, the accretion rate could be much higher than one would suspect from Eddington's formula, just as Lyttleton had hoped might be the case. Although we were to make a wrong immediate use of the discovery, which caused us some unpopularity in astronomical circles for a while, the technical details as they were developed from that time, especially in collaboration, a few years later, with Hermann Bondi, have proved their worth in later years. Today there is hardly an issue of any astronomical journal that does not contain a paper concerned with some aspect of the accretion problem.

This was a sideline, however, over the first six months of 1939. My Goldsmiths' exhibition would soon be running out, and, if I was to continue in research at Cambridge, I had to bestir myself once more, especially with regard to the possibility of securing a College fellowship. The time had come, therefore, for me to stitch the various bits of research I had done into an essay. I submitted my achievements, such as they were, for the annual fellowship competition at St. John's College. My tutor, P. W. Wood, wanted me to submit also for the fellowship awarded annually at Emmanuel. Although it might have seemed rather like turning my back on my old college, I decided, after some heart-searching, to restrict my application to St. John's, which was one of the few colleges whose fellowships were freely open to any graduate of Cambridge or Oxford (within certain age limits). I knew John Cockcroft personally from the Cavendish meetings, and I knew Dirac a little. Both were fellows of St. John's, whereas I knew no physicist who could speak for me at Emmanuel. John's gave three or four fellowships, against only one from Emmanuel, and competing for a single award against candidates in all subjects (arts as well as science) had to be a chancy business.

On the St. John's application form, there was a question asking if one were a candidate also at another college. I knew this, because I had already applied both at St. John's and at Emmanuel in the previous year, and I had been turned down for a fellowship by both colleges. In the case of St. John's, Cockcroft told me I had been turned down because it was thought that Emmanuel would elect me. In 1939, I didn't want to fall between those same two stools again, so I confined my application this time to St. John's, and, as it happened, I was appointed to a fellowship there for three years from May 1939.

I had also applied for a prestigious award offered by the Commission for the Exhibition of 1851, and, to my astonishment, a week or two after the fellowship, I was successful in obtaining a "senior exhibition," which paid £600 per year for two years. Since the St. John's fellowship was worth £250 per year, plus accommodation and dinners, I now had the enormous sum of £850 per year, with little in the way of living expenses to worry about. But let the reader not be envious of my apparent good fortune, because, in May 1939, the prospect of war was becoming overwhelming. Soon I would be postponing both the fellowship and the exhibition until the end of the 1939–1945 war. When I returned to Cambridge in 1945 to take them up again, the income of £850 a year would have declined to about one-half of its 1939 value, and by then I would have a wife and two children to support.

Shortly after inveigling me into the secretaryship of the Delta-Squared V Club, Maurice Pryce left Cambridge to take up a lectureship at Liverpool University. If I was to retain my student status with the Inland Revenue, another research supervisor was therefore needed. Pryce suggested Dirac, so I went to Dirac and explained the position. Although normally he didn't accept students, Dirac broke his rule on this occasion because he simply couldn't resist the circular counterlogic of a supervisor who didn't want a research student who didn't want a supervisor.

Dirac didn't seem too much amused by farce or repartee, but he liked this kind of mental contretemps. The time I saw him laugh most was a year or two later, when I told him the following story, after I had returned from my first visit to the United States. During the war, you had to have a priority (of which there were, I believe, four grades) in order to travel on domestic airlines in the United States. If you wanted to make a journey and you had a priority higher than someone else on a plane that was already full (they were always full), you simply took the fellow's place and he was bounced off. One day, a well-known scientist was traveling to give an important lecture, when, unfortunately for him, he was bounced by a general. It was unfortunate also for the general, because he was traveling to hear the lecture.

More than any other person I have known, Dirac raised the meaning of words and syntax to a level of precision that was mathematical in its accuracy. He had nothing at all of the irritating habit of attempting to read hidden significance into your remarks. He paid everybody the compliment of supposing they knew exactly what they were saying, a compliment he sometimes took to extreme lengths. There was a time, during the 1939–1945 war, when the British government was fencing with the American government to get as deeply as possible into the Manhattan Project (for making a nuclear bomb). The clarion call rang out in Whitehall to get Dirac involved as a bargaining counter with the Americans. I did not witness the occasion myself, but I have it on good authority that the minister concerned, Sir John Anderson, telephoned Dirac in Cambridge to ask if he would call at the ministerial office when next in London. Dirac said that he would. Sir John then went on to ask, as an afterthought, how often Dirac was in London, to which Dirac replied: "Oh, about once a year."

Those scientists who moved first into what was called "war work" quickly took up the most influential positions. The Cavendish began moving already in the summer of 1939, a few months ahead of the actual outbreak of war on 3 September 1939. Considering my close connections with the Cavendish six months earlier, it would have been natural for me to have gone along with such people as John Cockcroft and Philip Dee. The reasons I didn't do so were personal; they would lead to my marriage on 28 December 1939.

My life had now entered its second act, and to this point the actor, or rather the actress, who was to prove the most important character in the plot, had not appeared on stage. It happened quite without plan and quite without warning, as everything that turns out to be important always does. The point is that, if you try to plan your life, nothing important can happen because it is impossible to conceive of what turns out to be really important. The crucial aspects of our lives, it seems, can only come out of nowhere.

There was one undergraduate friend who had not passed into the shadows, because he was, and is still, an excellent correspondent. Richard Beetham had been one of the mathematics students in Emmanuel whom I had joined when I made the switch from natural sciences in October 1933. Now, in May 1939, he wrote from a northern school, where he had taken a post teaching mathematics, to say that he would be visiting Cambridge at his half-term and to ask if I would meet him at such-and-such a time in the Dorothy Café. The Dot was a place the size of a ballroom situated on the upper floor of Hawkins, a confectioner with a big shop on the first floor. It was a popular resort for undergraduates, with the Whim Café taking its place for us research students. Hawkins is everlastingly associated in my mind with the composer Johannes Brahms. Shameful to relate, I once had an argument with Ray Bell of the Treasury as to the relative merits of Mozart and Brahms, with Bell, as befitted a Treasury man, taking Mozart's side. Overhearing our argument, another chap supported Bell, saying the tune of the scherzo in the Brahms Fourth Symphony sounded to him as if Brahms could do no better than to say to himself: "Shall we go to Hawkins? No!" The remark did Brahms in for me, a position of which Tchaikovsky would have approved.

Anyway, I made my way to the Dot within a minute or two of the time Beetham had given me—I think it was 11:00 A.M. Surveying the many tables, I picked him out from the throng by the wave of his arm, and, as I approached his table, I saw two girls sitting there. They had been his pupils. Jeanne Clark was taking a teacher's training course at Homerton College, and her younger sister, Barbara, had come to Cambridge for an interview at Girton College. Instantly, I knew that the important actress in the play that constituted my life was suddenly, and wholly unexpectedly, on stage. There was little to be done about it there and then, for the solid reason that a chap whom everybody called Percy, the presiding genius of the Dot, with a big round face and big round spectacles, now began to thunder away on the piano and to boom out the latest jazz hits. But a little before that day I had bought my first car, a twelve-horsepower Rover of the 1936 vintage. It had cost £125, which was a lot to pay when you could pick up a tolerable secondhand car for £20, but my parents had always insisted that, unless you can afford a really good article, don't buy at all.

A while later, I had a license to drive alone, and the Rover somehow found its way north to Richard Beetham's school. Barbara of the long plaits seemed mostly astonished that I apparently thought so little of my well-being as to have holes in the soles of my shoes, just as I had much earlier. Even so, I managed to persuade her to make a trip in the second half of August to the Lake District. Two days after leaving her, on 30 July 1939, I wrote my first letter to my future wife:

10 de Freville Ave.,
Cambridge.

Dear Barbara,

I forgot to ask your postal address. Don't be surprised at this, I never do anything right.

After leaving you I decided to pay a short visit to an old friend

living in Sheffield who writes me periodically asking me to come
and stay with him.

I am sending this to the school, and would you send to my address giving yours, so that when I have the preparations made I
can let you know?

Fred

Having arranged to go on holiday with a girl, I had contrived not to discover her address, nor to leave her with mine.

Mid-August 1939 came round, and the two of us drove off for the Lake
District. We had intended to stop the first night at Buckden in Wharfe-
dale, but the inn there was full. We continued to Hubberholme and Ough-
tershaw Moss, by exactly the same tiny road that Fred Jackson, Edward
Foster, and I had walked some five years earlier. We stayed that night at
a quiet old inn beyond Hawes, where four months later we would spend
our honeymoon.

Barbara had arranged also to visit friends in Cheshire. We drove there
from the Lake District, after which I continued to Bingley. Four days later,
at the very end of August, I drove again to Cheshire to pick up Barbara
and take her back home. A few miles along the road, we stopped on a
grassy verge. The news on the radio had been consistently bad. However
much we wanted to believe otherwise, it was clear now that war was coming. War would change everything. It would destroy my comparative affluence. It would swallow my best creative period, just as I was finding
my feet in research. But it also made a nonsense out of the two or three
years of courtship that was considered proper in those days. On the grassy
bank somewhere in Cheshire—I was never able to find the exact spot
again—we decided to marry forthwith. More precisely, we would marry
as soon as Barbara's parents could be accommodated to the idea.

The wedding was held early on 28 December 1939. The after-wedding
ceremonies being over by 2:00 P.M., we set off to chase the light of the
short winter day, through Doncaster, Leeds, and Skipton. The roads became icy, and, trying to hurry, I had a bad skid between Skipton and Settle. Fortunately, there was no other traffic on the road. Crossing the moor
after Clapham, Ingleborough came in sight at last, glowing pink in the
last moment of sunshine, with snow on its upper slopes. Not long afterwards, we were at our destination, with a bright fire in the little sitting
room to welcome us.

The war had come everywhere to the towns and cities, but it had still
to reach up into those Yorkshire dales. The ten days we spent there were
to be the last surviving breath of the old prewar world. The inn was run
then by a middle-aged couple. It would be twenty years on before we

would stay in the same place again, by which time a younger generation of the family would have taken over. But the old man was still around. One day, he looked hard at us, saying: "I remember you now. You were the honeymoon couple who came just after the beginning of the war." We had tried to hide our honeymoon status. When I asked the old man how he'd known, he just hopped away along the passageway that led from the kitchen to the main outer door, chuckling to himself.

PART TWO

The Larger World of Science
1939–1958

The Portsmouth–Chichester Area

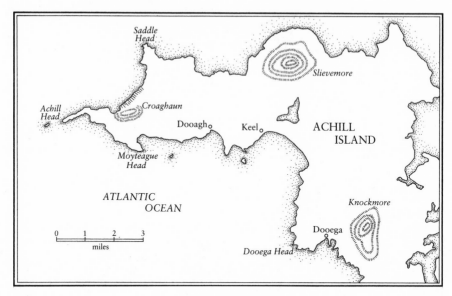

Achill Island

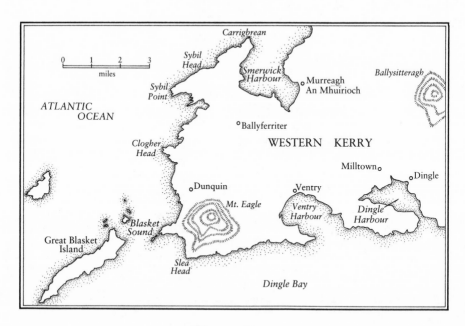

Western Kerry

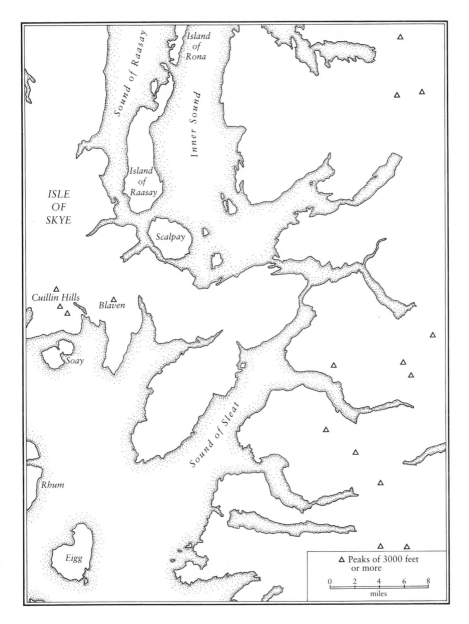

Part of the Scottish Highlands

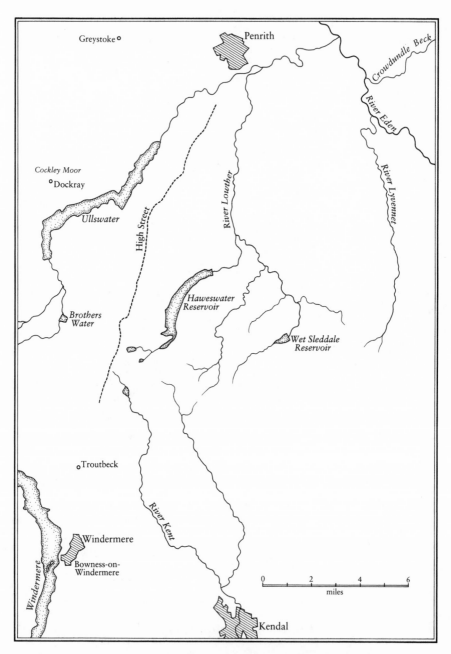

Greystoke ○

Penrith

Crowdundle Beck

River Eden

River Lyvennet

Cockley Moor
○Dockray

Ullswater

High Street

River Lowther

Haweswater Reservoir

Brothers Water

Wet Sleddale Reservoir

○Troutbeck

River Kent

Windermere

Windermere

Bowness-on-Windermere

0   2   4   6
miles

Kendal

The Eastern Portion of the Lake District, Showing Dockray, Ullswater, and Where the Romans Marched (High Street)

Chapter 11

Sir Arthur Eddington

EDDINGTON WAS a slender, nervous-looking man. More than most people do, he really looked like his photographic studio portrait, the one carrying the name of Ramsey and Muspratt, whose premises were in a narrow lane leading in an arc from St. Andrew's Street to Petty Cury, nowadays an artery of the notorious Lion Yard scheme, which changed the center of Cambridge in several ways, all of them bad. He was of middle height, about as tall as I am. He was also shy, like me, so it wasn't with any anticipation of sparkling conversation between us that, after lunching in St. John's College, I set off on a spring afternoon in 1940 to walk out along Madingley Road to the Observatories to see Eddington on a small matter of business.

Eddington contrived paradoxically to be both way down among the world's worst lecturers and yet to be one of the best. For a man who wrote with great clarity, he was astonishingly incoherent in speech. Yet he moved precisely. I have never heard of anyone moving pedantically, but that is what Eddington did. In my later days as a lecturer and professor at Cambridge, I conventionally wore a long black academic gown. Additionally, Eddington wore a mortarboard, which we called a square. For senior members of the university, like Eddington, the tassel was long and was worn dangling to the front, but for students, the tassel was short and was thrown carelessly to the back. It was a trick among undergraduates to go to a lot of trouble to pick their squares to pieces and then to reassemble the bits in distorted shapes. A common form of the trick was simply to remove the cardboard interior of the square itself, giving the outfit a gloomy, drooping appearance. Others cut the size of the squares by half or so (which is to say, cut the area by four). The point was that students, research students as well as undergraduates, were required to wear their squares in the streets after dark, the game being to see how far the modification of the squares could be taken without inviting a challenge from the progs, the university proctors. The problem for the proctors was to describe the offense, their verbal command usually not being up to it. On the few occasions when precedent demanded that, as a university officer, I should wear a square, I always put it on at a slight rake, the way service officers did and the way everybody else did, except Eddington. Eddington always wore his square precisely straight.

I am thankful that, throughout my teaching career at Cambridge, lectures in mathematics were given in what was called the Arts School in Bene't Street, hard by Barclays Bank, not in the aseptic rooms currently used in Mill Lane. In the academic year 1935–1936, I attended a lecture course of Eddington's. The beginning of every lecture was exactly the same. Eddington would come on the stroke of the hour, unlike the burly Ralph Fowler, who would appear five to ten minutes late, look up at the wall clock, and then roar with laughter. Eddington made for the standard rather high podium with which all lecture rooms were equipped. First he deposited his papers. Then he took off his square and laid it carefully beside a chalk box on a table to the right, a table that formed an integral part of the podium. Ultimately, he reached into an interior pocket, took out an empty spectacle case, carefully removed the pince-nez he had been wearing, carefully put it into the empty case, and carefully returned it to the inner pocket. Then he took out another pince-nez (presumably for reading), put it on, and returned the second case to his pocket. After looking for a moment or two at his notes, he at last set off in a rather high voice.

Which was where the trouble began. Charles Goodwin put it well: "Eddington gave a moan and then stopped for what seemed a very long time. He moaned again and stopped again for a very long time. Then he shook his head vigorously and said: 'No! That's wrong.'" The problem was that Eddington had no proper connection between his brain and his mouth. As far as I could tell, he began in midsentence and stopped at the end of the hour, without any full stops between. He drifted along from one subsidiary clause to another, never finishing anything. Nor did he write any too clearly on the blackboard.

Eddington's ghost will be glad to hear that, bad as his expositions were, those of the great physicist Niels Bohr were worse. There were the same convolutions, but, whereas Eddington could at least be heard, Bohr spoke in a largely inaudible mumble—or at least he did on the one occasion when I heard him speak, and Rudolf Peierls assured me that it was always the same. A friend who was also a friend of Bohr's told me he had once raised the matter of Bohr's mumbling. Why didn't he take a grip on himself and speak more clearly? After a minute or two of thought, Bohr replied with a gentle smile: "Because I don't like to speak more clearly than I think."

Ben Preston (1819–1902), my great-grandfather. He was said to have written the well-known hymn "Onward Christian Soldiers."

My great-grandmother Anne Preston (née Aitken), the wife of Ben Preston.

Mary Ellen (Polly) Pickard, my maternal grandmother, who, in later years, terrorized the village.

William Pickard, my maternal grandfather, who died when my mother was a small child.

My mother, in a photograph taken in the early 1900s.

My father, in 1915 (at the time of his recruitment into the British Machine Gun Corps).

A portrait of my mother and me, from a fragment of a photograph that my father carried in his pocketbook in the trenches of the First World War.

My parents, in the mid-1920s.

Here I am in early childhood, evidently persuaded in the mistaken belief that the world is a better place than I ever subsequently found it to be.

Approaching the age of seven, with my sister.

At the age of seven, at Mornington Road School, I am third from right in the back row. This picture is in a time capsule with nothing intervening between then and now, and all the faces are needle-sharp in my memory except my own. The class is in the playground where the older lad with the horsewhip roared around, clogs on his feet.

Approaching the age of ten, I have profited by my years of truancy and am beginning to experiment with chemicals. Notice the high wall of Milnerfield on the right, up which none of us could climb.

In this church group, at the time of my first scholarship, I am on the right of the back row of boys.

My grandmother Polly's house, where I always got a penny whenever I called. Notice the wooden door opening directly onto the village sidewalk. There is coal in there, which is why there is a sweeping brush leaning against the wall. There were few petty thieves in those days—because, in those days, petty thieves were whipped. Yes, Sir!

The Five-Rise Locks, opened to loud huzzas in 1774, were then one of the engineering wonders of the world. There is steam in the middle of the picture. During the continuing months of my truancy, I spent much time here, learning practical fluid mechanics and thinking it might not be a bad idea to become a bargee. Even today, when I see the big barges on European rivers, the idea doesn't seem too bad. (*Photograph by D. Burrows.*)

Milnerfield. This was where the footman in white gloves came from. Everybody who came to live here died prematurely, according to the old women of our village. It is easy to see why it was eventually abandoned and was left to fall down stone by stone. (*Photographs courtesy of A. Mirfield and B. Moore.*)

This photograph reveals that gardening at Eldwick School had got completely out of hand. Tommy Murgatroyd is standing toward the right. He is carefully dressed, as I always remember him. The school was generous in its teaching of a sort of earthy common sense. (*Photograph courtesy of A. Mirfield and B. Moore.*)

A sketch of Bingley Grammar School made in 1929, at the middle of my time as a pupil there.

Without the encouragement and determination of Alan Smailes,
I would never have reached Cambridge.

Paul Dirac has an undeserved reputation for being impractical—an absurd notion, for he was actually immensely practical. He wrote clearly on the blackboard, and every word he said was clearly audible. He invariably opened a lecture at immense volume (even louder than the formidable Geoffrey Burbidge in modern times). Within thirty seconds, Dirac's volume was lower; within a minute, it was down even further, to a level he maintained for the rest of the lecture. This was, of course, in the days when there were no microphones, speakers, and amplifiers in lecture rooms. I eventually graduated from being Dirac's student to being a fellow professor, to the point where I knew him better than most. When I asked him one day about his lecturing technique, especially about beginning, he said: "Every lecture room has its own acoustics, and you can't spend ten minutes learning about it. So I start at such a high volume that people in the back row jump in their seats. Then I lower the volume progressively until I see them ease back comfortably, and then I know that it is the right volume."

Paradoxically, Eddington was in a sense also a very good lecturer. You remembered and thought a lot about the big issues he raised, long after you'd forgotten apparently much better presented lectures from others. So this was the background, as I made my way on the spring day in 1940 along Madingley Road towards the Observatories and to where I knew Eddington would be. Or, rather, it was my part of the background. To allow the reader to understand the situation better still, I should also explain the best known of the reasons why Eddington was justly famous. The matter had to do with a difficulty first made clear in the nineteenth century by William Thomson (Lord Kelvin) and by the German scientist Hermann von Helmholtz. The difficulty was predicated on the strong nineteenth-century belief in the indestructibility of the atoms of which the chemical elements are composed. On this supposition, the energy that can be got out of an ordinary-sized lump of material is confined to the yield from joining its constituent atoms into molecules—which is to say, chemical energy of the kind that we are familiar with in a fire. From a human point of view, a raging fire might seem a pretty formidable thing, but, from the point of view of supplying the energy of the Sun, any conceivable supply of chemical energy would soon exhaust itself, in at most a few thousand years. Nowadays we laugh at Archbishop Ussher's assertion that the events described on the first page of the Book of Genesis occurred in 4004 B.C. But, if you stick to chemistry, even very good chemistry, the Archbishop's estimate can be criticized only for being a bit on the long side.

In the latter part of the nineteenth century, Kelvin and Helmholtz independently realized that gravitation, while negligible as an energy source for small quantities of material, could be far more important than chemical energy for large masses like the Sun. A heavy compact object falls with increasing speed to the ground because of the Earth's gravity. A comet falls with increasing speed towards the Sun because of the Sun's gravity. And a body like either the Earth or the Sun becomes hot in its interior if it shrinks because of gravity. So, by shrinking slowly, the Sun might develop the energy (heat) to compensate for what it is losing through the flood of radiation that it emits into space. Kelvin and Helmholtz independently calculated that gravitation, as an energy source, could provide for the Sun's radiation over a period of about 10 million years, which was so much longer than the historical time scale that it seemed to solve the problem. The solar system was 10 million years old, and that was that.

But, unfortunately for Kelvin, that wasn't that, for along came geologists with the claim that many features on the Earth's surface require times of 100 million years or more for their production. A good place for people in England to see for themselves is at High Force Waterfall in Teesdale. Below the waterfall is a gorge about a quarter of a mile long, which then eases gradually into gentler slopes over a total distance of about a mile, gentler slopes that have evidently been smoothed by weathering from an earlier gorgelike chasm. The entire formation has clearly been cut by the river, and the question is, How long did it take? If the rocks were cheesy soft, the time might have been fairly short, but the rock is the notoriously hard Whin Sill. When one looks at the waterfall and at the rock, it is hard to believe that the river could cut its way backwards at more than an inch or two per century. So High Force Waterfall alone eats up a fair fraction of the Kelvin-Helmholtz estimate of 10 million years for the age of the solar system.

Kelvin reacted in the way in which I think I would have reacted in the same circumstances. His estimate was a precise one based on seemingly secure foundations, whereas geology is not in its nature an exact science. But secure as Kelvin's assumption of the indestructibility of atoms might have seemed, there was an uneasy weight about the geologists' point of view. The position became still more uneasy with the discovery of radioactivity, in 1897, by Henri Becquerel. The story is told that, when a junior assistant reported to the chief assistant in Becquerel's laboratory that photographic plates stored in the same cupboard as a jar of uranium salts had become fogged, he was simply told to store them in different cupboards. But, when Becquerel heard the tale, he knew otherwise. It soon transpired that a small fraction of the uranium atoms were ejecting rapidly moving nuclei of helium, and it was these that were fogging the photographic plates. Similar experiments were tried in many laboratories to see what other naturally occurring elements would fog photographic plates in a similar way, but the haul was thin. Of the elements present in the Earth in any quantity, only thorium would do it.

Mass for mass, radioactivity is a million times more powerful than chemical energy, and a hundred times more powerful than gravitation in the Sun. So Kelvin's position was undermined on two fronts. Here was a potentially more effective energy source, and the doctrine of the indestructibility of atoms was wrong. Yet Kelvin's position was still defensible tactically if not strategically, for there could not possibly be enough uranium or thorium in the Sun to produce a significant effect. Even so, the stable door was opened, although it wouldn't be for another twenty years or so that the horse would make up its mind to bolt out of it.

In my mind's eye, I see Francis William Aston as a dignified gray-haired figure. He is always dressed in a sports jacket and is always walking unobtrusively across the Great Square of Trinity College. Aston lived deeply in the shadow of the Cavendish Professor, Lord Rutherford, and later of James Chadwick and John Cockroft in a generation younger than his own. Yet his contribution to physics was immense. Of the nearly 300 stable isotopes of the elements, Aston isolated and measured the relative masses of more than 200. Counting carbon as 12 units of mass, those of the rest were all nearly integers, showing that all nuclei were made up either of protons or of protons together with some neutral particle having a mass nearly equal to that of the proton, a discovery that, in effect, predicted the existence of the neutron. In Aston's system, hydrogen was nearly one unit and helium nearly four units. In more precise terms, helium was about one percent less than the sum of the masses of four hydrogen atoms. This meant that, if, in some way, four hydrogen atoms could be joined together to produce a helium atom, there would be an energy yield about a thousand times greater than the gravitational energy source of Kelvin and Helmholtz. Instead of the lifetime of the Sun being 10 million years, it would become 10,000 million years (10 billion years), thus accommodating the geologists by a wide margin.

I am not sure if it was Eddington or the physicist Jean Baptiste Perrin in France (or someone else) who first noticed that the fusion of hydrogen to helium would resolve the conflict between Kelvin and the geologists. More important, of course, than the mere resolution of a human conflict, it would explain the source of the energy that has provided for the emergence and sustenance of all life on the Earth, not just today but throughout the geological eons. Not many of Eddington's personal papers seem to have survived. We know, however, that, by 1920 or thereabouts, he was pressing the point of explaining the physical details of the hydrogen-to-helium fusion process, a problem that was not to be solved until 1938–1939, then by Hans Bethe and by Bethe and Charles L. Critchfield. The best advice Eddington could get in 1920 was that fusion at the temperatures present inside the Sun would be impossible. It was in this situation that he made his famous reply, advising his physicist colleagues in the Cavendish Laboratory to go and find a hotter place.

When, in 1939, I became interested in astronomy, I had the immense advantage of knowing what the relevant fusion reactions were and how to calculate them. I had a second advantage in having a big stumbling block to progress removed at Oxford in the early 1930s by Tom Cowling, a student of Edward A. Milne. The earlier situation would not have been so difficult if a suggestion of the Princeton astronomer Henry Norris Russell had been accepted. Russell proposed that the generation of nuclear energy inside stars happens suddenly at a critical temperature, the way water boils at a critical temperature. But work by the famous astronomer-physicist Sir James Jeans claimed to show that any such temperature sensitivity was impossible, for the reason that the star would be unstable. It would start to pulsate in–out, in–out, with ever-growing violence until, in a comparatively short time, it would come apart altogether. The fact that stars were not like this, except perhaps in rare cases, was interpreted to mean that Russell's idea was wrong. Actually, it was right, as Bethe showed in 1939. With the benefit of hindsight, it is possible to see that, had the idea been accepted, most of the stellar calculations made up to 1950 could have been done already in the 1920s. How in these seemingly unpromising circumstances it was possible for Eddington to obtain results with high numerical accuracy still remains something of a miracle.

From a spectroscopic analysis of the light of a star, it is possible to estimate its surface temperature with good accuracy. Eddington's achievement was to calculate a star's energy output (luminosity) in terms of its mass and surface temperature. He compiled a list of stars for which masses, surface temperatures, and luminosities were all known from observation and showed that, provided his result applied to any one star in the list (for example, the Sun), it applied to them all. The result could be displayed in the form of a curve relating mass and luminosity—the mass–luminosity relation, as it became called. With the curve passing through the point representing the Sun, it passed through (or very close to) the points representing all the others. And this despite the range of the luminosities being enormous—the brightest stars were about a million times more luminous than the faintest.

While I was still at school, there had been a long-running debate at meetings in London of the Royal Astronomical Society, between the three theoreticians who dominated the British astronomical establishment, with Eddington alone on one side and Jeans and Milne opposing him together. I was told by those who sat in breathlessly on the debate that Eddington always seemed to lose. Obviously, he would seem to lose, because of his inability to end a sentence. But, in writing, Eddington always won. His curve made certain of that. What his critics would never understand was that, by using surface temperatures as well as masses, Eddington was circumventing the difficulties they raised. This should have been clear from Eddington's published papers, but, after fifty years in the game, I have come to doubt whether scientists ever read each other's papers, especially if they disagree about something judged to be important.

Over the previous six months, Ray Lyttleton and I had managed to improve on the problem. Using Bethe's nuclear reactions, we had avoided the need to combine surface temperatures and masses together. Given only the mass, we could calculate surface temperatures as well as luminosities separately. Or, what was the same thing, we could calculate the luminosity of a star and its radius separately, provided it was of a type known as a main-sequence star. For stars evolving off the main sequence, something else was needed, and, in the case of one of Eddington's favorite stars in his list, Capella, we thought we knew what the something else should be. The method we used was, in essence, the method still used today.

Like all young people who make an improvement on the work of an older generation, we thought we were all-knowing and wise. But this was quite wrong. It is always the first investigation to succeed with an element of precision that really counts. Early vague prognostications are not the same thing. The issue is to have something that can be calculated more or less exactly and that can then be shown to agree with observation, signaling to a younger generation, like ours, that a profitable line is there to be followed, making further progress inevitable. The ancients understood the position when they made Janus the god of doorways. It was Eddington's investigation that was the doorway, making inevitable all that has happened since in the theory of stellar structure and evolution. If we had not made our improvement, someone else would quickly have done so. The trail started with relatively simple main-sequence stars. It progressed from there to giant stars, to variable stars, to esoteric stars nearing the end of their lives, and ultimately to exploding stars, supernovae, which can temporarily be as bright as a whole galaxy. It was a triumph of what was to become pretty well the most secure part of astrophysics when supernova 1987A was found observationally to have the highly complex properties that had been expected from calculations of the most exquisite precision. It was my privilege to have a little to do with some of the stages along this trail, stages I shall mention from time to time in later chapters.

My business with Eddington on the spring day in 1940 was simple. He was an officer (Foreign Secretary) of the Royal Astronomical Society, and I wanted him to sign a form for me, something that I expected to take less than a minute. In the event, my meeting with Eddington took the best part of two hours. So far as my architectural knowledge goes, I believe the main Observatories building to be Georgian. I noticed most its pleasant, mellow appearance. As Plumian Professor of Astronomy and Director of the Observatories, Eddington followed the time-honored custom, now unfortunately discarded, of living in the right-hand wing of the building—right-hand as you approached it along the drive from Madingley Road. Eddington's friends, such as the Dutch scientist Willem de Sitter, well known for the famous de Sitter model of the Universe, must surely have penetrated into Eddington's wing on his visits to Cambridge, but no one I knew ever had. A considerable door barred you from the rooms and corridors where you can walk freely nowadays, and beside the door there was a bell. I rang it and, after a moment or two, there was a patter of feet on the other side. Eddington's sister (Miss Eddington, as she was always called—nobody ever used her first name) acted as his housekeeper, and it was she who was standing there when the door opened. I explained the purpose of my visit. Miss Eddington nodded, closed the door, and pattered away on the other side. The door opened off the library, in which I was able to occupy myself for the several minutes in which it took Eddington to appear. As I had anticipated, signing the form took only the short side of a minute. The matter satisfactorily completed, Eddington then did an amazing thing. He asked me what I had been doing recently, which was amazing in that he should have overcome his shyness to such an extent.

I doubt that Eddington remembered me from attending his lectures in 1935–1936. I had answered his questions on general relativity in the final part of the mathematical tripos. Since I did well enough in the examination, it is a reasonable presumption that he marked my answers generously, and, in such circumstances, examiners sometimes remember a student's name. But, more likely, in the months since I was elected a fellow of St. John's, I had attended several colloquia (or seminars, in modern speech) at the Observatories and had commented on issues in the voluminous style to which young men of my age were given. Unlike Eddington, I never had any difficulty in communicating from the brain to the mouth, or indeed thence to the stomach.

It was easy to answer Eddington's question. I told him about being able to calculate stellar surface temperatures as well as luminosities. His interest quickened when I came to the fraught issue of evolution off the main sequence. His own model, what he liked to call the standard model, led to central temperatures, for stars like Capella, that were much too low for hydrogen-to-helium fusion to be significant. The Russian-born American physicist George Gamow had recently suggested changing the nuclear fuel for such stars from hydrogen to lithium, which would indeed "burn" adequately at Eddington's low temperature. But I said I thought there was too little lithium in the world for this to be a viable possibility. Eddington asked what I thought the solution might be, and I said a developing nonuniformity of composition as more and more helium accumulated in the central regions. Eddington said no, it had in his opinion to be something to do with the opacity formula. I countered by saying that Lyttleton and I had actually integrated an explicit nonuniform model and had obtained a measure of expansion off the main sequence that would explain a case like Capella. What he might have replied I shall never know, for at this moment there came the faint tinkle of a bell. The animation fell from Eddington's face, which became set like the Ramsey and Muspratt portrait. "My tea," he said, and was gone all in a moment behind the door into the forbidden wing beyond, exactly like a character in *Alice in Wonderland*.

There were two things that would have astonished us equally if we had been told of them by a shadowy bystander to whom all things are known. We both believed that the Sun was made mostly of iron, two parts iron to one part of hydrogen, more or less. The spectrum of sunlight, chock-a-block with lines of iron, had made this belief seem natural to astronomers for more than fifty years. And there really is a solution to the problem of calculating the Sun's luminosity that is based on the notion that the Sun is 35 percent hydrogen and 65 percent iron. But there is also a second solution that requires only 1 percent iron and 99 percent hydrogen and helium (we didn't worry much about helium in those days). Nobody, to my knowledge, believed in the second solution, although I understand there are papers at Harvard Observatory that show Cecilia Payne (later Payne-Gaposchkin) had the composition of the second solution in stellar atmospheres as early as 1926. Those being the days of women's nonliberation, Cecilia's paper was never published—at least until Henry Norris Russell published a similar analysis a year or two later. These were for atmospheres, however, not the deep interiors of stars. The high-iron solution continued to reign supreme in the interim (at any rate, in the astronomical circles to which I was privy) until after the Second World War, when I was able to show, to my surprise, that the high-hydrogen, low-iron solution was to be preferred for interiors as well as for atmospheres. My paper on the matter confounded a doctrine of Lyttleton, who used to say there are three stages in the acceptance by the world of a new idea:

1. The idea is nonsense.

2. Somebody thought of it before you did.

3. We believed it all the time.

This matter of the high-hydrogen solution was the only occasion, in my experience, when the first and second of these stages were missing.

A further respect in which information from our shadowy omniscient bystander would have astonished both Eddington and me was that I was destined to be the next astronomer to occupy his Chair. While this would surely be a good line on which to end this present chapter, there is something further about Eddington that I must mention. There has been a fashion, in recent times and in some quarters, to represent him as a mean and vengeful person. This is not consistent with my experience, as the second of my encounters with him will make clear.

The topic is again rather technical, for which I apologize. Recalling the in-and-out pulsations that James Jeans thought would occur disruptively in stars with temperature-sensitive nuclear reactions, there are certainly stars that do pulsate in and out repeatedly. The Cepheid variables provide an example. Cepheid variables play a critical role in establishing the distance scale of the entire Universe and therefore have been much studied and discussed. In his book *The Internal Constitution of the Stars*, Eddington devoted a whole chapter to them. Lyttleton was perhaps the first person to realize that the explanation given there for the pulsatory motion cannot be correct. Think of the cycles of an ordinary internal-combustion engine in analogy with the cycles of a Cepheid variable: the engine is kept going through the energy released by fuel consumption at a particular phase in each cycle. Eddington's explanation was similar, with the release of nuclear energy in the compression phase of a Cepheid taking the place of the energy released in the explosion phase of an internal-combustion engine. But just as an ordinary engine fails to work if its valves cease to function, so a Cepheid would fail to work for a lack of suitable operating valves. This realization led Lyttleton to a quite different idea.

Just as the Earth moves around the Sun, so it is the case that many stellar systems consist of two components moving around each other; and just as different planets have different distances from the Sun, so the separation of the stellar components in their orbits around each other can differ from one case to another. In some binaries, as they are called, the components are widely separated; in others, they are closely separated. Just how close could this separation be? Could it be so small that effectively the two stellar components touch each other, such that the components share a common atmosphere, swirling around in the manner of egg yolks swirled by an eggbeater? Could this be a Cepheid variable?

After the summer of 1940, neither Lyttleton nor I had much time for astronomy. Yet we managed to work up the several investigations described above into papers suitable for publication by the Royal Astronomical Society. By 1943, our Cepheid paper was ready and submitted. In contrast to the work on nuclear reactions, which was received favorably by the society—and indeed commented on favorably by Tom Cowling and E. A. Milne—the Cepheid paper ran into trouble. With hindsight, I can see that this was fair enough, for the eggbeater model eventually turned out to be wrong. Yet the paper began by proposing a possible solution to another problem that turned out to be right. This was the part in which we extended an idea of Gerard P. Kuiper, an influential and forceful Dutch-born American astronomer, that it was possible for material to be exchanged between the two components of a binary system, a possibility that has become of great interest with the modern development of x-ray astronomy.

By 1943, I was not in a position to go to London any day I pleased. It had therefore been with considerable effort that I had arranged to describe the Cepheid paper at a meeting of the Royal Astronomical Society. However, a few days before the meeting, I was informed that the society would be rejecting the paper. I then wanted to cancel my presentation of it, but my wife insisted that, if I still believed in the work, I should go. So I went. Eddington was there, in the front row. Now Eddington had recently published what appeared to me a very strange notion indeed—namely, that a Cepheid operated like an engine with stuck valves to which a periodic suction is applied at the exhaust pipe. This was an incredible perception that eventually turned out to be right. But there was no way of knowing it for sure until it was shown, in the 1960s, by Robert Christy of the California Institute of Technology, using a fast computer, to be so. So the score, as it eventually turned out to be, was that Eddington was right and that Lyttleton and I were right in the first part of our paper but wrong in the second part.

After I had done my bit at the London meeting, Eddington was called on to speak, and a fairly considerable argument ensued between us. A few days later, I had a follow-up letter from the secretary of the society, saying it had been decided to reverse the earlier decision. Our paper would now be published, at the behest (as I learned later) of Eddington. So this was my experience, an experience very far from suggesting him to be a mean and vengeful person. I recall that, at the meeting, I noticed he had dark circles under the eyes. I was too inexperienced to guess what this meant, and it was only with the news of his death the following year that the likely truth hit home to me. The lifetime score between us was one goal each, for I was to be proved right about nonuniformity of composition causing stellar giantism and Eddington was right about the Cepheids. I am glad I didn't lose the contest zero to two, but, at this distance in time, I am also glad I didn't win it two to zero.

In looking back, I now feel that Eddington's death was of much greater consequence for me than I realized at the time. In a strange way, a rapport had become established between us, despite our very different degrees of seniority. We were both basically shy, and we both came from pretty much the same part of the country. If events had been permitted to develop naturally in postwar years, I think my relations with the British astronomical establishment might well have been greatly changed.

Chapter 12

At War with Germany

I BEGAN with a dissertation on men's hats. Now it is the turn of women's hats, the tight-fitting helmetlike cloches that were considered smart beyond belief in the 1920s. What went on in women's heads that caused them to want so desperately to look that way? Well, something very different from what goes on in women's heads today. In a like fashion, what went on in the heads of the millions who clustered at 11:00 A.M. on 3 September 1939 around radio sets of antediluvian design to listen to the Prime Minister, Neville Chamberlain, was very different from what goes on today. As we heard his gravely delivered words—"This country is now at war with Germany. We are ready"—there can have been few who expected British losses would turn out to be smaller than they had been in the First World War, a total said to be roughly a million dead. Weapons, aircraft especially, had increased in efficiency over the intervening decades, so it seemed natural to think that the consequences of war would be worse. Yet, instead of everybody demanding in a frenzy that the war be stopped at all costs, the general mood was that it was about time. Indeed, there was an undercurrent of anger in the country because it had taken so long to come to it, and, consonant with this mood, Chamberlain's government has gone down in history as a deplorable group of weak appeasers. *Guilty Men* was published during the war by Victor Gollanoz (a publisher who bound his books in bilious yellow), a book that was later said to have had a big influence on the postwar general election of July 1945. Can one imagine a government that wanted so much to save a million British lives being denigrated today as a despicable crew of appeasers?

In the event, there were about a quarter of a million British deaths in the armed services, and about 100,000 British civilians lost their lives. Day in, day out, month in, month out, over six years, civilians lost their lives at an average rate of about forty per day. This would surely seem unacceptable to the modern generation, whereas, in 1945, my generation thought we had got away remarkably lightly. It can hardly be contested that an immense difference exists between the way I was conditioned to think and the way the media represent people as thinking today. To give the media their due, I think they could have prevented the outbreak of the First World War. I think too that the media played a significant role in exposing the bankruptcy of the socialist regimes of Eastern Europe, on the positive side of the ledger. But I do not think the media could have stopped the Second World War. By perpetual harping on the weakness of the British position in 1939, I suspect they would have undermined our resolve to stand fast, with the consequence that the confident faces we see so frequently nowadays on television screens would have grown up as Hitler's children—and bad luck to them, some of my generation might say.

I often wondered what Chamberlain meant by his last sentence: "We are ready." He might have meant the Spitfires and Hurricanes that were then coming into production, or he might have meant the protective radar screen that was nearing completion; but I doubt it. Almost certainly, he meant the blizzard of regulations that would overtake us in the next two months. Registration, evacuation of children from city centers, blacking out of light from windows at night, ration books, the slowing of trains to a crawl, and the cosmetic commitment of an expeditionary force to France, a force equipped with old-fashioned rifles, machine guns, and cannon. The inadequacy of it all was at least a proof to history that the appetite for war had not been strong in prewar Britain, whatever it became afterwards.

The registration to which all males in my age group were subjected shortly after the beginning of the war consisted in visiting a local labor exchange on a specified date. One was required to report name, address, and technical qualifications. Unlike the First World War, in which scientists were drafted into the armed services regardless of their qualifications, in the Second World War, scientists with research experience were not drafted at all. Some tried hard to get themselves into uniform, but it was just as hard for a research scientist to join the armed services in the Second World War as it had been hard to avoid it in the first.

Nor, so far as I could tell, was any scientist of appreciable ability or seniority ever instructed to take up work of a particular kind. Things were done by personal contact, as the Cavendish Laboratory had moved by special arrangement in the summer of 1939, mostly to join the Air Force radar establishment known as TRE (Telecommunications Research Establishment). If I had been around the Cavendish at that time, as I had been six months earlier, probably I would have moved with the others, first to Swanage, on the south coast, and then later to Malvern, in Worcestershire. But, because I had just been married, I was not in Cambridge at the time, and I therefore found myself, early in 1940, in the position of a forgotten man, which explains how I was able, in the spring of 1940, to discuss problems of astronomy with Eddington.

Lyttleton also found himself a forgotten man. Since we were under instructions to stay put until called upon, and since we were not called upon, the sensible thing was to continue working in astronomy. It was during this interregnum of ten months, as it fell out, that I decided to shift my research from theoretical physics to astronomy—a critical decision, as it was eventually to emerge. There were two good reasons for the switch. Surges in human activity are usually followed by troughs. Physics, from 1925 to 1935, had passed through a golden age. A pause for recuperation had become inevitable. Moreover, the effective creative lifetime for a theoretical physicist is usually reckoned to be not much more than ten years. With the war likely to eat up a considerable fraction of my own effective creative lifetime, there was an evident need to diversify my interests. The choice lay between astronomy, with Ray Lyttleton as mentor, and biology, with my friend George Carson. The decision in favor of astronomy was simply because the problems happened to fall that way.

There was one piece of work we did at that time, additional to the investigations mentioned in the previous chapter, that turned out to fit very well with later developments. It was already known that the diffuse gas existing between the stars tends to occur in clouds, but not until then had these clouds been thought to be relatively dense ones. Instead of a tenfold concentration of the gas, we had a concentration of a thousandfold or more, and we realized then what became accepted among astronomers only many years later, that hydrogen in dense clouds would exist largely in molecular form. This was another case in which Lyttleton and I experienced difficulties in gaining publication for the work, a situation quite unlike anything I had experienced in physics. Refusing each other's papers seemed to be endemic among astronomers, unless they were dull repetitions of what was known already. For the next fifty years, I never found anything different. In the late 1960s, at an international conference, a remark sometimes quoted was wrung in anguish from me: "Astronomers seem to live in terror that someday they will discover something important." According to the Oxford scientist Victor Clube, there are profound sociological reasons why this happens to be true, reasons I will discuss eventually in a later chapter. However, a third referee commented favorably on our paper, pouring a fair measure of scorn on the comments of the first two. From the ebullient style of the third referee, Lyttleton was convinced it had to be Henry Norris Russell.

Referees are permitted by editors and learned societies to remain anonymous, a practice that has always seemed to me objectionable, if not indeed corrupt. Corrupt it certainly is in some cases. It is wrong that an unknown person or persons should have access to new work several months in advance of anybody else, and the more important the work, the greater is the scope for shenanigans. It is not unknown for a referee to contrive the rejection of a paper and then to make use of what he has been privileged to read. On the other hand, a scrupulous person may be inhibited from following up his own independent ideas as a result of being asked to comment on similar ideas in a paper by someone else. I am told that Wolfgang Pauli was inhibited, in essentially this way, from publishing what today we call the Schrödinger equation.

On 3 September 1939, it was generally believed that, in a few years, the old way of life would all come back again, but of course it never did. The end of the war in 1945 brought nothing different except freedom from physical danger. The same gray, wartime existence was to continue in Britain for six years more, for a total of twelve years. It was to be a chasm in the lives of everyone.

In the Chesterton region of Cambridge, two small new houses had been built in de Freville Avenue, not far from where George Carson and I had shared lodgings. George was also married now, and he had rented a house nearby in Chesterton Hall Crescent, a house with a small garden in which stood a magnificent walnut tree that bore the sweetest thin-shelled nuts I have ever tasted. My wife and I returned from our honeymoon in Yorkshire to one of those houses in de Freville Avenue, which we had succeeded in renting. But the call away from Cambridge came eventually, from a chap at the Admiralty in Whitehall whose name was Fred Brundrett. I learned eventually that Maurice Pryce had followed the same road a month or two earlier, and I suppose that Brundrett learned of my existence through Pryce. I have never seen government papers describing how scientific recruitment was done, but my feeling is that those who did it, like Brundrett, were instructed to take care in fitting people to jobs. Exactly how someone like me (who had not worked in a laboratory since his schooldays, and whose background was in quantum mechanics and relativity) was to be fitted had presumably baffled the scientific civil service, which explains why I had been a forgotten man for so long. In an interview at the Admiralty, Brundrett asked me what I knew about radar. When I said I knew nothing at all about it, he didn't flinch, I can only think because he had already gone through the same thing with Maurice Pryce. A few months earlier, Ray Lyttleton and I had taken the initiative for ourselves by approaching the Meteorological Office to see if they had any use for our services. We had been asked what we knew of radiosonde. When we asked "Radio *what?*" we were quickly shown the door. Brundrett didn't show me the door. Instead he told me, with remarkable frankness, that eight out of ten of the high-ups with whom he unfortunately had to deal were dead from the neck up. Then he offered me a salary that was less than a third of my current emoluments. I could see that Brundrett, an apparently small and ineffective chap, was really a clever fellow, and I admired him for it.

So it came about, in the autumn of 1940, that my wife packed my suitcase and I walked out of the bright little house in de Freville Avenue for the last time. We promised each other that I would soon find a house to rent, wherever it was that the Admiralty intended to send me—somewhere near Portsmouth, it seemed, since I had been told to report that afternoon at Signal School, Portsmouth. This was exactly the period in which the Luftwaffe, after failing to bomb London and the south-coast ports by day, was doing so by night. This was the period that contributed most heavily to the eventual average of forty civilian deaths per day for the six-year duration of the war. Naturally, our goodbyes were emotionally tinged; and, in truth, there were few nights, over the next two years, when enemy bombers were not overheard.

I have, by now, had the sad experience of walking out of a house for the last time on eight occasions—about the same as the number of teeth I have lost. Only one of the eight houses do I still actively regret having left (although, through the sixteen years we lived there, it was a devil of a place to cope with); but I look back on the last moments before leaving each of those houses with some sadness, whereas I look back on the moments when a dentist yanked out my teeth (their positions being such that their removal had no serious cosmetic consequences) with complete indifference. Maybe there is something important in that. It suggests that what we are is a mixture of hardware and software, and what really matters is the software. When a piece of the hardware gives trouble, we are glad to be rid of it. Were there "garages" in the human body into which new components could be fitted—so that, for example, a worn-out disk in the spine might easily be replaced by a new disk—I think we would look back on the old discarded component with a similar indifference. Indeed, we should be delighted to have an aching back replaced by a new back that doesn't ache. But, for houses, the sadness remains, and this is because the days on which we leave houses for the last time serve as the ends of chapters in our lives.

The chapter that closed on that autumn day in 1940 had begun also on an autumn day, in early October 1933, when I had carried my green truck with wooden hoops around it from the Cambridge railway station to my assigned digs in Mill Road, at the beginning of my first undergraduate year. In those seven years (almost to a day), I had passed gradually from a highly tentative beginning to a more assured conclusion in the final undergraduate examinations, and thence from the tentative beginnings of research to my college fellowship in 1939 and, at last, both to my marriage and to work that had a measure of lasting value. There is a cliché to the effect that the undergraduate years are the best of our lives, and, as one grows older, the truth of the cliché becomes apparent. Those were the years of joy, never to be repeated. As things fell out, I was to return to Cambridge for far longer than I had yet spent there. But it would be a return to responsibility, to some measure of achievement, and (less happily) to a confrontation with the more petty aspects of human nature. After 1972, there would be sixteen magical years, but magic is not quite the same as joy.

In the interests of conserving coal supplies, the number of trains from Cambridge to London had been much reduced, which meant that those there were were immensely long. Engines groaned and wheezed with much emission of steam at the need to drag such enormous weights. Starting up was a big production, with slipbacks occurring whenever the gradient went slightly the wrong way. Because the trains stopped at every station, the slipbacks became a way of life. In short, the journey was ultraslow, giving me time to reflect. One reflection was of a train journey from Cambridge to London in 1937, in connection with a meeting on beta decay at the Royal Society. The return trip cost six shillings (thirty pence), and the time into Liverpool Street was just three minutes over the hour. Repeatedly, in the second half of the twentieth century, British Rail advertised its intention to make the journey from Cambridge to Liverpool Street in "under the hour." I can't say it never did, but, if it did, I never experienced it. The fastest journey I ever did, and the cheapest, was the one in 1937. It has been just the same with the driving time for the eighty-odd miles from Cambridge to Oxford. In 1939, Lyttleton and I made it in just two hours in Lyttleton's little Morris 7, and on poor narrow roads. Today I make it in just two hours in a six-cylinder job that the manufacturers claim has a top speed of 140 miles per hour. It is the same with traffic in London. The horse-drawn cabs of 1900 managed 11 miles per hour. The modern taxi also manages an average of 11 miles per hour, the private motorist usually less. All of which goes to show that the world has its own way of doing things, a way that makes a mockery of our human wishes and desires.

The enterprise on which I was setting out did not seem too well chosen, nor was it. Radar depends on knowledge and ideas that I had last encountered in my second undergraduate year. There was little I had been involved with in the last five years that would be useful, and this seemed a waste. Then, if radar it had to be, I would surely have been better placed with the Royal Air Force than the Navy. The reason has to do with logistics. My best hope of contributing to what was called the war effort would be in ideas, not in hardware—that was for the engineer and the experimental physicist to construct. A plane returns to base in a few hours, but a fighting ship may not return to port, after a refit, for one to two years. A new idea can thus be tried out in planes on a day-by-day basis, whereas anything new in a ship had to be completed and carefully engineered before it could be made useful. This necessarily made the development of new ideas a much slower-moving business for the Navy than it was for the Air Force.

There was a small nuclear-bomb-making project that I knew about, centered at Cambridge, but I hadn't been attracted by it and had not tried to get into it, despite the fact that being able to remain in Cambridge would have been far more convenient. Right up to the successful bomb test by the American Manhattan Project, I remained hopeful that the thing wouldn't work.

Then there were the deciphering chaps centered at Bletchley. Perhaps if I had known as much about codes and the like as I did about the nuclear situation, I would have tried to get into the symbol-chasing race and so might have contributed to really winning the war, according to some. The claim has been made that the German master code operated by the machine called Enigma was cracked so completely that the British government knew all the time exactly what the German high command intended to do. Without in any way doubting that the decipherment chaps at Bletchley played a key role on specific occasions, I find it difficult to accept the breadth of the claim that has been thus expressed.

The shape of the claim is wrong, rather as Father Brown said the shape of the murder weapon was wrong. Consider the formidable extent of Germany's military achievement. Almost single-handed, Germany took on not only both superpowers, but also Poland, Belgium, Holland, Denmark, Norway, France, Britain, Canada, Australia, South Africa, Greece, and Yugoslavia. Germany carried on the struggle for six years, maintaining the upper hand for a fair fraction of the time. To claim that all this was done with the Allies knowing in advance what each major move was going to be surely crosses the boundary of the preposterous.

As a riposte to the argument that the Allies never used the information obtained from Enigma messages because the Germans might otherwise guess that their messages were being deciphered, there are many examples of situations in which information, if it had been available, could have been used without the Germans suspecting in the least that their code had been cracked. In September 1943, Italy gave up the war on Germany's side. Allied prisoners of war in northern Italy could, for a while, simply walk out of the camps to which, until this point, they had been confined. On the assumption by the Allied command that the Germans would retreat rapidly to a defensive position at the Alps, the prisoners of war were advised to stay where they were, it being thought they could then be reached more quickly than if they were first scattered into the local population. The Germans did not retreat to the Alps, however. They elected to fight every inch north from Capua to the Alps. They took all control away from the Italians, and prisoners of war who had stayed put on the advice they had received unfortunately remained prisoners of war, but under German rather than Italian control. If the Allies had known from Enigma that the Germans were going to behave as they did, the prisoners of war could have been told to scatter, without there being any danger at all of Enigma's cover being blown.

Because of the underground (the subway system), at no time in the war was London a difficult city to get around in. The Inner Circle took me from Liverpool Street to Victoria Station, where I caught the next train to Portsmouth. On account of its being electrified, the Southern Railway ran more smartly than the other systems. There were fast trains stopping only two or three times, and there were slow ones stopping at every station. Somehow, typically, I ended up on a slow train.

Portsmouth was gray in a thin rain, more burned and wrecked than any city I'd yet seen, but not to compare with Cologne and other German cities in 1945. (Of all those I saw, I think Aachen in 1945 was the worst.) I took a taxi to the naval docks, and, as it drove through the dilapidated streets, I didn't like what I saw. In only a few hours, I had traveled psychologically a very long way from the bright little house in de Freville Avenue.

The boss at Signal School was named C. E. Horton. I don't think he ever told us what the initials stood for. You often didn't, in those days, and it was a convention that you never asked anyone to divulge a first name. I became friendly with a chap called MacIver, C. C. MacIver. On weekends, we went walking together, but I don't recall that I ever discovered what the C. C. stood for. We simply called him Mac, nicknames being fine but not real first names. Horton looked pretty old to my youthful eye, but he was probably about forty. Chunky in build with prominent eyebrows and spectacles that gave him an owlish look, Horton had a splendid habit of minuting on contentious issues: "Important. *If* true." He asked me what I knew about radar. I said that, beyond the very little

I had been told when interviewed at the Admiralty in London, I knew nothing. He then suggested that I look over a document while he attended to some matter or other.

The document was concerned with determining the speed of an approaching aircraft by measuring the Doppler shift of the reflected radar pulses. Actually, it couldn't be done in those days, because frequencies at the transmitter and in the local oscillator at the receiver were not sufficiently stable. I didn't know this, but I could see the author had the frequency formula wrong. He had used the usual one-way Doppler formula, whereas in radar, with two-way transmission, the shift is doubled. I suppose Horton thought my comment wasn't too bad for someone who knew nothing about radar. At all events, he gave me a naval car from the docks to a hotel that was much the worse for wear. I spent a miserable evening reflecting once again on the bright little house in de Freville Avenue, staring sourly at a poster in the hotel lounge that read: "Careless Talk Costs Lives."

The following morning, they packed me off to an outstation in the countryside to the northeast of the town of Chichester. "Nutbourne" was a large field on which all the antennae (aerials, we called them) for naval radar were designed. The nearby villages were East Ashling, West Ashling, and Funtington. I took a local train from Portsmouth, leaving it some way before Chichester, I think at a stop called Hambrooke. There were no signposts on the country roads, ostensibly to trouble the Germans, but I was the one who was troubled, as I struggled along with my suitcase. Eventually, I arrived at a field gate with a guard, from whose presence I knew I had arrived, as one might say.

Beside the gate was a small office used by the boss, A. W. Ross, a Cambridge physicist a year or so older than me. This was on the eastern side of the gate. On the western side were workshops where prototype aerials were machined. Straight ahead was a largish square hut, and in the distance to the right there was a second, similar hut. The best feature of the place was a fine view of distant downs to the northeast. From a secretary in Ross's office, I learned that a work panic was out. She told me I had been assigned a table in the distant square hut to the right. In addition to the huts, there were sundry horse boxes scattered over the field—not with horses in them, but with aerial systems, in various stages of development, drooped over them. But it was on one particular horse box that the current activity was focused, a horse box not far from the gate. Chaps— "bods," as air-force slang had them—were swarming over it, some with soldering irons in their hands, others with soldering irons (switched off, I supposed) tucked into their belts. An immense amount of cabling radiated in all directions from the box. Every now and then, somebody would switch on the main electrical power. For a few moments, there would be a loud humming, but every such session ended in a bang and a shower of sparks, making me fancy that I smelled burning flesh. Then somebody would take bits of equipment to the workshops, where they did God knows what with it. I staggered with my suitcase to the more distant square hut, feeling as a zoologist might feel who has just come upon a real live dinosaur.

Chapter 13

The Nutbourne Saga

I TOOK FROM my suitcase a few books, including Abraham and Becker's *Classical Theory of Electricity and Magnetism*, unopened since my undergraduate days. I had already noted that this bucolic limbo of a place seemed to have no library. Abraham and Becker appeared to be my only rational link with the real world outside. If the shadowy figure who knows all things had been standing there and had told me I was destined to spend two years on that field, I would have been mightily depressed. If the ghostly figure had also told me I would be sorry when the day came for me to move on to larger fields, I would have been mightily astonished.

There is nothing to tell you today exactly where Nutbourne really was. I have looked for it several times. I know the spot roughly but not exactly; the fencing has been changed, and trees have grown up. Forty years on, I heard that the Admiralty still had a station in the area; so, the next time my wife and I were in West Sussex, we made a detour to find the field, but we found only grass and cows. Enquiry in the nearby village of Funtington revealed that the station had been moved two or three miles in a northerly direction. We eventually found another gate, another guard, and another name. What had started as an outstation of the Admiralty Signal School had become the Admiralty Signals Establishment in 1942, or thereabouts, and was now the Admiralty Weapons Establishment. I explained my business to the guard—or rather my lack of business, for we were there only in memory of times long past. The guard went away to telephone, and, within a few minutes, my wife and I were sitting drinking coffee in the station head's office. It brought a catch to the throat to find that those of us who had been there in the bad years of the war, 1940 through 1942, were still remembered.

I have, however, a mild complaint to make about the present-day situation. I feel the Admiralty and the owner of the Nutbourne field should agree to allowing a small plaque to be erected, a kind of X to mark the spot. Otherwise, how is the passerby to know where George Owen fell off the roof of one of the square huts? George was a slightly rotund mathematician from Cambridge, younger than me and so more susceptible to pressure from above. The pressure at Nutbourne was to have a soldering iron in one's hand for most of the time. A mystique of the soldering iron had grown up at Nutbourne: as long as you were soldering something, no matter what, you were seen to be doing your bit for the war effort. Having succumbed to the pressure, George Owen was working, soldering iron in hand, on the roof of the nearer square hut, when he rather naturally fell off balance. It wasn't the event itself that was noteworthy but the manner of it. People were always falling off things or putting their fingers across the main power lines. We had one chap with particularly dry hands whose idea of testing to see whether the electric power supply was restored after an air raid was to put his fingers deliberately across the 220-volt lines, using himself as a voltmeter. What George Owen contrived to do as he fell, a feat an expert tumbler would be hard pressed to emulate, was to land with each foot in a fire bucket filled with sand, one foot in one bucket, the other foot in another—astride a horse, as it were.

I had learned that I was to be billeted with a Mr. and Mrs. R. Murray in the village of East Ashling, some two miles away. With my books unpacked, and with another curious inspection of the dinosaur, from which flurries of sparks were still emerging, I set off with my suitcase to walk, first down a hill and past a duck pond in West Ashling, then along a narrow path through fields to East Ashling. The Murrays lived in an enclave opening off the Chichester-to-Funtington road. Their house, I found, was bright and cheerful, like the one in de Freville Avenue, except bigger and better. Ron Murray was a squarely built man in his forties. He was a keen gardener, so that his house and garden were two orders of magnitude superior to what I had left the previous morning. With the prospect of being able to rent such a house as this, my spirits rose for the first time since I had left Cambridge. What the prospects of renting any house at all really were I shall describe soon enough.

The phenomenon of billeting deserves comment, for this again is an illustration of the difference between our present-day self-centeredness and the way it was from 1940 through 1943. Billeting was then a way of life. Either you were billeted with someone or someone was billeted with you. A spinster living in a bijou cottage in a Cotswold village would have a couple of runny-nosed kids from the East End of London billeted with her—after a few years, with favorable results for all three, mostly. Between 1940 and 1943, Britain came as close to being a classless society as it had been at any time since William the Conqueror. We even had a vision of the moment in Shaw's *Arms and the Man*, when Captain

Bluntschli says: "I have the highest rank known. I am a free citizen of Switzerland." Had the vision become a reality, Britain would never have declined in the postwar years—quite the reverse. But, by 1944, the knives were out again, at first in the highest places, and then, after the election of July 1945, in the lowest. After a brief moment of promise, we were returned to the self-canceling rivalries that have persisted ever since.

People with whom others were billeted had the feeling, for the most part, that they were "doing their bit," and I was, in any case, warmly welcomed by Ron and Betty Murray. Ron Murray was a dentist in the nearby town of Chichester. He was an excellent dentist, a judgment well proved by a difficult filling he did for me a few months later that has stood the test of time for nearly half a century. On my first evening with the Murrays, I wrote a letter to my wife, saying that I was fine and that it would not be long before I had found some place to rent. Hope springs eternal, they say.

As the weeks and months passed, my interpretation of the poster that said "Careless Talk Costs Lives" was put increasingly to the test. Ron Murray was enthusiastic about the prosecution of the war. A small victory would keep him happy for days, while a defeat like the repulse in Greece and Crete would depress him greatly. With me working at a secret establishment, and therefore supposedly in the know about most things, it was certain he would be looking for morsels of information, especially if they led in a hopeful direction. To begin with, while I still knew next to nothing, the position was straightforward, but, as I began to read reports marked "Secret" and "Most Secret," I had a decision to make. Would I sit at dinner in the evenings like an eleven-stone lump of putty, or would I divulge an occasional morsel? I arrived at what I still believe to be a proper balance. I saw no harm in divulging what the enemy knew already. It is, and was, the instinct of intelligence to keep absolutely everything secret. Thus, the nature of radar was considered secret, which I saw as ridiculous, since the Germans knew very fully what radar was. When the Battle of Cape Matapan came along in March 1941, I saw no reason to hide that it was radar that made the Royal Navy's victory possible, since the Italians and the Germans knew it to their cost already. What I considered wrong, however, was to say anything about the technical details of the new form of radar that made that victory possible. An official history (from the *Encyclopaedia Britannica*) runs thus:

> At the end of March 1941 a sea battle was fought to the northwest of Crete, off Cape Matapan on the Peloponnesian mainland. In scattered fighting through the daylight hours of 28 March, the Italian fleet evaded a group of heavy ships under the command of Admiral

Cunningham. With the coming of darkness the Italians paused to regroup, thinking that safety had come with the fading of the early-spring twilight. But owing to new radar equipment [from Signal School], operating on the hitherto unprecedentedly short wavelength of 1 metre, the Italians took such heavy losses during the ensuing night that never again did the Italian Navy attempt any surface ventures in the Mediterranean.

What I did hide was the bit about the wavelength being 1 meter.

There were differences between scientists and professional intelligence at a higher level than mine. Richard Feynman's tussles with security at the Los Alamos laboratory of the Manhattan Project have become widely known. Paul Dirac's sense of humor had about an average score, but it was curiously constituted. I doubt that a typical sitcom would have amused Dirac, although the TV series "Yes, Prime Minister" probably would have done so. What Dirac particularly liked were situations in which the practices of bureaucracies contained the logical seeds of their own destruction, as in the proposition that the greatest threat to national security comes from the security agencies themselves. I would like to make a loop in my narrative, at this point, simply to illustrate this question of security a little more closely.

Where precision of definition was important, it was generally the case that, the shorter the radar wavelength, the better. It was the unprecedentedly short wavelength of 1 meter that made the action at Cape Matapan possible. Owing to limitations set by the speed of light, it was generally thought, in 1941, that a wavelength of 1 meter was nearly as low as it was possible to go. I think the great physicist Ernest Rutherford deserves credit for breaking this belief. Rutherford thought anything was possible, provided it was not contradicted directly by the laws of physics. He also believed that discoveries, even very fundamental discoveries, could be achieved by simple means—moreover, by simple means not costing the immense sums of money that scientists today think essential to success. Rutherford conveyed his views emphatically to his students, among whom was an Australian called Mark Oliphant. By 1941, Oliphant was head of physics at Birmingham University, and it was in Oliphant's laboratory that the break was made—and by simple means, just as Rutherford would have predicted.

To understand the solution, think of a cylinder of copper—say, three inches long and two inches in diameter. Cut out the interior of the cylinder so that you are left with a cylindrical shell half an inch thick. Now drill perhaps eight cylindrical holes along the length of the shell, arranging them uniformly and with slots connecting them to the inner empty space.

Next, put along the axis of the cylinder a wire of a suitable material to emit electrons in large number when heated—that is, a cathode. Next, evacuate and seal the device. If the cathode is heated, and if a voltage is put between the cathode and the outer copper shell, an electric current flows across the inner space—nothing remarkable yet. To complete the program, arrange for a strong magnetic field to be directed parallel to the length of the cylinder. The electrons, instead of flowing smoothly, break now into bunches that emit radio waves of remarkably short wavelengths, much shorter than can be obtained from a normal transmitter, and the power output can be commendably high. It is not hard, for example, to obtain megawatts of power at a wavelength of 10 centimeters, which is only one-tenth the wavelength used at Cape Matapan (1 meter). And there is further good news: If this device (a magnetron) is inserted in a suitable way into a metal pipe (a waveguide), the radio output travels along the pipe. And, if the pipe opens up in a suitable way into a horn at the focus of a metallic dish of appropriate shape, the radio waves will be broadcast out into space.

All this was known before the war, and it was particularly well known in Germany, where the properties of the magnetron were closely studied. The bad news was that such a magnetron produces a chaotic set of radio wavelengths—far too chaotic to be of any practical use, the Germans decided. A reasonably well-defined wavelength is crucial in radar; otherwise, efficiency declines badly at every stage of the process. The waves tend to get choked in the pipe to the aerial—mismatched, as we said—causing sparking and feedback that damage the equipment. The antenna performance is bad twice over, both at emission and at reception, and receiver tuning is also inefficient.

H. A. Boot and J. T. Randall, working in Mark Oliphant's laboratory at Birmingham, found that, by interconnecting the cylindrical holes by wires—"straps," as they came to be called—all these many disasters could be avoided. The electron bunches became much more ordered, and the radio waves they produced, although by no means confined to a particular wavelength, were sufficiently well defined for a glittering range of new possibilities to be opened up. For the Air Force, there was the possibility of obtaining a detailed radar map of the ground to assist planes flying at night (the technique being similar to that used in later years for determining the surface terrain of the planet Venus and for obtaining distance scales within the solar system). For the Army and Navy, there was the possibility of accurately controlled gunfire, and, from the point of view of those of us at Signal School, there was the possibility of detecting U-boats at long range when they came up to the ocean surface during the night hours to recharge their batteries.

To produce the magnetron in large numbers, to equip ships and aircraft so widely with equipment featuring it on so great a scale, hundreds and probably thousands of bods, manufacturers, airmen, radar operators, and officers had to be in the know. Yet a knowledge of the significance of the magnetron's "straps" never reached the Germans, even though, in the late stages of the war, a magnetron from a wrecked plane actually fell into their hands. The ability of lots of ordinary people to keep a secret is the opposite of what spy-story writers would have us believe. They did so a great deal better than some supposedly important people working in intelligence agencies were able to manage.

The morning following my arrival at the Murrays', I determined (as I walked the two miles back to the Nutbourne field) that I must somehow acquire a bicycle—*somehow* because, by then, new bicycles were simply unobtainable, and everyone was on the lookout for secondhand bicycles for sale. Eventually, I did acquire one secondhand, from a chap called Davies, who had a garage (with a dark, mysterious workshop attached to it) in the village of Funtington. He charged 5 shillings for it, and he said he was robbing me—which, to a degree, he was. The tires were a seriously weak spot. Of course, you couldn't buy tires either, so repairing punctures was to be a way of life for several years. But the worst things about that bike were the chain and the main sprocket wheel. Whenever any real pressure was put on it, the chain jumped several sprockets. I was destined to ride thousands of miles on that bike—at 10 miles per day on average, roughly 10,000 miles. The chain must have jumped at least 30,000 times. Every time it jumped, I made a vow that, come the end of the war, I would never again ride a bicycle, a vow I have kept religiously.

With the first seriously rainy days, it also became clear that I needed protective apparel to wear on my way to work, in order to avoid sitting wet and miserable in the more distant of the two square huts on the Nutbourne field. Go and buy something, you say. But I had blown the last of my clothing coupons to buy a pair of shoes, also to keep out the wet. I managed to acquire a sort of tweed hat in a jumble sale, as well as a rubbery cape that had originally seen service as a ground sheet for a tent. Dressed in this outfit, and with the bicycle chain persistently jumping its sprockets, I looked, as Maurice Pryce used to say, as if I were playing a part—which, in a manner of speaking, I was.

We were kept out of uniform for sensible reasons. I am sure that, without the freedom to stand up for what we thought might be correct, many of the scientific ideas that emerged would never have bubbled to the surface. The negative side of being kept out of uniform was that we received no assistance in the minutiae of daily living, no issues of shoes from naval stores, and so on. Politically, to the bureaucrats of Whitehall, the situation might have seemed desirable as a denial of elitism and privilege. On occasion, however, it went beyond what could reasonably be considered fair. In December 1944, I had to visit a radar research station in Canada, wearing only light clothing, when the air temperature was -30°C (-22°F). I remember walking two miles after dark and seriously wondering if I could ward off hypothermia. On that visit, I had to climb 50 feet up a vertical steel ladder in the middle of a field, without gloves. I still remember my fingers sticking to the metal. There would have been a scandal in Parliament if service personnel had been required to suffer such conditions. New radar equipment didn't leap magically from the manufacturers to the ships to which it was meant to be fitted. Someone who knew the equipment had to supervise the fitting, and someone had to instruct ships' radar officers about the details of its use. Signal School (and, later, Admiralty Signal Establishment) had a whole section that undertook this work. They were off traveling a good part of the time, frequently at night. Yet, because they wore civilian clothes, it would be quite awkward to get even a cup of tea at a NAAFI (Navy Army Air Force Institute) canteen. Nothing was done to obviate these problems. In 1944, when I had to cross the Atlantic in a boat with a top speed no greater than that of a surfaced U-boat, an occasion not without a fair amount of risk, I found that no provision could be made for my wife to receive a pension if risk became reality. By then, we had two young children, which made me feel keenly that the situation had been poorly thought through.

There was one notable exception to all this. In later years, when we had to travel quite a bit about the country, we used an official Royal Navy car. To save the need for a driver, we drove ourselves. It proved possible to stop at any Army camp or Royal Air Force station to fill the tank with petrol and to get a meal at the officers' mess. On one occasion, when the front wheel bearings of my car seized, the nearest RAF station repaired the problem, requiring only my signature. After that, we came to realize that we could get all manner of equipment from station adjutants, provided only that we signed for what we received. Although acquiring necessities from the Navy remained difficult, it seemed that the key to obtaining almost anything from the Army and the Air Force was simply to sign for it. An exuberant and personable member of my section used to say you could even sleep with a WAAF, provided you signed for her. These interservice courtesies had come a long way since Neville Chamberlain's "We are ready" on 3 September 1939.

On the morning following my arrival at the Murrays', the bods were still at it on the Nutbourne field, swarming like bees around the horse box. More humming, more bangs, more sparks, more soldering. There came a day when the fanciful equipment blew up once too often and, like a forlorn colony of animals, the bods abandoned it. But not before I had ample time to read Abraham and Becker. I found a few books belonging to the others, and I read them also. Then I found a cache of Admiralty reports in Ross's office, and I read them too.

I began to appreciate that things were not quite what they were supposed to be. Bombs, shells, and high explosives generally have a mystique that is hard to quash. Bombs and shells acquire emphatic notice because the blast they create is felt, at least to a minor degree, over a wide area. From the first of the air raids I experienced, a problem with this mystique had been at the back of my mind. You were awakened maybe at 2:30 A.M. by a siren whose loud wail had been modeled on the hooters of the textile factories of my youth, hooters that, at 6:00 A.M., awakened the ill-paid laborers, who proceeded to pour darkly down from my native village to the satanic mills below. You heeded the air-raid siren by pouring down from your bed to whatever place in the house you had decided provided the best shelter from broken glass and from the roof falling in. My own favorite place was under a stout table. The government's favorite place was a corrugated-iron affair called an Anderson shelter. This was another of the things that Neville Chamberlain meant when he said "We are ready." You were supposed to dig a hole in your garden, to concrete the area you'd excavated, and then to erect the Anderson according to specs. It gave no more protection against a direct hit than a stout table, and, if the hit was not direct, the table was just as good. Water soon accumulated on the floor of the Anderson so that, if you used one, pretty soon you were back in bed, but this time with pneumonia. Fortunately, people didn't use them. If you happened to be in a southern town, such as Portsmouth, the arrival of enemy bombers was signaled by the opening up of antiaircraft fire, between whose salvoes would come the steady patter of falling shrapnel. I must admit to being really scared by falling shrapnel. There was so much of it, and being killed by a couple of inches of jagged metal tearing into your skull was just as fatal as being blown to smithereens in the direct hit of a bomb. Once, I had to cross London in the middle of the night when bombing was going on. We weren't issued tin hats. So what you did was to find a dustbin with a lid to it. Then you made your way hopefully through the streets under the purloined lid.

When the enemy bombers arrived, they made a curious rhythmic noise with their engines that is hard to describe but that became so familiar that everybody knew it for what it was. After the noise had gone on for a while, there came the shrill whine of falling bombs, each whine followed by a distant *crump*, or by a louder, sharper noise, if they fell close. People generally learned to take a philosophical view. They learned that, short of a direct hit, they were not in much danger—and, if you were directly hit, you knew nothing more of it. This led to the saying that it was all a question of whether your name was written on the bomb. I verily believe that some people thought that this was literally true. Somebody back in Germany, or someone somewhere above the clouds, was occupied in writing British names on German bombs.

After going through this process of responding to an air-raid siren several times, you wondered, as you crouched under your table, what the Germans thought they were doing. They had occupied a fair fraction of the time of their best aeronautical and engineering brains to design the bombers that were now throbbing overhead. A big manufacturing effort had then been made to produce them. They were piloted by the cream of German youth, some of whom would not return from the present mission. And to what end? To the deaths of an average of forty ordinary British civilians per day. It seemed an extraordinarily convoluted and expensive way of killing people.

Those who remain anxious to support the high-explosive myth might argue that the Allied raids of later years were much more cost-effective in killing ordinary people. But this would be yet another illusion. The 90,000 who are said to have lost their lives in the British attack on Hamburg did so, not because of high explosives, but because the city of Hamburg contained a lot of wood and other combustible materials. Hamburg and other German cities were chemically susceptible to destruction by fire. The bombers were, in effect, simply so many matches. Cities made of sand or concrete or mud would not have experienced disasters of such magnitude.

Pursuit of the high-explosive myth is the surest way for a nation to beggar itself. Excessive expenditures on the Navy early in the twentieth century set off the decline of Britain. More recently, the Soviet Union was brought to a similar pass. And even the United States, which has the world's strongest economy, has been placed under considerable strain, running itself seriously into debt in pursuit of the myth.

A naval bombardment is an order of magnitude more effective than bombardment by land-based artillery. In a comparison between Nelson and Napoleon, Nelson's cannon weighed ten times more than Napoleon's and were ten times more accurate. They could be moved by sea from place to place at ten times the speed and at one-tenth the cost. Yet, even at sea, high explosives are not entirely overwhelming. The sinking of Germany's powerful battleship *Bismarck* required immense and sustained gunfire from the *King George V*, the *Rodney*, and the *Dorchester*, even after the

Bismarck had been disabled by a torpedo from a tiny swordfish plane from the *Ark Royal*. The contrast between the effectiveness of the torpedo and the relative inefficiency of traditional gunfire requires explanation.

Compared with the amount of energy released by a nuclear explosion, the amount released by detonating a quantity of a chemical explosive is very modest. It is about the same as that released in burning a lump of coal of similar size. The damage comes from the speed of the release, which causes a destructive pressure wave to spread out from the explosion point. In air, the strength of the blast falls off rapidly with distance from the explosion point, essentially as the square of the distance. It is this rapid falloff that explains why bombardment, to be effective in war, must be heavy enough for the explosion points of shells and bombs nearly to overlap each other. An explosion in water is different, however. In water, the blast wave falls off less rapidly, more as the distance than its square. Thus, explosions in water are a whole order of magnitude more damaging than explosions in air, which explains the efficiency of the torpedo that crippled the *Bismarck*, and it explains also the grim sense of danger everybody felt in response to the threat of the U-boats. Where U-boats were concerned, however, what was sauce for the goose was sauce for the gander. Just as torpedos from U-boats delivered their explosive energy into water, so did the depth charges with which the Allies hoped to destroy them. It is said that, of those Germans who were drafted into the U-boat service, only a quarter survived.

As I studied the Admiralty reports during my first weeks at Nutbourne, it was gradually borne in on me how extraordinarily fragile ships are and how vulnerable they are to underwater explosions. I could see that everything possible was being done already to protect ships against submarine attack, but could anything be done about attacks from the air? The *Bismarck* had not yet been sunk, but it was already apparent that torpedo attacks from the air could, on occasion, be even more serious than underwater attacks. This indeed was to become very apparent in the early spring of 1941, at the time of the Battle of Cape Matapan, when far too many British ships were sunk from the air in their efforts to assist army landings in Greece and Crete.

Our ships had been provided with early-warning radar—Type 79, it was called. A ship's captain could therefore know an air attack was coming and could greet incoming enemy planes with all the barrage his ship could throw up into the air. But, for the reason I have just given, such barrages were more impressive in appearance than in their results. Guided ship-to-air missiles were still decades away, although their possibilities were being contemplated even at that early date. The best defense against air attack was interception by planes launched from aircraft carriers, but aircraft carriers were large and ungainly in shape and were themselves seriously vulnerable to attack in the same way. In the autumn of 1940, the balance between the defensive capabilities of carriers and their exposure to attack lay in favor of the enemy—or so it seemed to me as I read the Admiralty reports during my first days at Nutbourne.

The reports made it clear that, if each of our intercepting planes could know the altitude of an attacker as well as its range, the balance of advantage might be changed. Knowing the altitude could make the difference between attackers being driven off and carriers being sunk. Unfortunately, altitude information had not been a requirement when Type 79 radar was designed, and the hardware could not be modified to provide it in a direct way. But could the information be provided indirectly? It was from a report on the aerial design of Type 79, written by Ross, that I got the idea that eventually solved this problem.

Ross had designed the aerial of Type 79, the wavelength of which, about 7 meters, was ideally chosen for its purposes. The sea, being a good conductor, had the effect of producing a mirror-image aerial, which, together with the actual aerial, generated an interference pattern. This meant that, along certain angles of elevation, an aircraft would be invisible; along other angles, the equipment behaved as if it had four times its actual power. At small angles above the sea, there was a principal lobe, the design of the equipment being such that incoming aircraft would, in those days, be detected first in the principal lobe, and the greater the altitude of the aircraft, the greater would be the range at which it would first be detected. If you invert this statement, you will have a potential answer to the problem: the greater the range of detection, the greater the altitude, so that (in logic, anyway), by measuring the range of detection, you can infer the altitude. It was easy to conceive of a curve drawn on graph paper relating altitude and range, from which a ship's radar officer, having measured the range of detection, could simply read off the altitude. This could be done in seconds without any important waste of time, such as there would be if the officer had to make a calculation himself.

But how was I to draw the curve? Its shape (it wasn't a straight line) I knew quite well, but its precise placing in relation to the altitude and range scales depended on many factors—transmitter power, receiver sensitivity, cable losses, antenna design and performance, and the reflecting capacity of the aircraft in question. This was a formidable list of unknowns—or at least things unknown to me as a beginner in radar.

I now had something purposive to do. Some of the data I needed were available at Nutbourne, but data about transmitters and receivers had to be obtained in Portsmouth, where I was able to renew my association with Maurice Pryce. We hadn't seen much of each other since Pryce left

Cambridge, circa 1938, to take up a lectureship at Liverpool University. We talked of this and that, with consequences that were eventually of some importance.

Unfortunately, the result of all this cogitation turned out to be a disaster. My curve, which I wrote up as an Admiralty report, gave altitude estimates that were wrong. Always quite seriously troubled by mistakes, I took to walking the downs disconsolately whenever I could. My system for walking the downs was to leave my bike in Funtington because, if it turned out I had a slow puncture, I had Davies there to repair it for me. Open fields led up to the crest of the downs, the flanks of which had not yet been ploughed. On the higher ground were clumps of bushes with sweet grass between. The soil, although thin, supported a profusion of wildflowers in spring and summer. I usually walked in a wide arc, descending down the eastern arm of Kingley Vale. But, with the disaster problem on my mind, I came directly down into Kingley Vale through the deeply dark yew forest that covers its narrow head, wood from which forest must surely have provided the raw material for longbows used at Agincourt. The way through the vale is rather longer than one suspects, and it was while I was in the grassy bottom that the explanation of the disaster occurred to me.

I had appreciated that the performance figures I had been given for the various components of Type 79 radar would contain errors. I had thought there would be an error of 30 percent for each component. With six links in the chain between transmitter and receiver output, each link with an error of 30 percent, it seemed likely that the total error would translate into an error of only 15 percent in detection range (which depends on the fourth root of the total performance factor), an error that seemed just about acceptable. The mistake was that I had supposed the errors to combine randomly. The lectures of Eddington I had attended in 1935–1936, "Combination of Errors," had predicated randomness throughout, from the first lecture to the last, and so it had been natural for me to start from this assumption. But, if the errors were systematic, if there were a 30 percent error six times and all in the same direction, the result would be a big 50 percent error in the range of detection, leading to a similar inaccuracy in the associated altitude estimate. And this was just what seemed to have happened.

Why were the errors not random? Because each person had included a measure of hope in his estimate, and the estimates, therefore, were all in the same optimistic direction. It was the first time I had run into human factors so strongly affecting a scientific calculation. It was not to be the last time, for indeed this was to be an ever-recurring feature of controversial issues in later years.

So what was to be done next? Go back to all the people I had been talking to and ask them to take a more realistic view? I decided not. When human factors are involved, there is no reliable way to get them out. Some other device was needed, and what it should be did not take long to work out. All the uncertainties could be lumped together into a single uncertainty for the performance factor of the Type 79 radar. Instead of trying to say what the performance factor should be, I drew up separate curves of the previous kind for a spread of values of the performance factor; about ten curves for ten assumed performance factors was sufficient. What a ship's radar officer had now to do was to run a trial with a plane at known altitude, a plane of about the same size as the expected enemy planes. When the detection range of the trial plane had been found, it was simply a matter of picking out from my set of curves the one giving the measured detection range for the known altitude. This, then, was the right curve to use thereafter. I felt certain now that the method would work, and it did. Indeed, it worked so well that, despite all the more sophisticated equipment that was suggested for solving the altitude-determination problem, the simple Type 79 calibration method was used right to the end of the war. It had all sorts of advantages. It avoided unknown errors, just as I have explained. By repeatedly recalibrating always against known aircraft at known altitudes, variations in the performance factor could be allowed for when major components of the Type 79 were replaced; and, by changing the sizes of the calibrating aircraft to match changes in the enemy's attacking planes, that further uncertainty could also be eliminated.

Meanwhile, my attempts to find a house to rent continued. The only married person at Nutbourne to have found rentable accommodations within two or three miles of the field was Chris Fenwick, our chief engineer. He lived with his wife Grace and his small daughter Phillipa in an old dark cottage in the witches-in-the-forest tradition. It had neither electricity nor running water. Their water source was a well so plentifully stocked with organisms that it is no surprise that Phillipa became a microbiologist in later life. Other married folk at Nutbourne lived in distant towns—Chichester, mostly—and they commuted daily by station bus. It was, however, more difficult for me to enquire in distant towns than it was locally. My procedure was to spiral outwards to greater and greater distances from Nutbourne, knocking on doors and seeking intelligence as best I could. The results of my efforts are best illustrated by two examples. I found a squarish house with rooms to rent. The problem was that the place was full of cats. At my first and only reconnaissance, I counted eighteen of them. The circumstance that I considered the matter seriously put me under strain, to the point where never since have I been able to abide the sickly sweet stench of untrained cat. The second example was farther afield, in a village across the downs to the north of Funtington, perhaps four miles away. It was an adjunct in the grounds of a large house, the occupant of which was a lady with somewhat *haut monde* tendencies. Yes, she kindly said, the adjunct might be rented by someone as deserving as I was; except—and she was determined on this—nothing was to be removed. I looked over the property and thought it had sufficient strong points to write to my wife that she should come to look it over for herself. She came from Cambridge with her fine green three-speed bicycle on which she rode ahead of me on my junk pile over the intervening ridges to the north of Funtington.

The adjunct had both electricity and running water—strong points, truly. It had a serviceable kitchen—another strong point—and an adequate bedroom. The sitting room was a long rectangle—a corridor, almost—and it was here where the problem lay. Above head height, mounted along both long walls of the rectangle, were the heads of thirty-three stags. To get the feel of it, think of a traditional military wedding: The bride and groom emerge from the church, the bride in white and the groom in uniform. The groom's fellow officers are there in line at the sides of the path along which the happy pair must walk to their waiting car. The officers are there with drawn swords, which they hold high, crossing above the middle of the path, and under which the pair must walk. Well, that was the way it was whenever we walked the length of the narrow rectangular room, the difference being only that we had stags' antlers instead of sword blades. My wife and I have always known what the other was thinking, without words needing to be spoken. So we simply nodded at each other.

The lady was surprised we didn't want to rent the adjunct, and she looked at me with a decidedly *haut monde* questioning look when I told her so. The imp in my mind burst forth: "The trouble, you see," I explained, "is that there's no room for *our* stags' heads."

Eventually, we did find a house to rent, and at no great distance. In East Ashling, there were two largish unoccupied houses on which I kept a constant eye. I was told that, because the tenants were away for the duration of the war, these cases were hopeless. Nevertheless, I kept up my constant eye, to be rewarded one day by distant smoke. The house was in a quiet lane that opens off the Chichester road, Lye Lane, then unsurfaced and well stocked with potholes. I knew the house had two parts to it, the main house and an end cottage. It was the cottage I was after. I rang the bell on the door of the main house with hope but without much confidence. Yet all bad things, like good things, seem to come to an end. In the Second World War, Tom Groves had been badly wounded in the first battle of the

Somme, and he had much the same opinion of the competence of his commanders as my father had of the competence of his. Once I mentioned that my father had been a machine gunner from 1916 to 1918, there was never any question about the cottage being for rent—and at a modest valuation. There was running water, but not electricity for power; there was electricity for lighting from a generator with batteries. There was a vegetable garden that an old chap with a large goiter helped us to plant. I can see him in my mind's eye, but his name is gone from memory. (My wife feels sure it was Clapham.) There was a field opposite on the other side of Lye Lane where, in September, there grew more mushrooms than I have ever seen, except on a mushroom farm. The cottage itself had a downstairs kitchen with a coal-fired range, and there were two small rooms upstairs. The cooking range consumed too much coal for our meager standard ration, so we used the coal to warm ourselves in an upstairs sitting room. Through the ten months we lived in Lye Lane, my wife used a primus stove for cooking. Not a big primus, but the small camping primus I had bought in 1935. It had been quite a while since I had left the house in de Freville Avenue, but (as I had learned while hitchhiking in my student days), if you persist, you get there in the end.

Chapter 14

The Saga Continues

BY THE END of 1941, Germany had lost the war. This isn't a judgment of hindsight. It was apparent at the time, although, to those involved in the immediate urgency of military planning and action, it may not have seemed as evident as it did to those of us with the leisure to contemplate the events of what must surely have been one of the most remarkable years in all of British history. We started at the beginning of 1941 in a nearly desperate position—which is to say, a lost position—and we ended it in a certainly won position. The most astonishing feature of an astonishing transformation was that it came about largely through defeats, not victories. There had been only four bright features in 1940: the successful evacuation of the British Expeditionary Force from Dunkirk, the holding off of the Luftwaffe in the Battle of Britain, the relatively easy confusion of the German electronic guidance system in what became known as the Battle of the Beams, and the pushing back of the Italian Army from Egypt to Tripoli. The dominant mood at the end of 1940 was that our achievement lay in continuing to exist, not in winning a war whose end was looking infinitely far away.

The historian who sees all things from afar cannot appreciate, it seems to me, the emotional ups and downs of those who live through the day-by-day course of events. The ease, in the early months of 1941, with which we were robbed by Rommel and the Afrika Corps of our little victory of 1940 in North Africa produced deep depression in those of us who were not occupied in the detailed logistics of everyday planning. Worse was to follow. At the end of October 1940, Italy attacked Greece but came off second best in the fighting, an upbeat situation that led Churchill to make a radio broadcast in which he compared the modern Greek soldier favorably with those of ancient Greece. It was a stirring concept that suited the Churchillian manner, to which we all listened with much appreciation. But Hitler was having none of it, and, at the beginning of March

1941, the Wehrmacht moved south through the Balkans. The old story was repeated. Just as Poland had fallen in 1939; the Low Countries, France, and Norway in 1940; so now it was the turn of Yugoslavia, Rumania, Bulgaria, and Greece. Perhaps it was the memory of his recent broadcast that caused Churchill to make his quixotic gesture of sending yet another expeditionary force, now to Greece and to the Greek Islands—to Crete, particularly. Churchill was a past master in knowing what stirs the hearts of men, and, in 1940–1941, this was what we needed most. But, as a military tactician, Churchill scarcely came up to his skills as a writer and artist. Although the Greek venture gave us the victory of Cape Matapan, it led to a flood of our men pouring into German prisoner-of-war camps, and it brought too many of our ships within range of German dive bombers. So many ships were lost that the ability of the Navy to traverse the Mediterranean remained in question for months to come. It seemed a heavy price to pay for a quixotic gesture.

In June 1941, it happened that the two thieves who then dominated continental Europe fell out. The irony of the millennium, surely, was that Stalin trusted Hitler. Partly because of Stalin's bad judgment in this regard and partly because of his self-defeating extermination of his own generals, the German attack on the Soviet Union in June 1941 was at first phenomenally successful. I heard the news of the German attack in disbelief, thinking, at first, that it must be some desperate British propaganda designed to create tension between Germany and the Soviet Union, or some rush of blood to the head of the BBC. But Churchill's evening broadcast cast such doubts away. There had been whispers repeatedly of Britain reaching some sort of an accommodation with Germany. Now was the time, it seemed, when this might happen. But Churchill's broadcast dispelled that too. Churchill welcomed the Soviet Union to the boat in which we must cross the stormy seas together. In view of his past dealings with the Soviet Union, and in view of Churchill's continuing inner anti-Soviet conviction, it was a welcome that must have cost him greatly.

My strong conviction that Germany had made a devastating strategic mistake was based, I must confess, merely on a picture in a school history book of Napoleon retreating from Moscow and on hearing Tchaikovsky's *1812 Overture* performed at a Henry Wood Promenade Concert. But correct it proved to be. Yet, as the weeks and months of the summer of 1941 passed by with the German Army penetrating ever more deeply into first the Ukraine and then Russia proper, it seemed as if successful tactics were destined to overcome bad strategy, as sometimes happens in a game of chess. However, in the autumn of 1941, with German communications and supply lines ever lengthening and with the Soviet lines ever shortening, a stable position a little to the west of Moscow was eventually reached, a position that the Germans could not break before the onset of winter.

From the German point of view, the situation could still have been saved. At the moment when it first became apparent that Moscow could not be reached, when the defenses also held at Leningrad, there should have been a wholesale retreat, back to the comparative warmth of the western winter, thereby preserving the Wehrmacht intact. The Soviet Union could be reattacked the following spring, if need be, and the Soviet Union had, in any case, been gravely damaged at lesser German cost. There was, moreover, not much the Soviet Union could do in winter to repair itself. A temporary retreat could have spelled eventual victory, whereas leaving the army in the field, poorly clad against the Russian winter, spelled eventual disaster. When one looks at the unwholesome gallery of the world's past dictators, a pattern can be determined. A deep and ferocious cunning is a necessary qualification for membership in this club, as is also a total absence of humor and common sense.

It is worth pausing a moment to try to understand what lay behind the Japanese attack on Pearl Harbor on 7 December 1941. Faced with a rapidly rising population, and without the world trade it now enjoys, Japan had engaged, in the 1930s, in a number of imperialistic ventures, notably in China. This policy had affronted the Americans, who, over the years (for reasons that a non-American might find obscure), had developed an avuncular attitude towards China, the U.S. State Department being well-populated by what were called Old China Hands. Unwilling to confront Japan directly in the 1930s, the United States instead took economic measures, especially to reduce Japan's access to energy supplies. This, I believe, formed the underlying motivation for the attack at Pearl Harbor, a situation that Japan saw as an increasing economic stranglehold. To this, add that violence in one part of the world tends to breed violence in other parts; add too that a serious miscalculation was made. Japanese diplomats in Washington must have noticed the reluctance of the American public to allow the United States to become involved in the war in Europe, despite provocation and despite cultural ties with German-occupied nations and with Britain. The easy deduction must have been that, with its fleet heavily mauled by the Pearl Harbor attack, and with few or no bases available over the vast area of the Pacific Ocean, the United States would be slow to mount a military response, by which time the whole Pacific rim would have fallen under Japanese hegemony. If such a miscalculation was indeed made, it can at least be said that it was more understandable than the astonishing German mistake within days of Pearl Harbor. There have been numerous miscalculations in history, but I can think of none so gross or so unnecessary as Germany's declaration of war on the United States. Had Germany played the matter calmly, it would have been difficult politically for the U.S. government to give its first priority to the war in Europe. The public demand would surely have been, even granted that the United States would still have entered the war on the Allied side, to have given first priority to operations in the Pacific.

In early January 1942, my wife, who was expecting our first child, went to stay with her parents. As the expected day neared, I obtained compassionate leave. On my way from Chichester north, I contrived to meet with Ray Lyttleton for a few hours—we were still finishing our work on the structure and evolution of stars. Opportunities for discussion between us were now rare, and so the work progressed very slowly. Outside our astronomical talk, we discussed the current war situation, with Lyttleton saying that the sudden transformation, following the American entry into the war, was almost too great to be believed. In January 1942, we thought it was all over, bar a winding down that would probably occupy from one to two years, making the Second World War last about as long as the First World War. That it lasted a year and a half longer than we expected was to a major degree due, I believe, to an Allied insistence on unconditional surrender, a term that aroused the darkest thoughts in the German mind, and which therefore led to German persistence to the very last moment of what was physically possible. At the time I thought insistence on unconditional surrender to be plain stupid, but today I am not so sure. If the Allies had asked for anything less, the habits of mind that had led to the disaster of the Second World War would probably have persisted in some measure, ready to regrow and to sow death and destruction for the third time in the twentieth century. Anything less, and a new way of life and prosperity for Germany would probably not have been possible.

Meanwhile, we continued in our own small world at Nutbourne, a small world certainly, but one that, on occasion, could become entangled with the larger world with what seemed a startling suddenness. In his book *Most Secret War*, R. V. Jones gives a stirring account of the Battle of the Beams. From mid-September 1940 onward, the Luftwaffe changed from day raids to attacks by night using a navigation system based on intersecting radio beams of long wavelengths. The center of Coventry was spectacularly destroyed. According to Jones, it became a political question at the highest level as to how far the Midlands cities could be devastated by the same navigational system. Also according to Jones, radio experts from the prewar world, BBC radio engineers in particular, gave vague, equivocal answers. In reading *Most Secret War*, I felt this criticism to be somewhat unfair, because the usual method for answering such questions involved an immense volume of calculation, requiring, for accuracy, much more time than was available in the urgency of the moment. In technical terms, the standard solution of the problem of diffraction of long waves around the Earth was very cumbersome to apply.

I had a call one day from Maurice Pryce, saying he had resolved the diffraction problem in a form that could be calculated with much less labor and asking if I would check his solution to verify that it was correct. We met the following day, and Maurice handed me about a dozen sheets written in his usual impeccable hand. The solution was, of course, correct, and the question now was how to use the formulae obtained so that particular solutions of the kind encountered in the Battle of the Beams could be read quickly from graphs, rather than calculating solutions from the formulae themselves. Pryce asked me if I could handle this stage of the job. But calculating machines of the old-fashioned type would be needed, together with tables of functions, neither of which we had at Nutbourne. I would have been glad to have gone to Cambridge to use the facilities there, but this was at a time when I was still immersed in the altitude-finding problem described in the preceding chapter. So I asked Pryce if he would agree to having a young mathematics graduate, Cyril Domb (who had just arrived from Cambridge), take on the work. Pryce did agree, and arrangements were made for Domb to work not in Cambridge, as it turned out, but at the computing department under J. C. P. Miller of Liverpool University. The eventual result was indeed a set of graphs from which the solution to any such problem could be determined in about half an hour.

The work took about six weeks, with Maurice champing on the bit for more speed over the last three weeks—to no avail, for it was characteristic of Cyril Domb that he always took his own time. Two years later, the redoubtable Hermann Bondi said of Domb that he always took six weeks to solve anything, whatever the problem. If the problem was easy, you thought him slow; if it was hard, you thought him fast. This was at a time when there was a fad for solving little problems of the following type. On a table is a pile of nine coins, one of which is counterfeit and of a weight different from that of the others. There is also a balance. With no more than three uses of the balance, find the counterfeit coin. It was par for the course for people with a little mathematical aptitude to solve this problem in half a morning—if indeed you had half a morning to spare. Cyril Domb, however, took what seemed an unconscionable time. When at last he came up with his solution, it was quite different from the solutions the rest of us had come up with. Ours answered the question, that and nothing else. But you might wonder, What is special about nine coins? Why not twenty-nine, or any other number? Cyril's solution solved the entire class of all such problems, which rather distinguished the real mathematician from the mathematical craftsmen. There is actually one case in which the problem can't be solved (interestingly, the simplest case): when there are only two coins.

A year later, I had another call from Pryce, a call whose consequences were to spread in a multitude of directions. It was to lead me personally to the beginning of my work on the chemical elements. It was to lead indirectly to developments that would have consequences for NATO during the Cold War period of the postwar years. And it was to lead to Tommy Gold (even in those days a controversial figure) and me being thrown, in a disheveled state at midnight, into the main street of Fishguard. Maurice appeared not to think much of our discovery, although, to my surprise, I heard him speak glowingly of it a few years after the war to no less a person than the famous theoretical physicist Wolfgang Pauli. When he'd heard the story, Pauli responded, with his well-known derisive cackle, "Ah ha! That was done by Debye in 1909."

When we met shortly after his call, Pryce told me that the operators of the first of our radar types working at a short wavelength close to 10 centimeters had reported occasional days on which radar detection ranges for ships were far greater than they should have been, which meant the electromagnetic waves were traveling, and at considerable strength, far beyond the optical horizon. In a sense, this was like the diffraction problem that Pryce and Cyril Domb had solved. But diffraction goes the opposite way. It is important at long wavelengths, but it should have been almost negligible at such short wavelengths as 10 centimeters. So the reports, if they were true, had to be due to some mysterious kind of refraction by the Earth's atmosphere. But what should be causing the refraction? Not air, which was obviously too feeble. Pryce went on to say that he'd noticed that molecules of water had a very large dipole moment (this was the Debye part of the affair), which would make each molecule of water much more effective in refraction than a molecule of oxygen or nitrogen.

Unlike diffraction, the refraction effect became weaker at longer wavelengths. Therefore, to be meaningful, our report had to discuss wavelengths explicitly, but we were loath to put any useful information into enemy hands. The report, therefore, had to be classified MOST SECRET, a classification that always prevented wide circulation. And, to keep matters confidential from the beginning, I undertook to get the necessary typing done at Nutbourne, which turned out to be pretty awful, to Maurice's distress. Today, its scrubby production gives the report an attractive Old World charm. At the time, however, Maurice thought it would detract from its impact. But it didn't, and soon there were substantial repercussions.

Edward Appleton—a small, heavily built man who was then, I suppose, in his middle fifties—was a scientific battleship on the London
scene. Appleton decided to form an interservice propagation panel on
which various bodies were allotted a specified number of representatives.
The Admiralty got two and decided, at a rarified level, that one representative should be from Signal School and the other from the Hydrographer's staff, which meant a meteorologist. This turned out to be Frank
Westwater, who had been a year ahead of me studying mathematics at
Emmanuel College in the mid-1930s and to whom I had gone briefly for
supervision. To this point, Westwater had had an eventful war, having
been sunk twice. He had been senior surviving officer from a ship lost in
supporting the expeditionary landings in the eastern Mediterranean—
Churchill's quixotic gesture to the Greeks. Maurice Pryce insisted that I
should be the other member, whether from politeness or because he had
other fish to fry—I could never quite tell what Maurice was up to. It may
have been that he was heavily occupied in planning big changes at Signal
School, of which, to this point, I knew nothing. Or maybe he didn't like
Appleton—many people didn't. Or perhaps he thought the committee
would be an ineffective waste of time (which, in most respects, it was).

The terms of reference of Appleton's panel had cast the net wide. It was
to include propagation at all wavelengths, the very long as well as the very
short. This meant that all sorts had to be there, which was why we at the
Admiralty got so few representatives. But the basic facts and theory of
long-wave propagation were by now well known, so 80 percent of the
time of the committee was to be devoted to the short-wave effects of water
vapor, which Pryce and I had begun. There were mathematicians from
other establishments who reworked our models in what they implied
were new and more significant forms. Throughout my research career, I
have always disliked and despised this kind of disguised plagiarism. Unfortunately, it clutters research journals, and it explains to some degree
the insistent clamor for more government money. Governments remain
innocently unaware that plagiarism is a principal beneficiary of their munificence. Speaking personally, and perhaps a little self-righteously, never
once have I said to myself, "X has done something important. How can
I get in on the act?" Instead I say, "X has done something important. It's
a pity I was stupid enough to miss it."

There was one chap on the committee who moved in on the act in a
way I *didn't* dislike. His name was Cobbold. He was in his fifties, a florid-
faced man with a bit of a limp who found his métier by arranging propagation trials that he claimed were of importance to the Army. This
turned out eventually to be true, but in a way that even the Delphic oracle
would have been hard-pressed to prophesy. I wondered if Cobbold was a

scion of the famous brewing family, and the rather respectful way in which other committee members treated him suggested this might be true. I became convinced of it one day when I was walking with Cobbold in the London Strand. Suddenly he gasped out: "I must have a drink." We dived into some place and, to my surprise, he ordered two bottles of fizzy pop. When I told the Master of St. John's the story, he said, without hesitation, "Not a nice man to know."

Actually, Cobbold wasn't too bad a man to know, especially when he arranged one of his trials in East Anglia. The trials consisted in setting up a transmitter at a wavelength of 10 centimeters in one place and a receiver in another place about 100 miles away. The transmitter, in this case, was in the Sleaford area, and the receiver was outside Newmarket on the Cambridge side. Naturally, I was there, to Cobbold's delight, and I managed to learn two things, one at each end of the link. At the Sleaford end, we had an official army car driven by a woman, from which I learned that, for speed and safety, women could be at least as good at driving as men. At the receiving end, I stayed in Cambridge, borrowing George Carson's bicycle to ride out to the site each morning. One day, I bought a large bag of ripe plums. I ate the plums while riding. There being no decorous way in which to dispose of the stones, I learned to blow them out. There being many plums, I became an expert at blowing out plum stones into the wind, a facility I have never since lost.

With ever vaulting ambition, Cobbold sought to examine short-wave propagation at altitude. He arranged a trial between a transmitter at the top of Mount Snowdon in Wales and a receiver in the Mourne Mountains of Northern Ireland. I also made two discoveries on that occasion. One was that, by using the Snowdon Mountain Railway, heavy equipment could easily be got up to the hotel at the summit of Snowdon (which was closed for the duration, except for a special occasion like this). The other was that all the sitting around I was doing had made my legs weak. Instead of using the railway for the descent, I made my way down via the Crib Goch Ridge. Halfway along it, I cut away towards Llanberis, and I found my legs unsteady by the time I reached the road in the valley. It depressed me to find how much I had fallen off from the fitness of the prewar days in the Cuillin Hills of Skye, and indeed it wasn't until I was in my fifties, when I became engaged in that curious pursuit of climbing all the Scottish mountains known as "Munros," that I recovered something of my former freedom of movement in the mountains.

By 1944, the U-boats were presenting less of a target to our short-wave radar. We were running into difficulty in distinguishing reflections from the snorkel tubes they were putting up (to suck in air for their diesel engines) from reflections due to waves in the surrounding sea—sea-clutter, we called it. This difficulty made it seem imperative that we study the properties of sea-clutter in the hope that we could find some property that would enable us to distinguish it from the metallic reflection from a snorkel tube. By 1944 also, I had developed an aversion to the kind of formal trial in which you requested an actual submarine. Either there would be a flat calm and there would be nothing to discover, or there would be raging sea and I would be too seasick to do anything. My instinct was to find an ongoing situation in which we would have all the time we needed to gather all the data we needed. It was then that I remembered Cobbold and the Snowdon Mountain Railway, and the hotel there on top of the mountain. All we had to do was to set up a link, like Cobbold's, from the top of Mount Snowdon to a receiving station—in this case, on the south side of Cardigan Bay. We chose a site near Aberporth on the Welsh coast near Fishguard. We would put the receiver there, using an automatic recorder. The transmitter would be at the Snowdon end, where, unlike the situation in Cobbold's trial, we decided the aerial could be inside the hotel itself. With the power generators inside the railway-station shelter, all our equipment would then be under protection from the weather. It was high summer and the usually ebullient Hermann Bondi was suffering terribly from hay fever, for which reason he offered to man the Snowdon end, arguing that the rocky bulk of the mountain would reduce the concentration of pollen in the air that was tormenting him, an argument that turned out to be substantially true.

The southern receiver station gave us trouble of an unexpected kind. I should explain that, because only one-way transmission was involved, different from the out-and-back transmission of radar itself, high power at the transmitter was not needed. Low-power magnetrons were well known and had nothing much of a security rating, which was why all these investigations could be done without precautions beyond securely locking up the horse boxes when they were left unattended. The equipment used was supplied by the North Wembley laboratory of GEC— General Electric Company (U.K.)—and a Mr. Archer-Thompson from GEC (apparently a descendent of the pioneering Victorian rockclimber of the same name) came along with Tommy Gold and me to Aberayron when we set up the receiver end of our link.

Normally, we would have received logistic support from the naval unit at Fishguard Docks, but Tommy Gold had found the captain there uncooperative—it was, however, the sole example of scarcely veiled hostility to civilian bods that I can recall. The captain had his offices at the one local hotel. Although he was said to occupy only one-half of the hotel, we always found it impossible to obtain rooms in the other half. I could have raised a fuss, but raising fusses was never much to my liking. On the few occasions in my life when I tried it, nothing satisfactory ever seemed to happen. My system in such situations, therefore, is to adopt what Americans call an end-around play. In other words, Archer-Thompson, Tommy Gold, and I found ourselves holed up at a place in Fishguard called the King's Café. We had a bedroom with two doubles in it.

Tommy and I offered to occupy together one of the doubles, leaving the other double to Archer-Thompson, but he declined our generous offer. Instead, he produced, with the air of a man prepared for anything, a gadget called a lilo—actually a rubber mattress—which he spread on the strip of floor between the doubles. This was the first time I had seen a lilo, always having really slept on the ground when I slept on the ground. The idea of inflating the space between two ground sheets seemed a good one until Archer-Thompson started to blow the thing up with a little blue pump of minute volume. On and on he went, pumping away, with Tommy and me, meanwhile, repeatedly falling asleep and being reawakened by the struggles going on in the narrow space between us.

We were thankful, the following morning, to see the end of Archer-Thompson, his work on setting up the receiver system completed. We stayed an extra day to establish contact with Bondi on top of Mount Snowdon and so to check that things were good also at his end. Back in bed again that night—again, one in each double, Tommy sitting up and smoking—we reviewed the day and decided we would return to headquarters the following morning. Suddenly, the bedroom door burst open, and in came the proprietor of the King's Café, a virago of a woman. "Why are you sleeping in both beds?" she shrilled. "Because we don't care to sleep in the same bed," Tommy replied calmly, taking a long puff on his cigarette. "You must have slept in the same bed last night," the woman shot back venomously. "We didn't," Tommy explained. "Our friend slept on the floor." At this, the woman's voice soared upwards to a pinnacle of wrath: "Are my beds not good enough for you, then?" she cried out at high F. This was too much for us, and we both guffawed openly. In response, the woman retreated out of the open door and shouted downstairs to where several men were talking noisily and drinking: "Come up here, Dai," a request that was followed immediately by the pounding of heavy feet on the stairs. At this, I confess, I had the foresight to jump out of bed and slip on my shoes. I should explain that, although my feet have carried me a long way in life, even over all the Scottish Munros, they are quite unlike those of most robust people, who tend to burst their footware apart with their strong square feet. My feet are slender, and I have a particular aversion to having them stamped on by Welshmen in hobnail boots.

Thin rain was falling as, a number of blows later, we stepped at midnight, more or less undamaged, out of the King's Café and into the street. I remember acutely the blackness of it. In the war, there were, of course, no street lights. Normally, with a bit of the sky clear, there was enough starlight to allow one to get around, once the eyes were thoroughly adapted to the dark. In heavy cloud, however, there was nothing to be seen. We had all manner of sophisticated equipment in our horse box 15 miles away, but we didn't have a flashlight with us. Not wishing to sit shivering in the railway station for many hours, we groped our way slowly and stumblingly towards the naval dockyard, receiving aid from a little light as we neared it. Our Admiralty passes took us past the guard to the duty officer, in whose office we spent the time until 3:00 A.M. drinking the captain's whisky. Thereafter, we curled up under a blanket to sleep until the first train to London was due to depart.

I cannot claim that anything relevant to the problem of detecting submarines came out of the Snowdon–Aberayron endeavor, although we did learn a lot about sea-clutter. The trouble was that a U-boat showing only a snorkel pipe was simply too small a target to be tackled by 10-centimeter radar. To have a chance, we needed to reduce the wavelength still further—to 3 centimeters. But this was outside the time scale available to us during the war. Nevertheless, something important emerged over the long term, something even more important than U-boat detection would have been.

By 1944, I had moved from Nutbourne to the main ASE (Admiralty Signal Establishment) headquarters, which had been relocated from Portsmouth to King's School, Witley, in the county of Surrey. In addition to British scientists, we had scientists and engineers from Allied countries—notably, Poles in their resplendent uniforms and curious caps, and Norwegians in blue uniforms reminiscent of Scandinavian lakes on a bright day. I had been in Norway for six weeks in the spring of 1939, and I had taken a great liking to the country, where, despite the language difference, I had felt instantly at home, perhaps because the people of my native valley came originally from Saxony and Scandinavia.

I got into the habit of dropping in on the Norwegians, mainly to chat with their senior officer, Helmer Dahl. Like the Walrus and the Carpenter, we talked of many things. There had recently been an excellent production in London of Ibsen's *Peer Gynt*, with Ralph Richardson in the title role, and we talked about that. We talked about the Children's Red Cap Campaign against the occupying Germans. Many books have been written on the German armies of occupation, but the trouble with history is always that it is a dull gray affair. It lacks the brightness of actual existence. In books about Norwegian history, I daresay you will learn about

Quisling and the Norwegian underground, but I doubt you will find much about the Children's Red Cap Campaign. I understood Dahl completely when he told me that it was an age-old custom in Norway for all children to wear red woolen caps in winter and for them to begin doing so everywhere on the same day in the autumn. Probably because our English valleys were warmer than Norway, we had abandoned this practice, but we applied it to all our winter games: Everywhere in our valleys, the same game would start on exactly the same day.

But the German Army didn't understand. When children suddenly appeared in red caps all in the same moment all over Norway, high-ranking German officers thought it was some subtle move against them. Norwegian adults, sensing the occupying army's discomfort, took to laughing at officers whenever a child in a red cap appeared. This led to a ludicrous situation, in which children in red caps were chased through the streets by soldiers who had been ordered to impound all red caps, which they did, building up a huge, useless store of them. Then Norwegian men took to wearing safety pins in their lapels. This, too, led to a ludicrous situation: Norwegian men were arrested in the streets and ordered to remove the offending pins. These little examples illustrate the futility of attempting to bully a people who are sufficiently coherent to form essentially a single family, a truth that the Jews have proved many times over the two thousand years of their troubled history.

With the end of the war now plainly within sight, it was natural that we should also talk about what we intended to do thereafter. Dahl's project was one that would have attracted me if I hadn't been returning to Cambridge. His intention was to use short wavelengths for communicating between metal dishes directly across Norwegian fjords and hills for the purpose of providing telephone service in remote places where direct cabling was impracticable. This technique is very familiar nowadays, with dishes and repeater stations being common in many parts of the world, both for telephony and for television, but it was a new idea in 1944.

Something very remarkable that had happened in Cobbold's trial between Mount Snowdon and the Mountains of Mourne jumped into my mind as I talked with Helmer Dahl, something that, to this point, I have concealed from the reader. Cobbold was both a good organizer and a big fusspot, the two traits probably being connected. At whichever end of a communication link he happened to be, he fussed about what might be happening at the other end. This meant that he was perpetually on the telephone to the other end. But, between the summit of Mount Snowdon and the Mountains of Mourne, there was no telephone. So the chaps from North Wembley were asked to add speech modulation to their equipment. There was nothing particularly difficult about it. All our transmitters, whether for high-power radar or for Cobbold's communication links, were pulsed—which is to say, the transmitter would emit for a brief while, then would come a comparatively lengthy pause, then another brief emission, another pause, and so on. Typically, there would be 500 such pulses per second. A telephone works by first converting sound-induced vibrations of a diaphragm into corresponding variations of an electric current, which are then either sent directly along a wire or used to "modulate" radio waves of much higher frequency, as is the case with modern satellite-mediated reporting of events up and down the world. At reception, the process is reversed. The corresponding trick with Cobbold's link was simply to use the electric current from the telephone diaphragm either to speed up or to reduce the pulse rate of the transmitter, again doing the reverse at reception. When this system was used, and I heard on Mount Snowdon a voice speaking from Northern Ireland, I knew instantly that an important discovery had been made. The quality was immeasurably superior to any telephone I had yet experienced, and, moreover, there was a complete absence of background static. You heard the voice against nothing, against plain silence. The system never appeared again at any of Cobbold's trials, and it needed no great measure of Yorkshire cunning to understand why not. The chaps from North Wembley had realized, as well as I did, that a discovery had been made. If the Royal Army could have shown, after the war, that the discovery had emerged and been used repeatedly thereafter in their trials, and that they had paid for it, GEC would, I think, have had little title to the idea. Although I was friendly enough with the North Wembley chaps, I didn't like the situation. Cobbold, as might be expected, didn't even notice it.

Anyway, I thought it opportune to tell Helmer Dahl about it all. He came instantly on duty. When I also told him we were currently working the same system between Mount Snowdon and Aberayron, minus only the speech modulator, which could easily be added, he was keen to take a look for himself. We visited Bondi on Mount Snowdon and had a high old time there. Then we set off to drive south to Aberayron. Timing made it necessary to stop for the night along the way. It was a small place in the country, and there was only one bedroom available. I have a strong distaste for sleeping with the bedclothes tucked in, a distaste that I had never thought of as having a cultural connotation, until Dahl freed the blankets and sheets on his bed and said in a positive voice: "I dislike the envelope system."

The war ended, and it was many years before I saw Helmer Dahl again. He told me the Norwegian Post Office hadn't been able to obtain the financial resources for his project, so he had found it best to stay with the

Norwegian Army. The project had then expanded to provide a network of communications extending from the Norwegian frontier with the USSR over mountain and dale, eventually to NATO headquarters in Brussels. Because of heavy winter snows and rapid ice movements, they had been obliged to cut channels in the rock to accommodate power lines over the mountain tops. So, in the end, by a curiously tortuous route, the British Army got its just deserts, especially as I believe much of the training that led to victory in the Falkland Islands was done in northern Norway. Good old Tolly Cobbold, some might say, despite his two bottles of soda pop.

In the autumn of 1942, Signal School headquarters was moved from Portsmouth to King's School, Witley, and we were given the new name of Admiralty Signal Establishment (ASE). Nutbourne survived the reorganization, but Ross, Domb, and I were transferred to Witley, with Maurice Pryce taking over at Nutbourne—a curious piece of leapfroggery. ASE was to be divided into nine sections (I believe it was eight at first), and I was to run one of them, with Domb and Hermann Bondi as members. Within a short time, Tommy Gold would join us and Charles Goodwin would pry himself loose from a rather uninteresting ballistics unit in Cambridge. There would be a young graduate from Oxford, John Gillams, and an unexpected addition from Norway, S. Rosseland, an internationally known astronomer. Rosseland appears in the astronomical literature only as S. Rosseland. He was a small, slight man with bushy eyebrows. He was of Eddington's generation, and, like Eddington, hard to get to know. I bounced all manner of balls at him, but I never perceived one that rebounded. He would answer a direct question, but, as Samuel Johnson remarked, "Questioning is not the mode of conversation among gentlemen." In contrast, John Gillams was the only person I ever saw catch a wild squirrel barehanded. It was a remarkable performance for speed, both Gillams's exuberant yelping swoop and the dispatch with which the squirrel slit open one of his hands. It was a performance whose repetition I cannot recommend, except perhaps to those whom I quite seriously dislike.

Every officer in the services to whom I have spoken about the phenomenon has agreed that it is a mysterious rule of war that, as soon as you become comfortable in a posting, you are moved. It was exactly so with me. We had struggled for months to obtain the cottage in Lye Lane, with my wife cooking there on the pocket primus stove. However, in the spring of 1942, I learned from MacIver, who was billeted in Funtington, close-by the house of George Booth, Chairman of the Booth Shipping Line to South America, that the Booth family butler had left to join the services and that a modernized cottage had thus been freed in Funtington. We got it, and now the walking track up to the downs started immediately from our back door. The way to the Nutbourne field was by Watery Lane, which really did have a stream to the left, in the bed of which there were great quantities of watercress. It carried drainage from the downs to the duck pond in West Ashling. Where it has mostly gone nowadays I can't imagine, since drainage isn't something that can easily be stopped. Perhaps the local people use it for watering their gardens and orchards. Anyway, in the autumn of 1942, we decided that, having at last arrived at a reasonably comfortable accommodation, we were not going to restart the time-consuming and fatiguing business of house-hunting in the Witley area, especially as my wife now had a young baby to look after. So each day, no matter what the weather, I traveled the twenty-five miles from Funtington to Witley—four miles on my junk-pile bicycle to Rowland's Castle and thence by slow train to Witley, and then the reverse in the evenings, which were dark in wintertime.

This was now the time I could have done with the blue 12-horsepower Rover, which I had acquired so triumphantly in the spring of 1939. Unfortunately, I had sold it. On my small current salary, engineered by the ingenious Fred Brundrett, we couldn't afford to run it in 1941–1942. In those days, it was considered mandatory to store any decent car in a garage. All we had was George Carson's in Chesterton Hall Crescent, and I couldn't go on and on cluttering George's property, kind as he was about the matter. Happily, we got a good price for the Rover, almost as much as I had paid originally. The transaction served to put our finances in a sound condition. My wife has always paid our bills on the dot, a policy that has secured for us the most excellent service; and generally, wherever money was concerned, we had long since discovered the impeccable validity of Mr. Micawber's advice about not overspending one's income. So, regrettably, the blue Rover had to go. In the spring of 1943, Ross, who had also decided to remain in Funtington, bought a car, in which, each morning, he gave me a lift to Rowland's Castle. A day came, however, when Ross's car, being high-spirited, jumped off the road, returning me yet again to the execrable bicycle. Then George Booth came to the rescue. He sold me a 1928 Singer two-seater with dickey seat (or rumble seat) for £5. There wasn't much to go wrong on that car: no petrol pump, just gravity feed; no starting motor, just a hand crank in front. The Singer was to run sturdily for five years, never refusing to start except once, when shellac melted and caused the magneto shaft to seize rock-solid. At top speed with a following wind, it did 38 miles per hour. It was the best bargain I ever made. If I had only had the wit to maintain it, the Singer today would be worth many thousands of times what I paid George Booth for it.

Thereafter, in a spell of bad weather and in the cold of winter, I used to drive on Monday mornings to Witley, returning on Saturdays to Funtington. During the week, I stayed with Hermann Bondi and Tommy Gold, who had managed to rent a house at Dunsfold, a village about six miles southeast of Witley. The house was sited on a little eminence in line with the main runway of an airfield manned by Canadians, which was why nobody else had wanted to rent it. The trouble was that, in the heavy raids then being mounted into Europe, planes would sometimes crash on takeoff—perhaps one a week, on the average—so that it was only a matter of time before a plane actually landed on the house itself, just as various planes had done already on other similarly situated houses. The curious thing, in retrospect, is that we were not greatly concerned about the danger, or even much aware of it, except at moments when planes really did crash nearby.

Although Witley had a Brave New World atmosphere, I doubt that much we did from the end of 1942 had an impact on the war—except, of course, for those sections that were concerned with fitting currently available radar into ships. We were now going through the process of designing equipment the Royal Navy would have liked to have had in 1939 but that would not go into production before 1945. It was really the inevitable process of refighting the war the way one would have liked to have done it in the first place.

A main component of my job had become the giving of advice, which I discovered is not the pleasing activity many people suppose it to be, especially when it becomes clear eventually that the advice you gave was bad. In late 1944, I advised C. E. Horton that our plans for the future were becoming too complicated, with the likelihood that they would not work out well in some cases. My doubts were overridden, the projects were continued, and, in the years after the war, they were indeed successful. The reason for my bad advice was that I had failed to foresee the major revolution in electronics that occurred in the years after the war, really the biggest revolution in electronic technology of the whole century—namely, the vast improvement in the reliability of basic components that occurred between 1945 and the early 1960s. The modern generation, which now takes electronic reliability for granted, can hardly imagine the extensive breakdowns that were once considered normal, everyday occurrences. What would have been overcomplicated for the unreliable components of 1944 became attainable, and even fairly straightforward, in postwar times.

A better situation occurred in the spring of 1945. I was called to a meeting at the Admiralty to discuss future plans for the Royal Navy, a meeting in which more gold braid was in evidence than I had ever seen in one place. The heavy braid was gathered together at one end of a long table, and, as a technical bod from ASE, I sat below the mast at the other end. With the meeting well advanced, the Admiral in the Chair went round the table asking what each of us thought. When it came to my turn, I said that extending the present striking range of normal ship's gunnery from 15 miles to 20 miles, or even to 25 miles, might be made possible through the use of aerial torpedoes with aerodynamic lift. The missiles could be continuously guided to their targets. I hadn't thought of a lock-on to the target's own radar or of using radar in the missile itself—otherwise, my exposition would have amounted to a fair approximation of the modern Exocet. Or it could be thought of as a Jules Verne version of the cruise missile. I recall that the chairman looked down the long table with a wry grin and said, "What a horrible idea," which, in 1982 in the Falklands, it eventually turned out to be.

A second tale with an unsatisfactory outcome is next. Because an explosion in water can be far more damaging than one in air, it seems odd that we were not told to give a high priority to helping with the problem of torpedo guidance. Nevertheless, we did have relatively informal requests for help from naval officers responsible for torpedo design. The aim was to provide a running torpedo with a midcourse correction to its target. Unfortunately, we became fixated with providing the torpedo with a simple aerial to which a message could be transmitted. The snag that entirely defeated us was the high electrical conductivity of seawater. If the torpedo had run at the surface, the problem would have been easy, but we could think of no satisfactory way of getting a signal to a torpedo running under the surface.

It was one of the better features of my wartime experience that almost everyone with some special knowledge, outside of our daily activities, would be able to make an important contribution at an appropriate moment. A sense that this was so had the effect of making everybody feel wanted. A far higher proportion of British people felt wanted during the Second World War than they have done since. Because "feeling wanted" is the most important contributor to what we call the quality of life, this should be an arresting thought to those who are concerned with solving the social problems of the modern inner city. These problems will not be solved economically. They will be solved only if people feel wanted. What we needed in order to solve the running-torpedo problem was to have had a keen rod-and-line fisherman in our group. A word about playing a racing heavy fish would have done it. The solution we never found was simply to have a line run out with the torpedo—a line with a metal core, along which correcting signals could be transmitted. Maybe that was what the Japanese did, and maybe that is why their torpedo attacks were so accurate.

My last story of this chapter is a story of success, however, which is more pleasant to speak about at length than failure, and it provides something of a lesson to governments on how not to support science. One day, C. E. Horton, the senior scientist of our group, asked me into his office to read three sheets of pink paper, which indicated the highest level of security, Top Secret. The concept was so simple that I had some difficulty in understanding its high classification. I felt any of us could have thought of it in half a morning, had we been in the business of helping planes to attack our ships instead of the reverse. It was the suggestion that our planes throw out quantities of aluminum strips in order to confuse defensive radar, a project with the code name "Window" to us, "Chaff" to the Americans. The notion that such an obvious idea could be classified Top Secret seemed absurd—and it was, for the secrecy was a self-deception. The idea had been thought of repeatedly and suppressed repeatedly by both sides, as far back as 1937. Defensive commands had argued vociferously against allowing planes on offensive missions to use it, their fear being that its existence would thus become known to the enemy, when, in fact, the enemy was going through exactly the same argument. By 1943, however, attack on the British side had become so dominant that it seemed Window would be used at last. Naturally, there was the wish to eat one's cake and have it too. So the call had gone out to the defensive sides of which we were a part, to consider if there could be an effective countermeasure to the German use of the Window idea. This was why Horton had shown me the three pink sheets. In short, could we think of a remedy for the thing? In the outcome I shall now describe, we did think of one. The idea we hit upon was never used during the Second World War, but it was used for many years thereafter.

It was a main thread of our work to match service requirements to the hardware available at Admiralty Signal Establishment. Our first step on examining a requirement was to attach what might be called a figure of merit to the hardware that would be needed to satisfy the requirement. If the figure of merit was beyond what we could achieve, we had to say so immediately, thereby saving fruitless work. If the figure of merit fell within range of what could be managed, the next question was to find the most economical way to give expression to the requirement. For this work, we had to know the kind of hardware performance I had encountered, when I first arrived at Nutbourne, in connection with finding the altitude of incoming enemy aircraft.

In 1944, we were years down the road from those early days, and we knew a great deal more about the necessary electronic details. But there was still a major unknown: the precise manner in which radio waves were reflected by target aircraft. At long wavelengths, it had been good enough to regard a plane as a reflecting sphere, with each type of plane represented by an equivalent reflecting sphere, larger planes having equivalent spheres of larger radius than small ones. At short wavelengths, however, things were not as simple as that. The reflected radar signal at short wavelengths, instead of being steady, as a reflecting sphere would cause it to be, varied up and down rapidly, seemingly in times from one-fifth to one-third second. This led, in operation, to some curious results, which we were frequently asked about by Admiralty officers. So the day came when I decided that the moment had come to find out what really happens when a plane reflects a single radar pulse. This could not be done with typical radar, which displayed reflections on some form of screen analogous to a television screen, and, just as what we see on a television screen is the average of a number of photographic frames, so our radar screens averaged a considerable number of pulses—typically, about 30 pulses. A new form of equipment would therefore be needed.

We needed an adjustable device by which we could pick out only a particular target in a small specified distance range, which would be the plane we wished to study. Receiver output from this "gate" would then be displayed on a second screen that simply showed a spot of light that moved up and down according to the voltage output of the receiver. Thus, an individual reflection pulse from a plane in the gate would cause the spot of light to move up at the beginning of the pulse and then to move down at the end. At our usual pulse rate, this would happen 500 times each second, which, of course, would be impossible to follow by eye. The trick was to draw photographic film past the screen at a rate of several feet per second, when individual pulses would become separated by a few millimeters on the film. Tommy Gold made up the camera and film-moving equipment himself. The gate was fabricated, however, by the receiver division at Witley. I thought I had managed a good stroke of business in securing this assistance, but it turned out to have been a mistake. When a department is asked to do an unusual special job, either someone pretty good or someone just the opposite is put on to it. Here, it was the opposite. The equipment we were supplied with gave us trouble with maddening persistence, and it had been supplied with inadequate circuit diagrams. It was in trying to repair the thing that Tommy and I came near to electrocution, Tommy many times.

With the equipment apparently ready in a horse box (equipment with a high-power magnetron and, therefore, with a considerable security rating on it), the final step was to find a plane. The normal practice was to request that one be flown specially on your behalf at a stated time and place. By now, however, I had learned that such "trials" were next to being useless. If the weather was good, your equipment wouldn't work; or, if your equipment worked, the weather would be bad, or the pilot would fly a wrong route, leading to a lot of unproductive argument. Such a practice, bad enough normally, was clearly out of the question for the dubious equipment I had mistakenly acquired.

Two years earlier, my wife and I had enjoyed a brief holiday at a farm we knew from prewar days, Bedruthan Farm on the north Cornish coast about midway between Newquay and Padstow. Thence I knew that a big coastal command station had been established at the nearby village of St. Eval; and within five miles was a naval station at St. Merryn, which could supply a necessary guard for our horse box, if the CO there were willing to approve our request. It occurred to me that the Liberators flown round the clock from St. Eval would supply all the planes we needed. With the horse box and its contents at the ready in the grounds of King's School, I told Tommy Gold to drive off into the blue. If he found a guard from St. Merryn at the spot we had picked out on the map, well and good. If not, he was to take the horse box with its secret contents to the St. Merryn station. Then I went to our administrative officer, J. D. Rawlinson, and asked if he would fix the guard. J. D. was naturally furious, but, in the end, he had to agree that an attempt to arrange for a guard through the usual channels would have taken weeks, quite likely. Besides, all he had to do, I pointed out, was to tell the CO at St. Merryn that he had two mad scientists on his hands, a concept that was always readily accepted, in my experience, throughout the services. Anyway, a guard was there all right when Tommy arrived, and excellent accommodations were available at a nearby farm (as far as my observations went, food rationing was unknown among the Cornish). Everything seemed set fair, and, if either the equipment had worked or we had been supplied with adequate circuit diagrams, it would have been.

The weeks passed by. As the heat of summer passed into the storms of September, Tommy became mightily fed up. There was no way I could call off the enterprise, unfortunately. After the fuss over the guard, I just had to have something I could represent as a success—for me, a rare venture into politics. It didn't need to be much, but there had to be a little straw to make a few bricks. So we drew up a roster, and one or another of us, according to the roster, would go down to Cornwall to keep Tommy company for a few days. My own memories are of waiting endlessly for a train at Par Junction and of riding the local bus, which took the twisting road from Newquay to Padstow. Then there was the pretty girl on the train to whom, for better or for worse, I gave up my seat. Then there was St. Peter's Hostel in Plymouth. I arrived there from London at 6:00 A.M., desperate for a cup of tea. There were no pearly gates at St. Peter's Hostel. Compared to it, the King's Café in Fishguard had been like the Ritz Hotel. Poor St. Peter.

My entire section risked electrocution in this performance. You may wonder, Why not simply switch off the power? Yes, if you have circuit diagrams. Then you switch off the power, think a bit, make a change, and switch the power on again. Without circuit diagrams, however, there is nothing to be done but to keep the power on and to see what happens when you change things in a kind of electronic random walk. *Bang!*— and you crash into the wall of the horse box. So it goes on.

Eventually, the steady supply of planes did its job. Tommy and I studied the first long rolls of film, processed in laboratories at Dagenham, with astonishment. We had expected fluctuations in the reflected pulses over times as brief as one-tenth second. What we had not expected—indeed, what amazed us—was to find that the radar reflection from a plane could change enormously in a time less than a hundredth of a second. Sometimes, there would even be a big change between consecutive pulses. We savored the situation for several days, when suddenly, all in a moment (as is always the case), the key to the solution of the Window problem burst upon us. The solution would have been impossible to suspect, had we not gone through this whole process. Unlike the rapid variation of the reflection from a plane, the reflection from aluminum strips would be very steady, one pulse nearly the same as another, whence a subtraction between adjacent pulses or adjacent blocks of pulses would give zero for aluminum strips but would give nonzero for a plane. The idea was simple in concept but hard to achieve in practice by the techniques of 1944–1945. The difficulty then was to store electrical information for as long as a few milliseconds. Tommy solved the problem in subsequent experiments by converting electrical information into sound, a development that was eventually used in the first British university digital computer, built in Cambridge immediately after the war.

To end, I will describe more or less exactly where we positioned our horse box. At Bedruthan, there were two quite different sets of steps down the high cliffs to the sea. In 1939, there was an easy, pleasant set of steps with Rundle's wooden tearoom at the top, but Shakespeare's "envious siege of watery Neptune" made havoc of those pleasant steps, and Rundle's tearoom fell down. For some years, you could see where the foundations had been. By now, even the traces of the foundations are disappearing. In addition to the rapid crumbling of the cliffs at this point, the bay below has also changed for the worse. Most of the sand is gone, and the beach is now far more stark and rocky than it used to be. In contrast to this, Pentire Steps, the second set of steps, half a mile away to the north, has not changed at all. Pentire Steps looks impossible from above, its essential feature being a little gully at the bottom that cannot be seen from a distance from any direction along the cliff. This tells you that Pentire Steps was a smuggler's and a wrecker's route. Go from Pentire Steps a hundred yards or so towards the south, to a spot opposite Diggery's Island. This was surely where the wreckers put their decoy lamp. It was also where we put the horse box from which the Window problem was solved.

Chapter 15

The Aftermath

THE WAR AGAINST Germany ended officially on 8 May 1945. On 13 May, huge crowds gathered to rejoice in the streets of London, with church bells ringing throughout the land and with a sense everywhere of indescribable relief. An appreciation of the cost in lives came later. Society finds it relatively easy to remember its dead in quantity 400,000 in Britain, 2,500,000 in Germany. It becomes harder with individuals. At first, family and friends suffer sharp memories of the person lost. Then (and this is the cruelest aspect of it all) even those memories fade, and what is left, in the end, is little more than an old photograph to show what might have been. If you climb to the summit of Great Gable in the Lake District on Remembrance Sunday at, say, ten in the morning, perhaps in mist and rain, it seems much like any other winter day in a remote place. But, as the minutes pass by, people begin to arrive there, at first in ones and twos, and then, as eleven o'clock approaches, they begin to appear in great number. On the stroke of eleven, a few words of remembrance are said, whereon the people disappear as they came, descending back to their valleys on all sides of the mountain. This simplest of ceremonies is the most moving I have attended, especially now that it is more than forty years on.

It has been said that, in the future, the twentieth century will be described as the century in which melody disappeared from music and art from art. But do not blame the musician and the artist. They deliver what society has asked them for. If they produce no melody, it is because our society has no melody in its heart. When one contemplates the two world wars, it is easy to see why this should be so. But, for the acute emotion of sadness, opportunity has existed for the poet, well taken in these lines from Laurence Binyon's poem "For the Fallen":

> They shall grow not old, as we that are left grow old:
> Age shall not weary them, nor the years condemn.

> At the going down of the Sun and in the morning
> We will remember them.

Even Shakespeare didn't say it better than that.

My last duty for the Admiralty was to visit Germany in the latter half of May 1945. I was supposed to discover what Germany had done in the way of naval radar, but the party with which I traveled had such diverse interests that it became impossible to discover anything, especially in view of the destruction to be seen everywhere. The destruction was indeed so bad that I thought a generation would be needed to clear it up. By 1948, however, when I visited Germany again at the request of the Foreign Office, I was surprised to find the wreckage largely gone. Of course, the returning prosperity of Germany is now very well known. The vigor of Germany's recovery was, for a while, a mystery to me. Eventually I saw that, unlike physical systems, biological systems, unless they are totally destroyed, will always regrow with little impairment, the cause being that crops regrow every year and flocks of animals double themselves—at any rate, they do with good husbandry. For these reasons, a nation apparently defeated in war is never really defeated. Turn your back on it for a while, and things will be back where they started, more or less.

It is peculiar how very poor we are at judging which current events are to prove important for the future. Television is choked at times nowadays, and newspapers spill oceans of ink, over issues that five years hence will turn out to have left hardly any lasting effect behind them. Yet other events of eventual consequence steal past us like ships in the night. Hardly a person who joined the rejoicing crowds in London on the night of 13 May 1945 realized that the British age of Empire was over.

In July, there was a general election. To the world's astonishment, Winston Churchill and the Tory party were roundly defeated. As one among many who voted against Churchill, I may be able to shed some light on why this happened. In the 1920s, Britain was stronger than Germany. In 1939, Germany was stronger than Britain. Therefore, at some moment, the curves crossed over, probably at a time not far from the German reoccupation of the Rhineland in 1936. This was not a matter of subtlety; it was apparent already in 1936. How, then, could one vote for a party that had permitted such a catastrophic inversion to take place? We know now that Stanley Baldwin, the Tory prime minister in 1936, had judged that the people, enjoying the first bit of prosperity after the depression years, would react unfavorably to the necessary taxes being imposed upon them. But a political party that thus put its own interests ahead of the country's interests could hardly expect to be electorally successful. Churchill himself, of course, was not tarred with this brush. But Churchill as a war leader and Churchill as a front for the Tory party were altogether different things. Besides which, Churchill's record in peacetime was not rated highly. His reversion to the gold standard in 1926 had deepened the depression, a circumstance that was still not forgotten. For me, however, there was a small point that cost him my vote. The Welfare State had been born, in 1942, in a report by the social reformer William Beveridge. Enabling legislation for the new form that education should take was passed by Parliament in 1944, but Parliament had voted down one clause of the bill. Parliament had wanted men and women teachers to be paid equally, but the original bill required them to be paid unequally. Instead of accepting the change as an amendment, Churchill forced the original clause through on a vote of confidence. This petty dictatorial action seemed, in a microcosm, to be what we were fighting the war against. Many people had similar small points of irritation, leading to a large majority for Labour in the July election.

The first Attlee government set about the dismemberment of the British Empire with zest, in contradiction to Churchill's often quoted remark: "What we have, we hold." Unfortunately, it is impossible to know what might have happened if the July election had gone the other way, but quite likely the Tories would have hung on as stubbornly as terriers, resisting the insistent American pressure for the dismemberment of the empire. For more than a century, in order to defend its territory, Britain had been involved in an effectively never-ending sequence of small wars, the strain of which had increased with time. With weapons of destruction ever more efficient—in particular, with a big jump in efficiency during the 1939–1945 war—the strain would surely have reached breakpoint soon, had the likely Tory intention to continue as before been put into effect.

The major achievement of the long-lived Attlee government was to devolve a huge empire at speed and in such a way as to leave very little bitterness behind. Today, the British Empire still holds together in the loose organization known as the Commonwealth, about which other members, notably Canada and Australia, sometimes seem more enthusiastic than Britain. And it holds together in all manner of minute ways—as, for instance, with an elderly Indian man in Bombay who had been mugged and slashed with knives, and who said to my wife, as he drew himself up, "It would never have happened if your chaps had been here." Or a banker from Nigeria, who was wandering about like a lost soul, apparently with nobody to talk to, at a European economic conference, whose face broke into an instant smile as soon as he realized I was from Britain, as if some bond still remained between us. No other empire has passed away quite in that way, except perhaps the Roman. Ireland, of course, was not conquered by the Romans. I was dining once with an Irish cleric at the Hibernian Hotel in Dublin. When it came to the cheese, he apologized for its being local and added, "There's no really good cheese outside the Roman Empire."

Like most intellectuals, I thought the Attlee government's plunge into state corporatism would be a success. The argument that everybody working for the general benefit would lead to a better condition of life than people working for narrow personal profit went back to Karl Marx. It had been widely held by what today would be called the opinion-forming class—university professors, economists, and playwrights such as George Bernard Shaw, who somehow managed to combine the avid collection of royalties on his works with the eager dissemination of socialist propaganda. J. B. Priestley had been a young writer when Shaw was at the height of his popularity. Although socialistically inclined himself, Jack Priestley told me that he thought the old man a humbug: "He went to Russia [circa 1925] and came back lauding the Soviet system to the skies, although he must have known quite well that without VIP treatment he couldn't have survived there for a week."

The years after the war were incredibly gray and drab—worse, in many respects, than the war years had been. By 1951, after six years of state corporatism, the British weekly meat ration was still a mere 8*d* (3⅓p) whereas in Amsterdam at that time you could buy any number of sandwiches at any moment of the day with more meat in each one than eight pennies' worth. I was genuinely puzzled by what had gone wrong, and even wrote a book, *Decade of Decision*, casting around for an explanation. Holland had been devastated by the war far worse than Britain. How had Holland managed to recover so much better than Britain? That state corporatism was the culprit was demonstrably obvious, just as it is today from the sorry condition of Eastern Europe and what was once the USSR.

In the summer of 1945, I could have done with a little capital myself. The circumstances were these. The Faculty of Mathematics in Cambridge had experienced a considerable loss of personnel, caused only partly by natural retirements over the six years of the war. Many lecturers had preferred, in prewar years, to stay in Cambridge rather than to accept professorships elsewhere. The circumstance that they had now broken the old Cambridge routine through being away in one capacity or another made them more likely to move elsewhere permanently, thus creating further vacancies. A decision was made to fill only about half of the vacancies with permanent appointments. Maurice Pryce and Ray Lyttleton, both several years older than me, were given such positions. For the rest, a number of temporary junior lectureships were specially created. Hermann Bondi and I each got one of those junior posts. Each was for three years, and each paid the not very glowing sum of £200 per year. My fellowship at St. John's, which still had a year and a half to run, paid £250. There was a good chance that, with a university appointment behind me, the fellowship would be extended for a further year. But it was standard policy at St. John's that research fellowships were not prolonged beyond a total of four years—otherwise, there would not have been a proper flow of young research people through the college. So the bottom line was that I had £450 per year for approximately three years. Additionally, there was some six months to run on an award from 1939, and this, we reckoned, would finance the move from Funtington to a house in the Cambridge area. It was a far cry from the £850 I had been favored with so spectacularly in 1939, especially as money had inflated during the war years, perhaps by a factor of two. Yet, as I started up the little Singer two-seater on a July morning, leaving Admiralty Signal Establishment for the last time, it was with some confidence that I contemplated the future. I felt I had a kind of passport to success in my pocket, rather as the young Richard Wagner had toured the courts of Europe with the score of the overture to *Rienzi* in his pocket. In my case, it was the key to the origin of the chemical elements that I had luckily stumbled upon.

In 1940, I had shared the rooms at I8, New Court, in St. John's with Lyttleton, the rooms where I first met Lyttleton in early 1939. I was to occupy these rooms for the next four months. There were three things to be done: to get ahead in preparing the paper with which I intended to appeal to the courts of Europe, to prepare for my first lecture course in October, and to find a suitable house at a suitable price. Unfortunately, I quickly discovered that neither the college nor the university was willing to help in the last regard.

I had been living in a fool's paradise, in which I thought a decent house in Cambridge would cost about £2000, as it had when I left in 1940. But no longer. Cambridge had acquired a considerable wartime population whose activities did not end with the war. Several government departments had become settled permanently in the town, located primarily in Brooklyn's Avenue, close by the University Botanical Gardens. The stock of decent houses had never been large, and demand had now outrun supply, with the consequence that what might have been bought for £2000 in 1940 was now going for £6000. Even if I had been able to raise the required fraction of £6000 demanded as a deposit by building societies, interest plus repayment of mortgage would just about have consumed all of my prospective £450 per annum. In those days, interest was about 5 percent per annum, and repayment was 3 percent. To fit my economic situation, I would therefore need to come down to a house selling for around £2000. In Cambridge, this meant a small house in a long street on the eastern side of town, where the quality of life would be notably inferior to what we enjoyed in Funtington. Besides which, my wife and I had by now developed a liking for country life. So it was to the villages around Cambridge that I turned my attention.

A friend in London, who had some reputation for having his ear to the ground, told me that my name was being mentioned in connection with the Plumian Professorship, which had been vacant since Eddington's death in the previous year. Now I understood acutely what it meant to be taken up the mountain by the devil and shown the Kingdoms of the Earth. How wonderful it would be to jump to a professor's salary. How wonderful to have the nearly unique advantage of a university house, the wing of the Observatories, to which Eddington had retreated when Miss Eddington's little bell sounded for tea. But, luckily for me, I had too much common sense to take the possibility seriously. My friend's story might well have been true, for (as I was to discover in later years) it is quite normal, at the first meeting of an electoral committee, to mention the names of everybody under the Sun. In the event, Harold Jeffreys, the St. John's mathematician, was elected. This was good, since Harold was a friend. Over the years to come, I was not to suffer hostility from above— not in Cambridge, at any rate. Indeed there was no contentiousness or rivalry or difference of opinion on astronomical matters between Jeffreys and me, for the good reason that he was a geophysicist. Jeffreys was one of the few persons I have met who really was modest to a fault. Lyttleton used to say that, whatever your distance from him might be, Jeffreys always contrived to pitch his voice so you couldn't quite hear him. You couldn't quite hear him if you sat on the front row at one of his lectures, and you couldn't quite hear him if you sat next to him at dinner, especially when he told a joke, of which he had an immense fund. I once heard a bombast in the Faculty of Mathematics say he wouldn't give an examination candidate a first-class mark unless he reproduced "properly" what had been said in his lectures: "Oh," Jeffreys replied immediately, "if anybody reproduced exactly what I'd said, I'd give him a second." The story is told that Harold was called in as a consultant by a wartime committee concerned with some grave issue. Through the morning session, Jeffreys sat silently, his legs crossed, smoking from a cigarette holder. When his silence became unbearable, the chairman at last took the situation in hand by asking what he thought, to which Jeffreys replied: "That it's time for lunch." The same thing happened in the afternoon, the same crossed legs, the same smoke curling from the cigarette holder, and the same stare into distant space. Utterly determined now, the chairman turned with a steely resolve to discover his views. To the demand that he state them, Jeffreys replied: "My view is that I'm glad it's your problem and not mine."

In one respect, Jeffreys' disinclination to involve himself in astronomical matters proved unfortunate. If he had taken over the directorship of the Observatories and had occupied the house there, these necessary adjuncts to the Plumian Professorship would have fallen my way when eventually I was appointed to that chair. Instead, Jeffreys disavowed the traditional associations, so that, when things finally fell my way, I succeeded to a title without an estate.

The idea of applying for a post in some other university crossed my mind. I decided that, if I could secure a really senior post elsewhere, I would abandon my prospects in Cambridge. To this end, I applied for the professorship of physics in a far northern university, but I didn't get it. Many years later, I spent a while in the same university, being entertained at lunch one day by members of the Physics Department, when I was asked shyly: "Is it true you once applied for the physics professorship here?" My unsuccessful application being remembered after forty or more years was almost better than getting the job would have been. It was actually the only time in my life when I applied overtly for a job, excluding the early applications for research grants in my student days.

The war really came to an end for me in September 1945, when my wife and I took a month's holiday in the west of Ireland. We journeyed in the Singer to Fishguard, which continued to be ill-omened for me, since it was in Fishguard where the Singer's engine seized solid, owing to shellac melting in the car's magneto. From Fishguard, we took the ferry to Rosslare, and thence by train to Cork, where we were met by Cormac O'Ceallaigh, a friend from my research days in the 1930s. We stayed for two weeks with Cormac and his wife Millie at their cottage on the beach some 20 miles south of Cork. Then we took a bus from Cork north to Galway, a second bus through Connemara to Westport, and then a third bus to Achill Island, past the local businessman Joe Sweeney's improbable shark factory, and on to the remote village of Dooagh, where the driver set us down at O'Connor's Guest House, where another of the guests was a Protestant canon from Tuam who told the best ghost stories I ever heard. There was a mad major who had a big house high up the bog behind. The major, who I was told had succeeded to a fortune in India, put on an excellent tea once a week, to which all were invited, and after which, under the major's tutelage, there was much dancing in the Irish fashion. We climbed Croaghaun near Achill Head, and it was there that, for me, the war was suddenly blown away by a wind from the Atlantic— or perhaps it was banished by the view from the cliffs, which fall to the sea more than 2000 feet below.

In 1934, my enthusiasm for walking in the mountains began. Edward Foster is on the right, Herbert Lee on the left.

With George Carson, the blue Rover, and the little house in de Freville Avenue. When photographed, George always had the impulse to act the tough Irishman from the 1920s.

Easter 1936, at Dale Head, Duddon Valley. Harry Marshall, standing at left, was killed in the invasion of Malaya. Charles (E. T.) Goodwin is standing at right, Ivan Inksetter is seated at left, and next to him is the agile Bill Sessions. The summer of 1936 was a high point of my life. Soon the clouds of war would gather, and, after that, youth would be gone.

Barbara Clark: The actress in the play that constituted my life was suddenly on stage.

Son Geoffrey, circa 1950.

Geoffrey with daughter Elizabeth,
circa 1950.

With my daughter Elizabeth, circa 1960.

Barbara, circa 1950.

These are the windows at Orchard House where the geese came in the wintertime.

Ivy Lodge, Great Abington.

With Ray Lyttleton, Moscow, 1958. (*Photograph by T. Gold.*)

In Moscow, 1963, with V. L. Ginzburg (in cap) and I. S. Shklovsky (leading the way).

With Tommy Gold and Hermann Bondi, circa 1960.

A difficult problem. With Tommy Gold and Don Clayton, 1976.
(*Photograph by D. D. Clayton.*)

With Prime Minister Harold Wilson.
(*Photograph courtesy of University of Bradford.*)

This is the problem: Why are Wilson and I shifting the table?
(*Photograph courtesy of University of Bradford.*)

With the formidable Geoffrey Burbidge.
(*Photograph by B. Gallagher.*)

The dedication of the Anglo-Australian Telescope, with Prince Charles, 1974. (*Photograph by M. Jensen.*)

A working party at Herstmonceux Castle, including many notabilities, with Dick Woolley fourth from the right in the front row.

[P-18]

A summer group at the Institute of Theoretical Astronomy. Few photographs show so many who were destined to influence the course of astronomy in the years after 1970.

[P-19]

Arthur Stanley Eddington (1882–1944).
(*AIP Emilio Segrè Visual Archives.*)

Paul Adrien Maurice Dirac (1902–1984). (*AIP Emilio Segrè Visual Archives.*)

Wilhelm Henrich Walter Baade (1893–1960).
(*AIP Emilio Segrè Visual Archives, Physics Today Collection.*)

In my late forties. An early back injury makes it difficult for me to work comfortably at a table.

On a visit to Stonehenge, at the age of sixty. (*Photograph by Philip Daly.*)

F. HOYLE   H. C. van de HULST   A. R. SANDAGE   J. A. WHEELER   H. ZANSTRA   L. LEDOUX

O. S. KLEIN   W. W. MORGAN   B. V. KUKARKIN   M. FIERZ   W. BAADE   H. BONDI   T. GOLD   L. ROSENFELD   A. C. B. LOVELL   J. GÉHÉNIAU

V. A. AMBARZUMIAN   E. SCHATZMAN

W. H. McCREA   J. H. OORT   G. LEMAÎTRE   C. J. GORTER   W. PAULI   W. L. BRAGG   J. R. OPPENHEIMER   C. MØLLER   H. SHAPLEY   O. HECKMANN

At the Solvay Conference, 1958. Many of those mentioned in the text are in this photograph. Wolfgang Pauli, in the front row, appears to be waiting to pounce on somebody. (*Photograph by G. Coopmans.*)

[*P-24*]

We were also privileged to see no less a person than the Devil. It was on the long beach at Keel, otherwise deserted, on a cold windy day. He was dressed in obligatory black. He was thin, and he walked with a limp. He appeared about half a mile off, as we willed ourselves to go bathing in gray waves, which, at the moment of our immersion, were no more than a couple of feet high. My wife suddenly let out a scream, pointing wildly away to sea. I hadn't noticed it before, because, without spectacles, I am next door to being blind. An immense wall of water was fast approaching. As it came over us, it was twice our height. The thing knocked us both head over heels. When, some minutes later, we had recovered a little, the waves were again only a couple of feet high, and the Devil had disappeared, from which it will be clear that the canon from Tuam had been in a measure predictive in his ghostly phantasmagoria. But the wall of water was real enough. In years of sea bathing, I have often experienced waves that exceeded the average by a few standard deviations, but I have never seen anything like that monster. It had the aspects of a tsunami, which perhaps was what it was.

In November, halfway through my first term of teaching, I heard (from an estate agent in the little crescent that goes from Gonville and Cains College to the Market Square) of a house to let in the village of Quendon, 25 miles to the south of Cambridge, the rent being £250 per year. That we should take such a possibility seriously—a house far out of bus range from Cambridge, a house whose rent would consume half my salary over the next two years, a house that had been occupied by the Army and whose decor was consequently in a bad state, a house, moreover, without electricity—shows the pass to which we had come in our attempts to find anything at all suitable in the Cambridge area itself. But, when we visited Orchard House on a weekend, it was immediately apparent what a potentially superb place it was. In its size and in the extent of its grounds, with violets all the way down a long drive, it was almost infinitely removed from the small cottages in which we had spent the last five years. At all events, we took it, because it was the sort of gamble we both enjoy.

I persuaded the estate agent in the little crescent to allow me to charge against rent the materials needed to redecorate the house. Determined not to stint the matter, Tommy Gold and I—Tommy came to stay with us over the Christmas vacation—put on 30 large tubs of a kind of colored whitewash called distemper and 10 gallons of cream paint. The plumbing at Orchard House was on the wrong side of insane, the cold-water supply outside running exposed high above ground level. Then there was a cold snap, and the water froze in the pipes. At the first thaw, burst pipes sprayed water everywhere, which then refroze at the next snap, converting large areas into an ice rink. In this sequence of crises, Tommy did sterling work with insulating tape and a black, sticky, venomous material that went under the equally repellent trade name of Bostick. We managed to get the big drawing room into shape in time for Christmas, when, with plenty of fallen wood from the many trees around the house, we enjoyed a blazing fire throughout the holidays.

Because nothing much grew during the months of winter, the big garden at Orchard House seemed at first to be no threat. But, when spring came, it was immediately clear that yet another monster had burst upon us. As always, Tommy gave good abstract advice—to have the entire garden concreted over—but, since we didn't own it, another end run was needed. My wife and I hit on the sort of smart scheme you eventually wish you hadn't. We would have a flock of geese. They would serve the dual purpose of cropping the extensive lawns and of providing sustenance for our table. Neither of these ideas worked. To start with, we grossly underestimated the mess the creatures would make. Nor had we anticipated their blank refusal to crop the coarser parts of the lawn, which, in the Army's tenure at Orchard House, had been permitted to get quite out of hand. Geese simply do not comprehend that, if they begin by facing up to cropping coarse grass, the sweeter grasses that they crave will gradually take over. Instead, they all go to the few best patches of the moment, which they crop and recrop until the ground is essentially bare, so that you end up with a piebald lawn of bare patches and rough patches.

It didn't take too long either for my wife and me to know that it was out of the question to have the creatures slaughtered. There was a study with windows coming down nearly to the ground and a fireplace in which, in cold weather, we often lit a fire. Almost as if they were in need of our company, the geese would parade on the outer side of the windows while, on the inner side, our small daughter gesticulated in joy, crowing with laughter. No, there was no possibility of the geese being slaughtered, or of our taking the hypocritical route of selling them, and thus slaughtering them indirectly. Since my youth—when, at the age of ten, I saw a farmer's wife erupt into a yard, seize a hen, and wring its neck in short order—I have experienced a developing distaste for killing animals. There is nothing consistent about it. I kill midges and mosquitoes without regret. My wife pounces on any bluebottle that starts buzzing in the kitchen. In later years, we had a house that had been built piecemeal over two centuries. We found it impossible, with so many junctions in the structure, to keep a tiny species of mouse from invading the house each autumn. Some of the mice got among our clothes, and it is amazing what damage mice can do if they get into your clothing. So I had to agree with my wife when she said: "It's either them or us."

Consistent in this inconsistency, I have developed a strong distaste for scientists who experiment on animals. This is argued for by some on moral grounds, and it is sanctioned by respectable scientific opinion, leaving me in a poorly supported minority. Yet I am glad to say this was the position of the scientist who, among all whom I have met, was the fastest in his thinking: Richard Feynman. Feynman was elected a member of the U.S. National Academy of Sciences at an unprecedentedly early age, the mid-thirties, I believe. When he realized the Academy was supporting the use of experimental animals, he resigned. Thereafter, the physics section of the Academy did all it could to conceal the resignation, and, for many years, Feynman was begged by the Academy to reconsider—which, as far as I am aware, he never did. Other explanations have been published for Feynman's resignation, but this is the way he personally told it to me.

But how to solve the problem of the geese? A villager told me to take good care to clip their wings, but only on one side, because, when lopsided, they cannot fly (which explained why the creatures were lopsided when we acquired them). So, of course, we didn't clip their wings, and—sure enough, on one fine day, the Sun shining on their backs—the geese took off, and they were last seen flying at a considerable height above the village.

Chapter 16

The Origin of the Chemical Elements

MOST SCIENTISTS agree that the highest attainments are contained in what are loosely called the laws of physics—loosely, because the use of the word "laws" suggests verbal statements, which the laws of physics are not. The laws are mathematical—and not just mathematical in an ordinary way, but subtly mathematical. A lot of hard work is necessary before it is possible even to understand the basic concepts. I can best illustrate this by an example. Beginning in a roundabout way, I was lucky, in the winter of 1947–1948, to see a fair bit of Werner Heisenberg, who had first conceived, in 1925, of the modern form of quantum mechanics. A short time later, Erwin Schrödinger published his famous wave equation, from which he obtained results similar to Heisenberg's from an apparently disconnected point of view.

The immediate mystery, in 1925–1926, was why two such apparently different points of view led to the same results. When he was staying several months at St. John's College in the winter of 1947–1948, Heisenberg told me how, in 1925, he had visited Cambridge and how, in lecturing to the physicists in the Cavendish Laboratory, he felt he was talking into a vacuum, with nobody there really understanding anything of the new quantum mechanics. Then, at the end, he was given a surprise. When everybody else had gone, up came a tall, thin, dark-haired young man of his own age. "As we talked," Heisenberg said, "I suddenly realized he knew more about it than I did.* But instead of giving me the solution there and then, he said he would send it in written form. After several weeks, I received a package with forty or so pages. Everything was precisely there, perfect down to the last comma and full stop, as he might

---

*The problem was the formulation of what later became known as transformation theory.

have sent it as a paper to the Royal Society." The dark-haired young man had been Dirac.

It was hard to get Paul Dirac to talk about any problem unless he had a perfect solution to it—in physics, that is to say. Dirac would talk freely about the best kind of machine for mowing grass, or whether putting a flock of geese to the task was a good idea, or what was the best system of central heating, or what he thought about the latest situation in Whitehall—but about physics, no. As an undergraduate, I once asked him a question at the end of a lecture. Instead of talking around the point, as most lecturers would have done, he simply said: "I will give the answer next time." At the next lecture, he began: "I have been asked a question. . . ." Then he gave the answer, taking about fifteen minutes over it, which was the best morale booster I had in my undergraduate years, for I reasoned that, if Dirac took fifteen minutes over a point that had baffled me, perhaps I hadn't been too stupid after all.

The big problem in 1926–1927 was that Schrödinger's equation was not consistent with special relativity, the mystery being that the natural extension of Schrödinger's equation to what is called the Klein-Gordon equation clearly could not be correct—it was inferior to Schrödinger's in important respects, whereas it should have been superior in all respects. Worse, there seemed to be no other possibility. It was in these circumstances, with physicists everywhere baffled, that Dirac broke through. He did it by an increase of mathematical sophistication, by introducing what are called spinor fields in addition to the more usual vector and tensor fields. At the risk of seeming incomprehensible, I would like to write down two equations, just to show how short and simple science is, provided you know what it means! The first defines a spinor field χ^β from a given spinor field ξ_α, namely,

$$m\chi^\beta = (\, i\partial^{\alpha\beta} + eA^{\alpha\beta} \,)\, \xi_\alpha.$$

Here, e and m are the charge and the mass of the electron, $\partial^{\alpha\beta}$ is the spinor form of the space-time derivatives, and $A^{\alpha\beta}$ is the spinor form of the electromagnetic field, while i is the square root of -1. No work has yet been done. But now comes the big statement, a statement that controls the structure of the world, even down to the behavior of the electronic gadgets we use routinely in our everyday lives. The statement is,

$$m\xi_\alpha = (\, i\partial_{\alpha\beta} + eA_{\alpha\beta} \,)\, \chi^\beta,$$

which is sophisticated but simple. You do the same thing again and get back to where you started. This equation, Dirac's equation, is not quite the whole of quantum mechanics. Certain symmetry relations must be added when many similar material particles are considered. The discovery of the appropriate symmetry relations was perhaps Wolfgang Pauli's greatest contribution to physics. The student is usually given the relations in a form that, although easily understood in principle, is awkward to apply. Again, more sophisticated mathematics produces simplification. Many particles are best studied by multiplying the spinor groups of the constituent particles, with the injunction that, from the irreducible groups so obtained, the unique, totally antisymmetric representation must be chosen. This leads to a situation, when one has labored to understand the mathematics, in which otherwise quite difficult problems, such as the term analysis in atomic spectroscopy of arrays in heavy atoms, can be determined with almost as little trouble as shelling peas.

These are refined examples of what James Jeans meant in his Reid Lecture at Cambridge in the 1920s, which was published by the Cambridge University Press under the title *The Mysterious Universe*, by the assertion that God is a mathematician. In my school days, there had been a big fuss about this book, especially among philosophers and religious believers. In the end, they mostly decided, to suit their peace of mind, that the book was a nut without a kernel. I never did take this view, and to this day I don't. I once felt impelled to work through a substantial fraction of the term analysis for the iron atom, the element that produces most of the numerous dark Fraunhofer spectrum lines in the spectrum of the Sun, and, after that experience, I feel that only the rabid should agree with Macbeth's view of the Universe: "It is a tale / Told by an idiot, full of sound and fury, / Signifying nothing."

Another remarkable thing is that the simple primary laws turn out to have immensely complicated secondary consequences. The explicit solution of the Dirac equation is even complicated for the apparently elementary case of the hydrogen atom, so complicated indeed that hardly any book on quantum mechanics tackles more than a part of it—the less difficult part. To explain all the details of the hydrogen atom in an undergraduate course of lectures would take up most of the course, which is why very few students today know such things. They are driven at so great a speed by the educational system that almost everything gets skimped, at no fault to the students themselves.

There is a great page in one of Richard Feynman's books on which he obtains the solution to all problems of a thermodynamic nature, compressing the result into a single, all-embracing formula. As he then goes on to remark, it is amazing that so much of what happens in the world can be summarized in so terse a way. But, of course, as soon as one tries to work out the formula, the details in even very simple systems quickly go beyond what can be managed, even with the aid of the fastest and most capacious of computers. The big lesson one learns from science is that the world is very simple in statement but very complex in execution (from which it can be said with certainty that any theory that is complex in statement and trivial in execution must be wrong).

Although the consequences of the laws of physics are exceedingly complex, they are not "full of sound and fury." There is also a high measure of subtlety and sophistication in the consequences. This shows itself already in the subject of the present chapter. Protons and neutrons are particles whose basic nature was explained by the mathematics of unitary symmetry, which developed from the mid-1960s onwards, ideas that, again, had the properties of sophisticated simplicity. The laws that were discovered in the mid-1960s lead, in a complex way, to clusters of protons and neutrons being able to hold together in stable ways, by what in earlier times were called nuclear forces, the clusters being the nuclei of the isotopes of the elements. But there is apparently nothing in the laws to say that all the possible stable clusters will be generated in the Universe, as they actually are. A Universe could readily be imagined in which some were realized and others not. Indeed, as we were to discover over the years, it is only through a glittering array of nuclear processes inside stars that all possibilities are, in fact, realized. Moreover, the proportions in which the chemical elements are realized has a critical relation to the existence of life. It is easy to see how slight changes could have made the emergence of life impossible, thereby preventing another remarkable possibility from being realized. To add to this consideration, using the basic laws alone, the Universe would become increasingly indeterminate, a situation that can, it seems, be avoided only through the intervention of consciousness, a situation about which the older generation of physicists was exceedingly uneasy, but that is generally admitted by the younger generation, a situation that caused Schrödinger to remark: "I don't like it, and I'm sorry to think I ever had anything to do with it."

I have a conviction that the large and small are interlinked, whereas the usual view is that the small (that is, the laws) are primary and that the large (that is, the Universe) is secondary. I suspect the two form a closed logical loop in which both exist together, without either being primary. There is a hint of this already in the Dirac equation. At the level of quantum mechanics, the mass m is the experimental mass of the electron, but, at the highest level of quantum electrodynamics, m is a so-called theoretical mass from which a quantity must be subtracted to obtain the mass measured in the laboratory, a procedure known as renormalization. The quantity to be subtracted is arbitrary, in that it depends on cutting off all electromagnetic fields at some arbitrarily chosen, very high frequency (that is, removing all field components with frequencies above some assigned value). The recommended scheme for avoiding this arbitrariness is to push the cutoff frequency ever higher and higher, until it goes to infinity. But then m in Dirac's equation goes down and down until, eventually, it goes to minus infinity—not a result to recommend the study of physics, you might think. Another way is to appeal to the universe to provide a definite cutoff, with m then having a definite unique value determined in part by the electron and in part by cosmology. This is the solution I favor, and I think it applies to a very wide range of physical problems. Its only demerit is that one has to see deeper to apply it, which makes things harder to work through. But there is no reason to suppose that the Universe exists to suit our convenience.

There can be few scientists who do not dream, in idle moments, that it might fall to their lot to discover a basic law, like Faraday's law of induction, or Maxwell's displacement current, or the equivalence of mass and energy, $E = mc^2$, or Einstein's of general relativity, or Dirac's of the electron. But, as the French physicist and mathematician Joseph Louis de Lagrange dryly said of Newton's discoveries: "There can be only one system of the world." In other words, it is essential to live at the right moment for every big discovery. When I was a student of Dirac in 1938–1939, he told me the time then for such discoveries was not ripe, which judgment proved correct. In a considerable measure, it had been Dirac's unfavorable prognostication in 1938–1939 that turned me to astronomy. Dirac was more careful in his precise choice of words than anybody else I have met, more careful even than my friends in the English Faculty in Cambridge. If there is a word used freely nowadays that would have caused him to grit his teeth, it is the word "researcher." Whenever he felt the need to shorten the cumbersome term "research worker," he described himself simply as a worker.

The idea of the chemical elements being formed in the stars had been held long before I came to it. Once Francis William Aston had done his mass-spectrograph measurements of the masses of the isotopes of the elements, it was an idea that came easily as a speculation. The stumbling block to the development of the idea was the enormous internal stellar temperatures that would be required. There even seemed great difficulty, in the 1920s, in seeing how the relatively easy building of hydrogen into helium could take place. The difficulty for the origin of the elements did not go away with Bethe's successful work on the hydrogen-to-helium problem, the trouble being that temperatures upwards of a hundred times greater than those used by Bethe would be necessary. The bearlike George Gamow, who had done almost more than anyone else to smooth the path with his solution of the quantum tunneling problem in the late 1920s, was led to argue that the best way to obtain the needed high temperatures was in the early moments of what today is called the origin of the Universe, in a "big bang." By the mid-1940s, George was pressing this idea, which, as it eventually turned out, had an important relevance to cosmology but was not the main solution to the problem. This could indeed already be seen in the 1940s, for there were insurmountable gaps in Gamow's attempted building chain at atomic masses 5 and 8. The alternative was to demonstrate, beyond reasonable doubt, that the required exceedingly high temperatures really did exist inside stars. It was this demonstration that I had in my pocket, like the overture to *Rienzi*, when I drove in the little £5 Singer in July 1945 from Funtington to Cambridge. I had arrived at this demonstration by a chain of accidents.

The chain started with the phone call from Maurice Pryce about the curious anomalous propagation of the new radar equipment at short wavelengths. It continued with Edward Appleton's committee on propagation and with my being a member of it. In the autumn of 1944, Appleton told his committee that a meeting on anomalous propagation was to be held at the end of November in Washington, D.C., mostly on account of the interest of the U.S. Navy in the anomalous-propagation problem. He went on to suggest that it would be appropriate for two Admiralty committee members (he suggested Frank Westwater and me) to attend the Washington meeting. E. C. S. Megaw, who had organized the large-scale production magnetrons at the General Electric Company (U.K.), was chosen as an outside third member. When the date approached, I was told to join the *Aquitania* at Greenock, an old tub from the 1920s that could hardly outrun a U-boat. I mentioned before that there would have been no pension prospect for my wife and family had the *Aquitania* been sunk. So I took out as much insurance at Lloyd's as I could afford. The odds I was given were sufficiently poor as to suggest I ought to contrive to see my parents, not, I hoped, for the last time. So I decided on a Royal Navy car to make the journey north. With 8:00 A.M. the following morning as the deadline for reporting in Greenock, I drove north to Leamington Spa, had lunch with Aunt Leila (my favorite), and then continued to my parents' home in Bingley. After staying there until the light had gone, I managed somehow to drive in the dark to Frank Westwater's parents' home in Edinburgh, a long, tormenting journey with no signposts to a destination for which I had only an address, tormenting because of the obligatory headlamp masks, which permitted only a little light to illuminate the road. I arrived eventually at some time after midnight, slept a little on a sofa, and left with Frank for Greenock three or four hours later, so that I never saw Frank's parents.

The voyage to New York took ten days. By a direct route, it would have taken six. There were 10,000 American troops on board, returning home

for Christmas leave. They manned the ship's inadequate guns with enthusiasm, and they perpetually practiced boat drill, which Westwater told
me was a waste of time. When a ship is torpedoed, he said, even a naval
ship with the strictest discipline, pandemonium reigns and nothing works
out as it is supposed to. The best plan, he advised, was to make for the
ship's stern and to slide into the sea down one of the many ropes you could
find there. Then you waited in the water until the ship went down, when
an immense quantity of jetsam came up. With luck, you might find something to cling to and thereby survive, which you never would if you joined
the press of people trying to get into the boats. So we examined the stern
of the ship, deciding which were the best ropes. I thought it favorable that
we all had life jackets, but, by the time we were in the middle of the Atlantic surrounded by thirty-foot waves, I obtained only marginal reas-
surance from my life jacket.

There were sumptuous meals three times a day, the service being practically continuous. Compared to the slenderness of the British wartime
diet, it seemed almost indecent, although I suppose the ship simply picked
up its provisions when in port on the American side. We spent a good deal
of time in working out the boat's secret course, which could be done with
tolerable accuracy by estimating the elevation of the Sun at noon, taken
with an assumed average speed in longitude of 15 knots. The course went
south towards the Azores. We also spent time in researches on the inci-
dence of seasickness, from which we both suffered. There was a place in
the ship, in an old ballroom that was still decorated in the style of the
1920s, where we experienced no seasickness, however much the boat
rolled or pitched ("It rolls like a pig," Westwater averred); but this was
the centroid of zero sickness. Then we experimented by going increasing
distances away from the centroid, finding that the bow, where the radar
officer had his cabin, was the worst. We would sit for a while, enjoying a
drink with the radar officer, when, in an instant, one or other of us would
rush at speed back to the centroid. Making it safely, it took 30 to 45 minutes for the chemical effects causing the sickness to disappear. The first
symptom is a tight feeling behind the ears, then a queasiness in the pit of
the stomach, at which stage you had better do something about it, I can
tell you. I had been through it all in acute form twice in crossing the North
Sea in the spring of 1939, and I had measured the recovery time, over the
most gut-bindingly strong coffee I have ever managed to swallow, in a
dark, dockside café in Stavanger. In Westwater's cabin—we traveled eight
to a prewar cabin of two—was a Fleet Air Arm pilot, who spent the entire
voyage reading very slowly a semipornographic book called *Cities of Sin*,
which he defended with a furious intensity from the many persons who
tried to get their hands on it. Westwater eventually discovered the inge-

nious place where the chap hid it at mealtimes, and he reported that *Cities of Sin*, in his opinion, was pretty tame stuff, which anybody could easily better for themselves over two days spent in Cairo.

How much pleasure can be got from a little deprivation! We had been five years now without lights in our towns. When we docked, I believe in the Hudson River at about Forty-second Street, it was coming up 6:00 P.M. of a short November afternoon, and the lights of the city made it seem like fairyland. We took a taxi to the hotel at which we had been told to "report." If in those days there was a rather upmarket place called the Barbizon Plaza, then that was the hotel. I don't believe either Westwater or I had any American money at that point, so, until we reached Washington, we must have borrowed from the efficient Megaw, who seemed to have spent the voyage commendably working on what he was going to say at the Washington meeting.

We picked up our train tickets to Washington from the hotel desk, and the next morning took a taxi to Pennsylvania Station on Thirty-third Street, again borrowing from the unfortunate Megaw. If you believe that technology improves with time, then you are wrong. Penn Station, I would guess, could swallow Waterloo three times over. There was a female announcer there with an indescribably fascinating voice that could be heard over speakers in every nook and cranny of the vast concourse. We listened, spellbound, for half an hour as announcements were made of trains departing for legendary places we had only heard of in geography lessons at school. Contrast this with the present annual prize offered by British Rail to the station with the greatest measure of electronic distortion, a prize currently held by Southampton, which somehow manages to combine ear-splitting volume with a total lack of intelligibility. Perhaps not quite that, because you can actually tell that a human voice is involved rather than some mixture of seagulls and a bellowing bull. But that is all you can tell.

We were shown to our seats in a Pullman car by a redcap. The seats were comfortable armchairs on swivels, so that you could turn around to converse with any one of a selection of people around you. With Westwater in Royal Navy uniform, we were never in need of people eager to converse. Cocktails were widely drunk on the journey, except in New Jersey, which was still dry. It is odd that my memories of that trip are all sharp and clear, except for the time we spent in Washington, our supposed destination. I remember clearly the hotel in Boston with the exceedingly hot tiled bathroom floor, on which Frank Westwater danced and shouted: "The deck's too hot to stand on." But I can't remember the hotel in Washington or the conference hall or what was said there. Perhaps this anomaly was due to my first impressions of Washington being so much at variance with later memories. I had the impression of a rather small, friendly place, much rooted in the history of the past. Perhaps the difference was due to an almost complete absence of traffic.

Our first port of call was the British Embassy, where a program had been drawn up for us. Notably, we had invitations to visit propagation units at MIT and at the naval headquarters in San Diego. The head of the unit at MIT was called Alvarez (which might suggest the physicist Luis Alvarez, but this was someone else) and the head at San Diego was named Smyth (which might suggest the author of the later Smyth Report, but this again was a different man). We drew a generous per diem allowance for the entire visit, repaid Megaw, and were now free men, ready to enjoy the three spare days before the conference was due to begin. The total allowance per person for the visit was $400 or thereabouts, which, in those days, was a mighty sum. In addition, we had travel tickets and hotel expenses. With such largess flying around in diplomatic circles, no wonder a second secretary was able to give us lunch at the Mayflower Restaurant.

I decided I would make use of the three spare days by returning north on the railway as far as Princeton Junction and then into Princeton to visit Henry Norris Russell. Frank Westwater decided to go too because, in the mathematical tripos of 1935, he had been awarded the Tyson Medal for astronomy. We stayed at the Nassau Tavern in Princeton, bought a gallon of wine for $1, drank it out of tumblers, and consumed large bunches of grapes, wishing our families were there to share the experience. We called on Russell the following morning. After two hours of the overwhelmingly ebullient talk for which Russell was famous, he invited us to lunch, when we found that his wife (a slight woman, as I recall) was not too far behind in the matter of talking herself. When Russell heard we were to go out west, he went into a long account of the languages, the archaeological history, and the physical appearance of the North American Indian, which could quite well have gone straight into the pages of the *Encyclopaedia Britannica*. On the matter of the San Diego visit, Russell urged me to call, if possible, at the Mount Wilson Observatory offices on Santa Barbara Street in Pasadena. There was a free weekend in our itinerary that might be devoted to such a visit, and Russell immediately offered to write to Walter Adams, the director of the observatory, on our behalf. A further link in the chain of accidents was now in place.

The conference that was the ostensible purpose of our trip came and went, and I remember little or nothing of it. What I can remember clearly was arriving at Washington National Airport for the evening flight to San Diego. Our tickets had been provided, but, if we had had to buy them, the cost one-way would have been $110, which I suppose is only about twice what you would pay today. (The rail fare from Cambridge to London has, in contrast, gone up thirtyfold.) The plane was a DC-3. There were stops at Knoxville, Tennessee; at Little Rock, Arkansas; at Amarillo (I believe), Texas; at Albuquerque, New Mexico; at Phoenix, Arizona; and so into San Diego by 11:00 A.M. the next day. The plane flew generally at a height of about 10,000 feet, rising an additional 5000 feet over the higher mountains. Each passenger had a big window, making the morning views spectacular. There was no food on the plane, the general practice being to order a hamburger or apple pie and coffee on the ground at each stop, which we did on each occasion. Smyth met us at the airport in San Diego. He took us to his home where we met his wife, a good-looking young woman with Irish dark hair and deep blue eyes. She met us with soil on her hands, for which she apologized, saying she had been digging for gophers, an activity unknown to me then. Mrs. Smyth, after washing off the soil, offered us pecan pie and coffee. Since leaving Washington, we had consumed a lot of pie of one sort and another. That evening saw us booked into the Valencia Hotel in La Jolla. After dinner, we walked the La Jolla beach under a moon, the beach quite deserted. Floating delicately on the wings of memory, I remember walking the picturesque streets of La Jolla's old town the following morning, finding a barber shop and stopping for a haircut. Yes, Geoffrey and Margaret Burbidge, I was there before you, at a time when there was scarcely a car in the streets. In later years, I ran into Jacob Bronowski on the steps of the La Jolla Post Office. I was there before him too, and before Francis Crick and Leslie Orgel. Jonas Salk paid me a visit in Cambridge in the 1960s, at the time he had raised the money for his institute in La Jolla but before any building had been done. I think he would have liked me to have gone there, and, if I had, the later chapters of this book would surely have been different. La Jolla had considerable attractions even in the 1960s. For instance, there was Chuck's Steak House, not far from the Valencia Hotel. Chuck's served what they called a combo plate: half lobster, half steak. The French astronomer Marie-Hélène Arp was in La Jolla in the 1960s. When I asked her recently, "Do you remember Chuck's combo plate?" she replied, with a nostalgic sigh, "Ah! It was the best thing in all America."

Two aspects of that visit to San Diego that were relevant to the time capsule of that moment were our first sight of massed naval ships in San Diego harbor and the long, open-air line of planes being assembled at the Convair Corporation plant. Westwater had seen far more warships in port than I had, but what he saw of the firepower in San Diego harbor staggered him. At school, I had learned much about the advantages enjoyed by the United States—the fifteen feet of topsoil in Iowa, the immense area, the forests, the great mineral wealth—but now I saw that even the quality of the light could also be a natural advantage. All the planes I had seen being made in Britain had been assembled by artificial light, for the reason that assembly could then be continued through the night. Here, in natural light, the speed of construction was much greater, the size of the factory floor being effectively infinite.

Smyth arranged for a Convair car to take us to Los Angeles. He had a southern manner of speech, allied to a western idiom. He never talked of leaving or arriving, but only of pulling in or pulling out. It was around 6:00 P.M. when we pulled out in the Convair car. There was, of course, no freeway to Los Angeles, not even a very large road. We made our way up the coast through a seemingly never-ending sequence of small townships. About halfway to Los Angeles, we stopped for the inevitable pie and coffee, arriving eventually at the Biltmore Hotel a few minutes after 9:00 P.M. We were told that a tram would take us along Huntington Drive to Pasadena. The fare was 25 cents, a real silver quarter, the world's most beautiful coin. For this we were deposited at the Huntington Hotel perhaps by 10:30 P.M., after journeying most of the way through orange groves. The surprise at the Huntington was to find food still being served. We paid $2 for a three-tier club sandwich. This was the equivalent of ten shillings, and we almost choked on it at the thought of paying such an absurd price. By this time, however, all the pie we had consumed had given us an appetite.

We walked the whole way, north up Lake Avenue from the Huntington Hotel to Santa Barbara Street, the mountains ahead crystal clear in the morning light. Walter Adams was there in his office. He parried my astronomical questions, telling us that a car would be leaving in an hour for the actual observatory on the summit of Mount Wilson, and that, if we wanted to spend the weekend there, we were welcome to do so. The driver of the car was Roscoe Sanford, who had delivered the killer blow to Harlow Shapley's pre-Copernican view of the Universe, a view in which Shapley had been supported by formidable establishment of Dutch astronomers. The story is this:

What we call spiral galaxies today were already noticed, in the early nineteenth century, in the sky surveys of William Herschel, but it was not until such examples as the Whirlpool galaxy, M51, were observed by the Leviathan of Parsonstown (a huge, unwieldy telescope built in the second half of the nineteenth century by the Earl of Rosse) that a few astronomers began to think seriously of spiral galaxies as independent objects outside our own galaxy. The majority remained unconvinced, however, preferring to maintain that our galaxy was all there was of the Universe, thereby setting up a pre-Copernican point of view, which was added to by Jacobus Cornelius Kapteyn, the leader of the Dutch astronomy establishment, who claimed that the Sun and our solar system lay close to the center of our galaxy. If you subscribed to this view, the Earth and we humans were situated essentially at the center of the Universe. Harlow Shapley did not go as far as Kapteyn, but he did become the acknowledged champion of what came to be called the island Universe theory. Heber D. Curtis took on the role of arguing for the dissenters, and the issue eventually boiled down to whether or not there is a fog of tiny absorbing particles existing in clouds between the stars. A photograph obtained by Roscoe Sanford settled the issue. It was of a flat spiral galaxy, seen nearly edge-on, with a dark band of absorbing clouds across it. This is how our galaxy would look if it were seen nearly edge-on from the outside.

There are two points of interest about this story. One is a demonstration of how almost the entire body of scientists can go wrong in a way that, in later years, seems absurd. To hear scientists talk today, you would think the first moment in human history in which nonsensical views are not widely held is now. Yet, when I contemplate the strong belief of many, that everything of real importance in the big-bang Universe happened in the first 10^{-43} seconds of its existence, I am compelled to wonder.

Roscoe Sanford drove us to the summit of Mount Wilson via the sentry box and by the back road. We were introduced to the chap in charge of "the Monastery," and to G. Stromberg and A. H. Joy. Then Sanford drove away, leaving us eventually to walk down the mountain back to Pasadena at the end of the weekend. The present smog-engulfed inhabitants of Los Angeles may be anguished to know that the air was so clear in those days that it seemed as if you could almost stretch out a hand and touch the snows of the distant Mount San Gorgonio and Mount San Jacinto. At dinner that night, Westwater committed a blunder. He produced two bottles of wine to liven up the proceedings. But the wine was shipped away in shocked silence by the chef. The two things most disapproved of at Mount Wilson were alcohol and women, but in what order of importance I never determined. The third important thing, but lower down on the priority list, was that everybody had to eat from tin plates.

I made a gaffe a year or two later to equal or even exceed Westwater's. It too was at a dinner, and it also arose from that visit to the United States in 1944. The dinner was at a London club where I was a guest. It wasn't until I arrived for the dinner that I learned I would be expected to say "a few words" afterwards, in the course of which, according to club rules, a joke should be told. Throughout the dinner, I sieved through my brain for a joke, and then I suddenly remembered an occurrence in the western United States in 1944 over which there had been merriment in the heavyweight eastern daily newspapers. I had told Dirac the story, and he had laughed uproariously, so I knew it was a good joke that should go down well at this exclusive club. There had been a national census. An official of one small town in Wyoming was puzzled by one of the questions— namely, "What is your death rate?" When I came out with the words "death rate," there was the same total silence there had been when Frank Westwater produced his two bottles of wine, for as I now saw, to my growing discomfort, all the members of the club were very old. There was no stopping, of course. You can't stop in the middle of a joke, especially after mentioning the words "death rate." In absolute silence, I ploughed ahead. The official of this small town in Wyoming replied, to the question that puzzled him, "Our death rate here is the same as everywhere else. One death for every inhabitant."

My shoes were well into a process of disintegration by the time we had walked down the front of Mount Wilson on the Monday morning, so I was delighted when we were met by a car in Altadena. Walter Baade was the driver. He had come from the Hamburg Observatory about a decade earlier to join the staff at Mount Wilson, ostensibly to settle an interne-cine conflict between Edwin Hubble and Adriaan van Maanen. Appar-ently, van Maanen had proved too slippery for Hubble to nail, and, as Baade told the story, van Maanen was even too slippery for him: "Vun day I thought I had him on the hook, but he got away." At home with his wife, Baade always spoke German. The German overtones he added to slick English idiom always sounded comical. When he arrived at one of the many parties given in later years by astronomers at Mount Wilson and Caltech, it was as if a gale had suddenly blown through the house. From the time I first met him in Altadena until his death sixteen years later, Walter Baade enjoyed the reputation for being the world's finest direct plate observer, a reputation I never heard challenged, either by his con-temporaries or by the later generation of astronomers. The mystery for me was how he did it, for Walter had no control over his hands, to speak of. His explanation was that he never made a move at the telescope or in the darkroom without thinking in detail beforehand exactly what he was going to do next. He never corrected the curve of a car as it went towards the edge of the road until almost at the disaster point. This penchant pro-voked trouble with the Pasadena police, with whom he had long-standing battles. For the last half decade of his stay in Pasadena, he simply drove without a license. You might think Walter was then a sitting duck, but he wasn't. He knew the police to be creatures of routine. What he did was to map the disposition and movements of prowl cars. Then he planned his routes along streets where he knew they wouldn't be.

After Baade had picked us up in Altadena, I spent the afternoon discussing astronomy with him. He gave me a number of reprints of papers I hadn't seen because of our lack of library facilities at Nutbourne and Witley, and the penultimate link in the chain of accidents was added. In the evening, Baade took us to an upbeat restaurant in Pasadena called the Stuffed Shirt. In a fit of exploratory madness, I ordered abalone, finding to my surprise that the tripelike stuff actually tasted very good.

After flying from Los Angeles to Boston, we crossed the Charles River to the other Cambridge to visit colleagues at MIT (and, in my case, at the Harvard Observatory in Garden Street). It was then that I first met Harlow Shapley. He was listening to Wagner when I called on him, but I don't think it was *Rienzi*. I also met Zdenek Kopal for the first time, and I renewed my friendship with Felix Cernushi from my days as a Cambridge research student. For a thoroughly amiable person, Felix somehow managed to get himself into an amazing lot of trouble, having begun the Students' Revolution back in his native Argentina and having somehow turned the Paris police upside down. After California, I was cold in Boston in my wartime utility clothes.

The route back home was via Montreal. The large new planes, Fortresses and Liberators, could be flown nonstop from Montreal to Prestwick, near Glasgow, in fourteen hours. This seemed infinitely better than ten days at sea in the *Aquitania*. The cargo of the planes was people, mostly officers bound for Europe. The system was that, after you arrived in Montreal, your name was put on a list. When eventually your name came near the top of the list, you were told to ready yourself a few hours ahead of a flight. We took the night train from Boston to Montreal, with a long border stop as Canadian Customs searched for contraband whisky, which, for some reason, American officers seemed impelled to smuggle. Frank and I parted in Montreal; we had lost Megaw long ago in Washington. I still had to visit radar research stations in the Ottawa region. Back in Montreal, I also hit stormy weather and had to wait around for several days. It was during those several days that the last link in the chain of accidents was to be fitted into place.

I was not short of people to see. There was Nick Kemmer. Nick had, I believe, been a student of Wolfgang Pauli. He had come to Britain circa 1938, and I had met him frequently in Cambridge. It was Nick who first classified the neutron and proton by a formalism analogous to electron spin, a formalism that later became known as isospin. At the time, I thought the formalism a bit arid, never dreaming that, with the addition of further quantum numbers, it would eventually yield the classification of the many subatomic particles to be discovered in subsequent years. Nick was tallish, blond, and with a gift for languages that made it hard to know what his native speech had been. I think it was Russian, although you would never have guessed it from his English pronunciation. Then there was who else? Maurice Pryce, of course. I knew Maurice had mysteriously disappeared from Admiralty Signal Establishment, but, as I said before, I could never quite make out what he was up to. The juxtaposition of Kemmer and Pryce already told me a great deal, however, about what was going on in Montreal—or, more accurately, at Chalk River near Montreal. The British nuclear-bomb-making project had been based, in 1940, in Cambridge, where it went under the code name of Tube Alloys, and Nick Kemmer had been a part of it. Although nobody told me so explicitly, it had been obvious that their goal was to separate the isotopes of uranium by gas diffusion, a project that was difficult at best and that I had hoped would prove to be impossible. It very nearly did. With much labor, the immense American effort at Oak Ridge, Tennessee, managed to produce enough uranium-235 for a single bomb, but that was about all. If centrifugation had been used instead of gas diffusion, the story might have been different.

Maurice Pryce's disappearance from ASE and his reappearance in Montreal told me that the bomb-making project must be nearing success, unfortunately. I couldn't believe that they had succeeded in separating the isotopes by gas diffusion, so it had to be some other way, and there wasn't too much mystery about what that other way had to be. The possibility of making a nuclear reactor (or a nuclear pile, as it was called in those days) had been clear since the publication in *Nature*, in the early spring of 1939, of the discovery of delayed neutrons from the fission of uranium. It was also clear that neutron capture by the common isotope of uranium in such a pile would produce plutonium-239, which, being of odd number, would be made fissile by fast neutrons. Moreover, because plutonium is chemically different from uranium, its separation from uranium would be less difficult than the separation of the uranium isotopes from each other. In 1939, Britain had not possessed the resources to follow up on both methods of bomb making. The uranium-separation method had been chosen by Tube Alloys. Before my visit to Montreal, it hadn't occurred to me how curious it was that the apparently more difficult route had been chosen. It could only be, I realized as I met Kemmer and Pryce socially, that there was a hidden snag somewhere in the plutonium route. I had heard, in a general way, of the immensely strong team of scientists the Americans were assembling somewhere in the West—at Los Alamos, New Mexico, as it turned out. It seemed odd that the Americans felt they needed such a team to make a bomb, which, to this point, I had believed to be nothing more complicated than firing a couple of pieces of uranium together at a comparatively low speed. It required no great power of divination to guess that it must have something to do with a greater difficulty in the plutonium method. It had to be that plutonium was more difficult to explode. This could only mean a problem with triggering. So how could the triggering be overcome, as Pryce's sudden transfer from Admiralty Signal Establishment suggested it had been? Whether I thought of it myself or picked it up from a hint, I have forgotten—it was probably the latter. It had to be a speed problem, with a solution that involved implosion. The idea of implosion being necessary to a nuclear explosion was the last link in the chain of accidents leading to my work on the origin of the elements.

I never heard a convincing explanation of what that strong team of British nuclear scientists at Chalk River at the end of 1944 was supposed to be doing. The explanation that it was in liaison with Fermi's group in Chicago, the group that produced the first operational reactor, seems unconvincing. The Americans surely would have thought it unwise to pass sensitive information across an international frontier, information that would likely fan out to be picked up by passing travelers, as it had a little in my case. Where the Americans really did agree to there being British scientists involved, the outstanding case being Bill Penney, they took good care to confine them at Los Alamos under the oppressive security described picturesquely in later years by Richard Feynman. Stories of the British government being miffed at American tightness in handing out information were rife at the time. Efforts were made to attract the Americans by involving such scientists as James Chadwick and Paul Dirac, whom the Americans respected. The obvious explanation of the presence of the strong British nuclear group in Montreal was that they were there as a listening post, to pick up whatever they could from the signs and portents, which was just what I had done myself.

It is a comment on the state of transport in those days that, although it took 14 hours from Montreal to Prestwick near Glasgow, it took 24 hours from Prestwick to my home at Funtington in western Sussex. Over the relatively quiet Christmas and New Year periods at the end of 1944, I was able to review the events of the past month. The best starting point might be my conversation with Walter Baade. The properties of novae had been a topic of avid discussion among prewar astronomers, the favored explanation being that they were caused by flares analogous to solar flares occurring on the otherwise cool surfaces of giant stars. But I had come on a paper by Milton Humason at Mount Wilson discussing his observations of the remnants of old novae. From the preponderance of blue light in their spectra, Humason concluded that the remnants must be very hot, with surface temperatures in the region of 100,000 kelvin.

This suggested to my mind the very reverse of the popular view, that novae were associated with small stars, not large ones—stars more like white dwarfs but with higher surface temperatures. This was the topic with which I opened my conversation with Baade. He quickly brushed it aside by saying it was clearly so, but then he went on to ask why, if I was interested in such things, I hadn't preferred supernovae to novae, supernovae being vastly more powerful in their outbursts. So we talked about supernovae rather than novae. Walter also gave me reprint copies of his papers, which had not been accessible either at Nutbourne or at Witley. These I read carefully over the holidays. Combining what I learned with my speculations about implosion being relevant to the plutonium bomb, I wondered if a supernova might not be like a nuclear weapon, with implosion as the cause of instability, leading eventually to an enormously violent outburst.

An idea leads nowhere unless it can be followed up either by an experiment or by a precise calculation. Experiment being out of the question, I therefore wondered next what form of precise calculation could be done. There was no hope then of tackling the complicated networks of nuclear reactions that eventually became possible, except in one case, the ultimate case. If a star were imagined to evolve on and on, with increasing internal temperature and rising central density, a stage would be reached at which protons and neutrons would come into statistical equilibrium. Given the density and the temperature, this evolution could be calculated. The form of the calculation was like that of an assembly of atoms and molecules, the problem of so-called mass action that had tormented me in my schooldays but that I had learned how to solve accurately when I had attended the lectures on statistical mechanics given by the burly Ralph Fowler in 1935–1936.

Protons and neutrons took the place of atoms, atomic nuclei took the place of molecules, with the equivalent of the molecular bindings being determined from a knowledge of nuclear masses—which, for the moment, I didn't have, but I knew I could acquire that knowledge eventually. One thing that could be done immediately, however, was to apply a general result from all such calculations—namely, at lower temperatures, nuclei with the strongest internal binding would be favored, whereas, at higher and higher temperatures, the heavier nuclei would come apart to yield mostly neutrons, protons, and helium nuclei (alpha particles). The latter, I saw, would be energy-absorbing rather than energy-yielding, inevitably causing the core of a highly evolved star to implode—the process I was looking for. Implosion would lead to the release of more than sufficient energy from gravitation to explain the eventual outburst of a supernova, provided some means could be found to transfer energy from the inner regions to the outer. There was the possibility of nuclear energy still being available in the outer regions and of this being made explosively available as the inner regions collapsed. It was to be many a year before the fine details of all the complex of processes that would be set in train were at last worked out, details that were almost exactly what actually happened in the outburst of supernova 1987A.

It was not until March 1945 that I found a reason to be in Cambridge. I was looking for information about nuclear masses in the Cavendish Library when I had the good fortune to run into Otto Frisch. I told him what I wanted. Leading me to his office, he produced extensive tables compiled by the German physicist W. Mattauch, which he was kind enough to lend me. I also obtained a book by V. Goldschmidt, the father of geochemistry, from the university library. From Goldschmidt, I was able to plot the relative abundances of the elements in the Earth, ranging in atomic mass from sodium and magnesium upwards. If the Earth were representative, these had to be the relative abundances elsewhere—in the Sun and in similar stars, for instance.

What I found was that, starting from a top level at the common elements magnesium and silicon, there was a drop to about one-tenth in the relative abundances of the elements through sulfur to calcium. If one then left out the range of masses from 45 to 60, there was a further drop, to about a hundredth, to the nonferrous metals copper and zinc. And so on, down through arsenic, bromine, selenium, silver, tin, barium, the rare earths, and eventually to the precious metals, with abundances about one-millionth those of magnesium and silicon. The big exception to the general rule of decreasing abundance with increasing nuclear mass was between 45 and 60, where there rose a peak of the same level as that of magnesium and silicon (or even higher, according to Goldschmidt). The peak rose from a very low value at scandium, up through titanium, vanadium, chromium, and magnesium, to a maximum at iron, thereafter declining, through cobalt and nickel, back to the generally falling curve at copper and zinc.

These were just the metals on which our civilization had been based, and the peak was just the statistical distribution of nuclei that I calculated using the nuclear masses I had obtained from Otto Frisch—or, at any rate, they were so within an acceptable margin of error. This could not be an accident. It implied that the physical conditions I used in the calculations, which were consistent with statistical equilibrium, really occurred in nature. The temperatures, in particular, seemed staggering: 2 billion to 5 billion degrees, more than a hundred times greater than Eddington had calculated for main-sequence stars. And if such enormously high temperatures could occur in a supernova, then surely all the intermediate temperatures between Eddington's 20 million and my 5000 million could also occur among the stars, where the complex of nuclear re-actions that must surely occur, not in statistical equilibrium, might explain all the other features of Goldschmidt's abundances.

While all this was going on, something rather unpleasant happened. Back at Admiralty Signal Establishment, I was one day called into C. E. Horton's office. He showed me papers that had come from the Foreign Office to the Admiralty and thence to him as the boss at Witley. The papers called for an explanation of my visit to the Mount Wilson Observatory, which had not been on my "tour of duty." I remembered I had heard of there being an astronomer on the staff of the British embassy in Washington, and I suppose it must have been that person who heard of my visit to Pasadena. It was my first indication of the malice that existed at no great distance below the surface of the British astronomical community. In later years, Dick Feynman once put his arm around my shoulder and said, "Remember, Fred—when you get real mad, always make certain your voice goes down, never up." Although he was three years the younger, Feynman always had a fatherly attitude to me. If I may say so, we had a lot in common. For instance, there would be problems in physics I would find interesting but would fail to solve, and then he would come along and solve them.

We also had in common that our fathers earned their living in textiles, Feynman's as a tailor and mine as a vendor. I once arrived to stay with Feynman in his Altadena home to find him putting the finishing touches to a patent application for cutting patterns in cloth. The idea was to punch a design onto a computer, which then instructed a powerful array of knives to do the cutting, quickly and without human intervention. He told me he wouldn't make much money out of the patent, because cutting was the easy part of tailoring. "Now if we could do the stitching, then we'd be rich," he said. So we thought for an hour or two about how to do the stitching. The trouble, of course, was that cutting is a two-dimensional problem, while stitching is three-dimensional, and we both knew enough topology to see the difficulties. We were both almost completely free from envy, and we were both capable of "getting real mad," and at more or less the same things.

It has been an unhappy feature of my life that, on almost all the occasions when I have been "real mad," some extraneous circumstance has prevented me from letting the hounds off the leash. This causes me to sit congealed, boiling internally, hardly able to speak, as Ray Lyttleton has graphically described it. On this occasion, when C. E. Horton called me into his office, the implied reprimand was of no consequence to me. The end of the war was close, and I would soon be leaving Admiralty service. Nothing, I thought, would be lost by unrestrained invective. But then I remembered that Frank Westwater would be continuing as a Royal Navy officer. So I wrote in explanation that southern California is well known for a strong temperature inversion, at which there is a density jump that could be of consequence for short-wave propagation. The matter was one on which the astronomers at the Mount Wilson Observatory were the best possible consultants. I thought this superior to making the excuse that our visit had been on a weekend. My explanation must have satisfied them. About a month later, I had a request from the Admiralty, asking if I would be willing to act as a consultant after the war. In view of the Mount Wilson incident, I refused.

Anyway, all this explains how it came about that I had the overture to *Rienzi* in my pocket (figuratively speaking) when I left for Cambridge in July 1945. Much water would still have to flow under various bridges before Geoffrey and Margaret Burbidge, Willy Fowler, and I arrived at our paper of 1957, and much more still before the theory was developed to the point where it became capable of predicting the behavior of Supernova 1987A, but I like to think that the direction in which the water had to flow was already decided.

Chapter 17

Brave New World

I MAKE LITTLE apology for beginning and ending this chapter with money, for money matters are universal among the young, unless they are sheltered by wealth or by family connections. The beginning, in the present case, is with a temporary job bringing in £450 per annum (say, with added small perks, £500 per annum) and, on the debit side, a big house in deplorable condition renting at £250 per annum. The house is 25 miles from Cambridge, with a £5 Singer two-seater of 1928 vintage as the means of transport to and from my place of work. We shall end, however, with a job for life and also, into the bargain, being rich. If you don't think the accumulation of £3000 in excess of salary was to be rich, then you have never studied the problem of finding £62-10-0 in time for the next quarter day, one of the days on which, four times per year, tithes used to be paid to the church. What I will not describe here are the beginnings of ideas in cosmology with connections to the modern state of affairs, which topic is better left for a later chapter on its own.

The credit for this transformation in an interval of four years belongs to the University of Cambridge—or, more accurately, to the University of Cambridge as it was then. The three teaching terms took up less than twenty-five weeks of the year, while the remaining twenty-seven were completely free—that is, free of committees and the like. The free twenty-seven were made up of six weeks at Christmas, six more at Easter, and fifteen in the summer. So clear-cut a division permitted one both to teach properly and to give wholehearted attention to research. During the teaching terms—eight weeks in the autumn, eight more before Easter, and another six after Easter—I gave maximum effort to my university lectures and to the detailed supervision of students for St. John's College. During the intervening vacation periods, I retired back home with books and papers. Even if the skies over Cambridge had fallen, I would not have been required to interrupt my research.

Einstein once said that the best job for someone in research would be that of a lighthouse keeper, because the essence of research is to have sufficient continuity in one's work to allow one to build up large patterns in the brain. Interruptions destroy such patterns, and repeated interruptions even destroy the will to reform them, in contrast to the way I felt at the end of each teaching term. I was always ready and eager to get started, and, until the beginning of the next teaching term, I never stopped, except for sleep.

I used to think that it would be better still to work in an institution entirely devoted to research. Now I am doubtful. I think conditions as they existed at Cambridge up to 1960, when committee proliferation first got into its stride, were pretty well ideal. Research does not always go well, and, in bad times, if research is all you have to do, it is easy to become depressed and to lose confidence, confidence being absolutely essential to success. However bad things may have gone during a vacation period devoted to research, the next teaching term always brought something that could be accomplished successfully. It is also easy, when doing research only, to become fixated by something and then to go steadily more and more wrong, whereas interruptions during the teaching terms provided the periods of reflection and reconsideration essential to maintaining balance. To achieve anything really worthwhile in research, it is necessary to go against the opinions of one's fellows. To do so successfully, not merely becoming a crackpot, requires fine judgment, especially on long-term issues that cannot be settled quickly. The trickiest present-day issue of this kind known to me is the problem of the distances of quasars. The popular opinion is that quasar distances are determined by their red shifts, in the same way that the red shifts of galaxies are related to their distances by the expansion of the Universe. But quasars have always obstinately failed to follow the criterion by which Edwin Hubble established the relation for galaxies. To hold popular opinion is cheap, costing nothing in reputation, whereas to accept that there is evidence pointing oppositely, and hence pointing to something very different as an explanation of quasar red shifts, is to risk scientific tar and feathers. Yet not to take the risk is to make certain that, if something new is really there, you won't be the one to find it.

Our teaching load, which would probably seem heavy to modern academics, was typically six lectures per week for the sixteen weeks up to Easter, and three lectures per week thereafter, for a total of 112 per year, to which six college supervision hours per week should be added. It was usual to reckon that three supervision hours equaled one lecturing hour, on which basis the annual load was equivalent to about 150 lecturing hours.

In contrast to the organizational frenzy that developed from 1960 onwards, under overdrive from George K. Batchelor and William V.D. Hodge, neither of them a Cambridge product (Batchelor came from Melbourne University and Hodge from Edinburgh), we managed the entire Faculty of Mathematics in the halcyon days of the 1940s and 1950s on one shorthand typist and a faculty secretary, whose duties were considered sufficiently light as to demand that no more than twenty-four hours of the usual lecturing load be remitted. The secretary's duties consisted in keeping the minutes of four faculty board meetings each term and of maintaining a chart of the lecturing list, which had to be communicated at appropriate moments to the University Press for printing and to the custodian at the Arts School in Bene't Street so that lecture rooms could be properly allocated. Lyttleton did the job for three years, and he always said, in the huge fuss that was made after 1960, largely by George Batchelor, that he couldn't understand what it was all about. It was done in the name of improved quality; but anyone who thinks that what Cambridge produced after 1960 exceeded, on the average, what it produced over the preceding forty years is surely out of contact with reality.

My load in 1945–1946, as a tyro junior lecturer, was 72 hours of university lectures and 144 hours of college supervision. There were 24 hours of geometry to science students and there were 48 hours of advanced statistical mechanics. Ralph Fowler had died suddenly, and I was asked to take over his prestigious course. Geometry was the last subject in which I could claim to have expertise, and after hearing one of Walter Baade's stories in later years, I know why. Baade started in research as a student of the great German mathematician Felix Klein. Klein set him some multiple quadric line-and-point problem, asking him one day how he visualized it, to which Baade replied by indicating how he would draw the problem on paper. "Quite wrong, Baade," Klein said. "You should never visualize things in front of the eyes, always in the back of the head." It is easy to verify for yourself that Klein was correct. Sit back, close your eyes, and try to visualize some simple geometrical figure—say, a triangle inscribed in a circle. It is possible only if you can manage to throw the image into the bulge at the back of the head—at least, this is the way it feels. If you try to visualize it in front of the eyes, the field is too bright, perhaps because of the flow of blood in the eyes. This story cleared up a bit of a mystery from my own undergraduate days. There were students who scored excellently in geometry but less well in other parts of mathematics. When asked how they did it, all of them said by not wasting time drawing figures on paper. So-called "pure" geometry depends on using visual perceptions to avoid a lot of tedious algebra. Having no taste for tedious algebra, and being a blood-in-the-eyes man myself, it is clear to me that geometry wasn't the best subject for me to cover in my first lecture course. Besides which, there was another snag.

The rate of entry of new students in mathematics was about 120 each year, which we considered much too large to take all together in a lecture course. So each course was given in duplicate, with about 60 students to a class. Most of the lecture rooms on the ground floor of the Arts School had a capacity of about 60 students. This was for students in their first and second years, when everything was common to all of them. In the third year, however, there were alternative subjects, making the classes smaller, 30 to 40, and for these we used the rooms on the first floor of the Arts School, like Room H, where Eddington had lectured. But the rate of entry of new students in what was called natural science was greater than that of students in mathematics. Their numbers were about twice as large—more than 100 rather than 60. Special rooms were needed, of which Room A on the ground floor of the Arts School was one. We all hated Room A because it seated 200 or more, and lecturing to a half-empty room has an uninspiring effect, especially if you have to do it three times a week for weeks on end. Somewhat naturally, I got Room A as well as geometry. Furthermore, immediately in front of the blackboard in Room A, there was a floorboard that squeaked loudly. The floorboards were rather narrow, and it took quite a while before you knew exactly which one it was that squeaked and a while longer before you became skilled at avoiding it.

In the summer of 1946, a major international conference of European physicists, but without representation from the USSR, was organized and held in Cambridge. Patrick Blackett, later Lord Blackett, then at Manchester, was a major organizer. Because of Blackett's interest in cosmic rays, the subject of cosmic rays formed a minor fraction of the topics discussed, the major fraction being the problem of infinities in quantum electrodynamics, the problem we had thought ourselves to be on the verge of solving when I had attended the lectures of Rudolf Peierls and Max Born in the term following Easter of 1936.

The visitor who enters Trinity College at the main gate will see, on the opposite side of Great Square, the Master's Lodge on the right and the raised Hall of Trinity on the left. There are long vertical windows in the Hall. It was there, behind one of those windows, that I met Max Planck briefly in 1946. He sat in a chair with sunlight coming in on him from the high window, looking old and (or so I thought) without energy. Only later did I learn that his son had been arrested in 1944 after the attempt on Hitler's life, and that, only a few days before the end of the war, he had been executed on Hitler's special order.

It was at this conference that I heard Niels Bohr defeat Eddington in the lecturing unintelligibility stakes. It was also where I first saw Wolfgang Pauli get up to his lecture-room tricks. On the positive side, with the possible exception of my colleague Jayant Narlikar, Pauli was the finest blackboard artist I have seen. There was a beauty about the way he wrote highly complex mathematical formulae and about his arrangement of them. Paul Dirac was too practical to be called a blackboard artist. Everything he wrote was in a plain up-and-down style designed for clarity. Accompanying his blackboard writing, Pauli had a wide range of extravagant gestures. He would march relentlessly back and forth in front of the blackboard, in contrast to Dirac, who nearly always stood still. As he marched, Pauli would gesticulate with his right hand, usually by wagging a finger. If you try wagging your finger, you will find it is the muscles of the forearm that are being employed, the muscles used for writing. Additionally, however, Pauli's gesture incorporated a subtle roll from the wrist, which converted what is usually an admonitory gesture into a comical one. By this time, too, his shape was becoming a fair approximation of a sphere.

Three or four years later, Nick Kemmer (who had become a mathematics lecturer) and I persuaded the Faculty Board of Mathematics to invite Pauli to give its annual specially named Rouse Ball Lecture. To hold the expected crowd, the infamous Room A in the Arts School was chosen. When eventually Pauli discovered the squeaking board, instead of avoiding it, he deliberately marched backwards and forwards on it, squeaking the whole time. Room A was packed to the chandeliers for the occasion, and the atmosphere steadily became more tense, with Nick Kemmer and me not daring to look at each other. Had there been even one cackle, the whole crowd would have gone off like fire in dry timber. Adding to the problem, Pauli was speaking on the resolution of the two-decades-old problem in quantum electrodynamics, which is a genuinely subtle matter and which most of the audience didn't understand at all. The worst bit was when Pauli went to the blackboard, added a minor symbol to an enormously long formula, retreated back to the squeaky board, marched back and forth a couple of times, stopped, wagging his finger the whole while, and then said: "That is conventional."

The contrast between the substance of Pauli's later lecture (with the big problem of quantum electrodynamics largely solved) and the exploratory state of the subject in the summer of 1946 was great. The details of how the transition came about illustrate the nature of scientific discovery at its highest level and are therefore worth commenting on a little. An example of the kind of idea Eddington was given to throwing out in his lectures was that a person imbued with adequate perception could guess the laws of physics without needing to experiment or observe the Universe at all. Physicists were amused at this notion because, of course, the laws had never in fact been arrived at in such a way. Eddington knew this very well, of course. His point was not that they actually had been but that they might have been discovered by pure introspection. It was the kind of remark that caused you to think about what Eddington had said for a long time after.

It is hard to believe the classical laws of electromagnetism could ever have been guessed in the form in which they were first discovered by Maxwell. But, in the later form, in which they were described by a so-called 4-potential, the issue looks less clear, and, if the situation is expressed in the elegant form of an interaction between particles, you might very well have the feeling that the mathematical structure of the theory is so inevitable that it could indeed have been guessed. The same holds for the special theory of relativity. The form in which the theory was expressed by Lorentz and Einstein could hardly have been obtained without the experiments of Michelson and Morley. But the eventual form given by Hermann Minkowski again looks almost as if you might have guessed it. To some students in my day, the Schrödinger equation had the same look. There was one, I recall, who went under the sobriquet of Nipple, who wanted to know what all the fuss over quantum mechanics was about. To my eye, however, the situation in quantum electrodynamics, which I am now going to describe, could not have been guessed. Perhaps an equivalent but simpler and more elegant exposition of quantum electrodynamics may be given in the future, when yet again things may look different.

The major difference between classical theory and quantum theory, in Feynman's way of looking at it, is that, in classical mechanics, a particle goes between spatial points either not at all or along a unique path determined by the forces acting on it, whereas, in quantum mechanics, the particle goes with a probability determined by a summation over all paths, each path contributing to the sum with a weight factor depending on the forces. For a particle with electric charge, the electromagnetic forces can be expressed, in an elegant form, as an interaction integral between the actual path of the particle in question and the possible paths of all other particles having electric charges. The outcome is in excellent agreement with experiments for a wide range of processes, including the way in which one electric particle deflects the motion of another, the phenomenon of scattering, and, with a change of semantics, the emission of light by one atom and its absorption by another. But now comes the difficult question. Does the path of an electric particle interact with itself in the same way it does with the paths of other particles? If you say yes, you run into immense mathematical difficulties. If you say no, you run into an impossible logical contradiction. If the paths of one particle were always separate from those of other particles, contradiction could be avoided, but electrons in the same atom can have paths in common. Such a common path must interact with itself because it has to be considered twice, once as belonging to one particle and once as belonging to the other. So, for consistency, paths have to interact with themselves, when every quantum calculation leads to infinite results, if carried through remorselessly. Only if parts of a calculation are retained and parts are thrown away is agreement with experiment possible.

The parts to be thrown away could be characterized in my student days by expressing the electromagnetic interaction in terms of waves by a Fourier analysis. First, terms in the Fourier analysis with frequencies above some arbitrarily assigned value were thrown away, yielding a highest frequency to be considered in the calculation, a so-called cutoff frequency. Then all those terms in the calculation that depended explicitly on the cutoff frequency were thrown away. The lectures of Rudolf Peierls in the term following Easter of 1936 had explained this procedure, while the lectures of Max Born had attempted to justify it. Born even gave us an estimate of what the cutoff frequency should be, about 10^{23} cycles per second. Ever one to believe that old ideas are not the nonsense the younger generation would have us believe, I think that something of the sort will eventually turn out to be right, except that the cutoff frequency is really much larger than Born supposed it to be. But I will come back to this point later. Nobody, as I recall, was worried in the 1930s about the introduction of a cutoff frequency, so long as it could be shown to work in an invariant way. The problem was to explain why the cutoff frequency appropriate to a particular observer should be different for another observer in motion relative to the first, as it could be shown to be. This was the problem as it was addressed by all who were interested in quantum electrodynamics, and it was the problem discussed by Bohr, Pauli, and Dirac at the conference in the summer of 1946, little progress having been made in the decade since I first learned of it. Reasonably, it can be suspected that the situation would have continued in the same way, had it not been for an experiment made in 1947 by Willis Lamb and his student R. C. Retherford, who found a slight difference of energy between the state of zero angular momentum and those of unit angular momentum in the first excited level of hydrogen, a small discrepancy from expectation by about one part in 3 million. Almost immediately, Hans Bethe and Victor Weisskopf showed that there would be no hope of explaining this unexpected result unless cutoff-dependent terms were included in the theory. So paths had to interact with themselves and terms that, it had been hoped, could be thrown away had somehow to be retained. The next step is associated with Freeman Dyson, a young British physicist who moved, at this time, from Cambridge to join Bethe at Cornell University. The idea was to do the calculations, not for the known experimental mass of the electron, but for a different "false" mass, later to be called the theoretical mass. This would produce a change, and the suggestion was that, for an appropriate choice of the false mass, it might be possible, in the case of a single electron, to make the change so as to combine with the cutoff-dependent terms in order to yield a result apparently without a cutoff frequency, and with the experimental mass of the electron replacing the false mass. Thus, the cutoff frequency would be eliminated completely for an electron taken by itself through a change from the false mass to the experimental mass. But, for an electron associated with a proton in the hydrogen atom, the effect would be to produce the small new term discovered by Lamb and Retherford. This program was shown to be correct by J. W. Schwinger, and Richard Feynman almost simultaneously produced a powerful mathematical method of widespread application that permitted the new extra term to be calculated with a far higher measure of precision than that of the experiments that had been done. So, by 1949–1950, the issue was inverted from what it had been in 1947. Instead of theoretical physicists having to follow along behind the experimental physicists, mathematics was now in control of the situation. It now became a case of the experimentalists refining their methods to verify the mathematical prediction. For a while, there was frenzied activity, but no matter how cleverly the experiments were performed, the mathematics always turned out to be ahead; and, when agreement between experiment and prediction had eventually been pressed to an immensely high measure of accuracy, the experimentalists at last lost heart. The field of battle was conceded to mathematical insight, at what is generally agreed to be the highest level of sophistication yet achieved in theoretical physics.

Eddington was wrong, in a direct sense, and yet he could be said to have been right indirectly. Without the experiments of 1947, it would have remained difficult to find the correct electrodynamic theory. But, with the correct route discovered, the insight of the human brain was little short of miraculous. The example adds spice to James Jeans's concept of God as a mathematician. It can at least be said that Jeans's view has more merit than what was said against it from the pulpits, many a sermon in the churches being preached about the affair.

The rest of the story of quantum electrodynamics has both good news and bad. How is the problem of the cutoff frequency being different for observers in relative motion to be solved, the problem that baffled everybody in the 1930s? Well, the cutoff frequency has disappeared in replacing the false mass with the experimental mass, a result that is said to characterize a "renormalizable" theory. So make the cutoff frequency higher and higher until it goes to infinity, a procedure that can be common to every observer. This is the good news. The bad news is that the mathematics now becomes a whirlwind of infinite quantities, a state of affairs that is like grit in the mouth of the pure mathematician. And the false mass, the theoretical mass, becomes negatively infinite, a state of affairs that is grit in the mouth of the experimental physicist.

There is a way to avoid the bad news, and this is to go back, in some degree, to how we used to think in the 1930s, when it seemed acceptable to use a real cutoff frequency rather than the fictitious cutoff frequency of the 1950s. The consequent lack of invariance can be dealt with by making the cutoff frequency of cosmological origin. With the cutoff frequency being of cosmological origin, it takes the same explicit value for all observers moving with the expansion of the Universe, in what astronomers call the Hubble flow of the galaxies. The latter defines a particular motion at every point—that is, for every galaxy. Observers who deviate from this flow have different cutoff values, and there is no reason why they should not. In other words, the invariance problem of the 1930s is an artifact of working in flat space. In the real expanding Universe, the problem disappears. If one takes this view, it is possible to work out an explicit cutoff value for observers in the Hubble flow. It is of the order of 10^{40} times greater than we thought in the 1930s.

Returning now from the sublime to my own small world, it was not until the spring of 1946 that the work described in the previous chapter was written up, typed, and sent for publication to the Royal Astronomical Society under the title "The Synthesis of the Elements from Hydrogen." The paper was long—far too long to have the impact I had hoped for—but accompanying the long paper was a short one: "Note on the Origin of Cosmic Rays." I had wondered what might be the fate of the residue from a supernova explosion. I thought rotation would probably prevent the formation of a black hole, and I had plumped for a neutron star having a rotation rate close to what in modern times is called a "millisecond pulsar." Walter Baade at the Mount Wilson Observatory and Fritz Zwicky at the California Institute of Technology had suggested that supernovae might be the source of cosmic rays, but, in their view, with the cosmic rays appearing immediately in the main outburst. My suggestion was that cosmic rays emerged later, their energy source being the rotation of the neutron star. What I didn't have in 1946 was the modern magnetic-slingshot effect, but otherwise the suggestion was the same as the suggestion that was made following the discovery of pulsars in 1967 and that holds the field today. I also predicted that heavy elements would be found in cosmic rays, a prediction that was confirmed a year later. So it was with some anticipation that I attended the meeting concerned with cosmic rays at the summer conference in Cambridge in 1946.

I talked to Patrick Blackett about the little paper. He asked if I could explain the energy distribution of cosmic rays, and, when I said I had no good explanation, that my process was one of cosmic-ray injection at the bottom of the spectrum, he said: "Then we wouldn't look at it." At that, I lost heart and turned my attention to other things. Only Marcus Fierz, the Swiss physicist who had been concerned with Pauli on the problem of wave equations of higher spin, remembered by writing to me when, the following year, heavy elements were actually discovered in cosmic rays.

There are lots of things I am quite bad at. Drawing is one and proselytizing is another. I never solicit to give lectures, nor do I seek out science reporters of newspapers and science magazines. When he was Director of Astronomy at Cornell, Tommy Gold had a story of a seminar he asked me to give that illustrates the way I feel. According to Tommy, at the end of the seminar, someone said: "I didn't find what you said about such-and-such to be very convincing," to which I apparently replied: "I wasn't trying to convince you. I was simply saying what I thought." Certainly, I have a strong feeling about publishing what I do, and for this I am willing to fight tenaciously. After publishing, however, an idea must look after itself. It is out in the world, and it must grow up if it can.

I feel about publishing the way I do about birds. In later years, we had a house with an outside conservatory. From time to time, small birds would get inside, either through a half-open door or through chinks in the stone wall that formed the base of the glass structure. Once in, only blue tits could get out. Other birds would dash themselves to death, battering away with their wings on the windows. To save them, you had to catch them. We learned that it is absolutely no good being tentative about catching birds. If you are, they flap their wings with increasing frenzy, and the result is the opposite to what you want. You have to pounce like a predator, the way my wartime assistant John Gillams caught the squirrel. But, unlike the squirrel, which slit Gillam's hand open, a captured bird lies placidly in your cupped hands. The transformation from wild terror to calmness seems entirely magical. And as you walk from the conservatory into the open air, the bird watches you with a bright eye. After attaining some little eminence, you throw up your arms and release the bird, and away it soars. I feel it should be like that with ideas.

The concept of credit for a scientific discovery is an uneasy one. I was in need of "credit" in 1946 because my job depended on my acquiring some of it. But acquiring credit in the frenetic way it is done by many members of the scientific community today is another thing. I feel, somehow, that science should not be used for establishing a pecking order, just as one might reasonably feel the same about religion. A modest modern joke expresses my feeling: A physics student, who is fortunate to meet an electron at a party, asks the electron, "Do you really satisfy the Dirac equation?" To this the electron replies, "Who is Dirac?" The story of quantum electrodynamics shows that scientific discovery is not really what we would like to think it is. What happens is that the Universe programs our brains. The success really belongs to the Universe, not to us, and this has been the way of it throughout human civilization.

Nevertheless, "credit" does play a major role in science. It gains one access to exclusive learned societies. I must confess to experiencing some irritation when, year after year in the 1950s, I was not elected to the Royal Society of London. My wife knew I was under some tension. She usually opens our letters in the morning, and, one morning in February 1957, she opened a tiny envelope addressed to me with a tiny letter inside. The writer was Mary Cartwright of Girton College, a colleague in the Cambridge Faculty of Mathematics. My wife burst into tears when she read the letter, which was to say that I had at last been so elected. Unfortunately, the same tension must darken many a household where science has come down to roost. Oddly enough, by the end of that week, the exhilaration I had felt initially was gone, and since then, I have never really thought anything much of it, although before the event I had thought quite a lot about it. Lyttleton used to say that such things are a device of society to maximize the pain-to-pleasure ratio.

An objectionable feature of "credit" is that, in actuality, it isn't at all what it is supposed to be. The fiction is that credit is awarded in proportion to discovery, whereas, in fact, credit is awarded to those who influence the world, irrespective of whether they were the real discoverers or not. The fiction gains credence because the two have happened to coincide in a number of spectacular cases, especially those involving discoveries relating to the laws of physics. The reason is that the discoverers, in those cases, have only to convince a small number of other people. I don't suppose Paul Dirac had to convince more than ten others, and Einstein had to convince even fewer for his general theory of relativity in 1915. From the small group thus impressed (the only group competent to understand the issue in the first place), an awareness of the general state of affairs then spreads out like waves in a pond when a stone has been dropped into it. Senior professors impress their students, who proceed, in turn, to impress their own students. All such cases are characterized by being highly mathematical and by simplicity of structure. Quantum electrodynamics was concerned with a single electron, or, at most, only a handful of electrons, not with the 10^{50} electrons in the Earth, the 10^{57}

electrons in the Sun, or the 10^{78} electrons in the visible Universe. As complexity increases in science, mathematical standards fall away, and, as they do so, the group that must become convinced that a discovery has been made widens greatly. It is then those who succeed in manipulating the larger group who are most honored. Whether these happen to be the real discoverers is then no more than a matter of chance.

When popular opinion indulges in what later is acknowledged to be a mistake, popular opinion rarely makes amends. Alfred Wegener's theory of continental drift was laughed about and scorned during his lifetime. When it was discovered, in the 1960s, to be essentially correct, some reattribution in reputation might have been expected—at any rate, in all supposedly responsible quarters. But there was still not even a short biography of Wegener in my 1980 edition of the *Encyclopaedia Britannica*.

It is interesting to compare the discovery of pulsars and that of the Mössbauer effect, both having been made by graduate students. It was during an interim period, when his doctorate work was delayed, that Rudolf Mössbauer discovered that the line emission of gamma rays from radioactive nuclei held bound in a crystal was far narrower than physicists believed it would be. His professor said he had had nothing to do with it, and he insisted that Mössbauer publish the discovery alone. So Mössbauer came to influence the world and to receive whatever credit there was to be had. For the discovery of pulsars, however, it was otherwise.

A week before the discovery of pulsars was announced in *Nature* by a large "team," an agent from the Cavendish Laboratory called at my office with the information that something of great importance was to be announced at the next Laboratory seminar, but that, being sworn to secrecy, he couldn't tell me what it was. The seminar was on the Wednesday before the usual Friday publication of *Nature*. There was nothing in the proceedings to tell me that a graduate student, Jocelyn Bell (as she was then known), had been the discoverer, nor was there any such attribution in the way the discovery of pulsars was presented to the world thereafter. Several years later, it happened to fall to me to explain, in somewhat brusque terms, what the truth had really been. By that time, with the world stood on its head, the truth was not welcomed in respectable quarters. The misattribution had become accepted. It was correcting the misattribution that was not accepted.

At a still later date, on the understanding that one good turn deserves another, Jocelyn Bell recalled what happened at the Wednesday afternoon seminar. After it had been said that pulsars were white dwarfs, it seems that I was asked for my opinion and that I said I thought they were more likely to be neutron stars. Jocelyn Bell added that it was the best example of fast thinking she had encountered. Honesty compels me to reveal otherwise. It was just that my thoughts had gone back to the little paper on cosmic rays and neutron stars that I had written in 1946, and so, if anything, it was a case of slow thinking.

If, in those days (in the late 1940s), you wanted to change one of your lecture courses, you simply spoke to the faculty secretary, and you advised him of your wish. Then he would look for someone agreeable to a swap. I wanted to opt, in this way, out of my geometry course, but I had the sense to realize that nobody would contemplate such a swap voluntarily. So I waited and, sure enough, the fates came to my rescue (as they usually do, provided you tip your cap at dawn in their direction). A. J. Ward, who was then the faculty secretary, actually came to see me to ask if I would take the very course I most would have wanted. It was called "Electricity and Magnetism III," and it dealt in such matters as Maxwell's equations, dynamos, and the like. Maurice Pryce was up to yet another of his moves, going as Professor of Theoretical Physics to Oxford, and this was a course he was vacating. Because I was only a junior lecturer, they couldn't ask me to take on more work than I was doing already, so I had to be relieved of the geometry. Some other unfortunate had to do it.

Pretty soon, there was another issue. Nobody, absolutely nobody, wanted the course on thermodynamics, and so there was pressure on me to do that as well. "Get Hoyle to do it," Leo Pars of Jesus College would have said. "He *knows* about thermodynamics." Pars was an older man of middle height, with spectacles, and a rather large bald head, usually thrust a little forward. He had exceptionally white skin, and he was always ill with something, although he lived into his nineties. James Jeans had used Pars, in his youth, as a dogsbody, getting Pars to do the hard work in connection with Jeans's textbook *Electricity and Magnetism*, in return for which, as Lyttleton said, Jeans couldn't even spell Pars's name correctly. In the preface to *Electricity and Magnetism*, he appears as "Mr. Pass." He used to track me down in the lecturer's common room, saying, "You must write a book on thermodynamics. Then we can all *know* about it." I pointed out there were enough books on thermodynamics to cobble dogs with, but he would say about their authors, "Yes, but they don't really *know* it."

Anyway, as a junior lecturer, I couldn't take on another 24 hours of lecturing, so either they had to find someone else to do the thermodynamics or they had to promote me. Forced by this choice, they decided to promote me, and, in the second of my three years, I became a full lecturer with tenure to retiring age, which, in those days, was 70.

At some point, the university administration—the General Bawd (Board), as many called it—decided to reduce the retiring age to 67, which was to invite an inevitable *non placet*. Any change proposed in the

University Ordinances had to be "graced." The phenomenon of "gracing" occurred in the University Senate House on known days in each term. You knew which days and times they were from a special little red diary that anybody could buy in Heffer's bookstore, and without which university business would soon have come to a halt, with nobody remembering where he was supposed to be or what he was supposed to be doing. What happened at a "gracing" was that the head proctor stood up beside the vice-chancellor at the far end of the Senate House as you go into it. Both were in full academic dress, and, if you wished to attend, you too had better be. The head prog raised his square and put it back on again. Then he read the new university regulation, after which he raised his square again and put it on again. If silence reigned, the regulation was law, and a ripple of relief ran through the higher levels of the university. But it was within the province of any member of Regent House who happened to be present and of evil intent to take off his square in the body of the senate and shout the dreaded words *non placet*. A member of Regent House was any university teaching officer or any fellow of any college, whence it will be realized that the potential number of malefactors was large—in my time, 800 or so.

What happened next was a big fuss. The *non placeter*, as he was loosely called, together with all the supporters he could muster, wrote a diatribe against the university for its proposed new regulation. This was taken to the University Press, which printed it up in what was known as a leaflet. The Press had runners who then took the leaflets at speed around Cambridge, dropping them in bundles at the various colleges. The General Board and its supporters followed by doing exactly the same, and, on really contentious issues, such as lowering the retiring age, the process would be repeated three or more times, with runners in the streets the whole time until voting day arrived. Voting occurred on specified Saturdays at 2:00 P.M., the weekends being a device of the university to keep all but those who felt strongly away with their families, or in London, or watching a rugby match at the university ground in Grange Road. On these occasions, if you turned up, you saw that the ayes were assembled on the right as you entered the Senate House and the sinister noes were on the left.

Everybody from 50 upwards was certain to vote against the reduction of the retiring age, so it was generally acknowledged that the vote would be close. I went along, not because I felt deeply about the matter, but because I was studying voting patterns at the time, what later came to be called psephology. Usually I made my way to the sinister side of the house, but, on this occasion, I voted with the angels on the right side. And a powerful lot of angels they were. Newnham College and Girton College had each put out a three-line whip, and essentially every woman entitled to vote was there. In contrast, the turnout on the other side had ragged features, and so, amazingly, they lost. It was my first experience of womanpower in politics, a long overdue revenge for being kept by the university in subsidiary status for so long.

One important thing I learned in my psephological studies was that, except in the unlikely event that the votes on the two sides differed by one or less, it didn't matter how you voted. It was William V. D. Hodge who got me into it. Throughout the university, it was said, "Vote with Hodge and you win." As far as I could tell, this was true, and I was puzzled to know how he did it. One day, on the Faculty Board of Mathematics, we were considering an item that, judging from the discussion, was going to be decided by a close vote. The chairman called for the ayes on the motion and began to count the number of raised arms. Towards the end of this process, Hodge's arm went up. At the next Faculty Board meeting, I watched as he took his seat, which essentially completed my investigation. Hodge sat where he could see the arms going up. If there was a shower of arms for aye, he put up his arm immediately. If there was only a sprinkling of ayes, he waited for the noes. But, if the issue was close, he had to count. He was a small, round, rubicund man, and I suppose it must have meant a lot to him to be on the winning side. Of course, the system wouldn't have worked on a card vote, which was the method we used in the Senate House, but then it would not be known how he voted, and his valued reputation would not be threatened (that is, not unless the registrar investigated the card votes, but the way in which registrars were chosen made that most unlikely).

I was given a fourth year on my college fellowship, but this really did seem to be the end of that line. St. John's was generous in its support of mathematics. It had four fellows permanently—college lecturers, they were called—whereas I was only a supervisor, a post for which turnover was generally rapid. We had two college lecturers on the "pure" side and two on the "applied" side. The latter two were Ray Lyttleton and Leslie Howarth, an expert in fluid mechanics, both of them senior to me. I could expect to retain the title of Supervisor, but the fellowship would go. Supervising undergraduates was the nearest you could come to a starvation situation. Hence I would lose, with the cessation of the fellowship, about as much as I had gained by the promotion to a full lectureship. As Lewis Carroll said (through the persona of the Queen in *Through the Looking Glass*), "it takes all the running you can do, to keep in the same place." Or, as it was expressed in a Hungarian cartoon showing a peasant flogging a horse dragging a heavy carriage, "You must work harder, now that the carriage belongs to you."

After two years at Orchard House, the strain of finding £62-10-0 on every quarter day had gradually run us bit by bit into debt. My wife continued to pay all bills promptly, the debt being entirely on Barclays Bank. In that way, we continued to enjoy good service from tradesmen, and we could not avoid knowing exactly where we stood financially. Eventually, we heard of a house in the village of Great Abington, eight miles south of Cambridge. The rent on Ivy Lodge was £150 per annum, which was more in line with my salary. Within bus range of Cambridge, I could now contemplate disposing of the £5 Singer, which, unfortunately, was reaching the end of its days, or, at any rate, the end of the days when I could drive it with an easy mind. The radiator was leaking badly now. Because radiators were essentially impossible to repair, I had taken to traveling always with a tub of water. To make matters worse, the exhaust system had fallen apart, so that the Singer sounded, in the confinement of village streets, like a powerful motor boat.

Nick Mayall, from the Lick Observatory, came to stay with us in Great Abington. Exactly how he came to be in Cambridge I have forgotten. What I do remember clearly, however, was that, on the last morning of his visit, I had to get him to the Cambridge railway station by 8:00 A.M., and the Singer's radiator was now leaking disastrously. Thumbing through a respectable motoring magazine, I found a recommendation for dealing with such situations. The author said that porridge was the thing. So, on the following morning, I poured a generous dollop of porridge into the radiator, fixed by Mayall's frankly incredulous gaze. But we made it over the Gog Magog Hills and into Station Road, boiling porridge and all. That, however, was the last of the Singer's exploits. Shortly thereafter, I gave it away to an enthusiast.

We would have no car, we decided, until our debt was paid off. The problem with using the bus was that, to have any chance of keeping my fellowship, I had to put in an appearance at dinner two or three times a week, which meant that I usually missed the last bus from Cambridge to Great Abington (the last bus left at about 8:30 P.M. in those days). There was still a bus from St. John's College to the southern boundary of Cambridge, but the next six miles of open country to Great Abington had to be walked. A bicycle would have been a possibility, but, given my wartime vow to eschew bicycles, I never contemplated it. The walk I remember most clearly was not at night, however. At the end of the autumn term in 1948, I got away immediately after lunch. Not wishing to wait through the afternoon for the school bus at 4:00 P.M., I decided to walk home along the ancient Roman road, which starts high on the Gogs and continues on and on past Great Abington and Linton almost to Haverhill. To reach Great Abington, one left the road along the edge of a large ploughed field above the village of Hildersham. The sky was clear, the air was exceedingly still, and the Sun was nearing the horizon as I reached the ploughed field. Smoke from the cottages in Hildersham was going absolutely vertically upwards, and there was a thin veil of smoke across the Sun, cutting down the glare from the solar disk. Then, so clear that I felt as if I could reach out and touch it, a large sunspot appeared.

Every unusual astronomical event I have witnessed has been like this, easy to see when the occasion was right. It was so with the planet Mercury, with comets, but not, I must add, with the so-called green flash that is supposed to occur on the western horizon immediately after sunset. I have looked for the green flash many times, but I have never seen it. The people with whom I have discussed the green flash fall into distinct categories. Some say they have seen it scores of times, while others, like me, say they have tried many times but have always failed. The explanation, I think, is that seeing the green flash may depend on possessing a physiological defect of the eye. But, of my sunspot over the village of Hildersham, there was no doubt. As far as I am aware, there are no medieval accounts of sunspot sightings, which strikes me as curious. One would think every countryman must have seen them once or twice in a lifetime—unless, in those days, there were no sunspots (which might be true, since solar activity is known to be quite variable over time).

Ivy Lodge in Great Abington was, like Orchard House in Quendon, a potentially beautiful and comfortable house. The poor state of the decoration when we moved in was quickly dealt with, but the other snags were not. The electric wiring had no power lines and the lighting circuits needed complete renewal. Add, too, that the kitchen and rear quarters generally needed gutting and refitting throughout. We could have done all this slowly if we had owned the house. Although we were to be there for a decade, we never approached the owner with a view to purchase—in the beginning, because we couldn't afford it, and in later years, because the Cambridge to Colchester traffic, which in those days ran immediately past the front door of the house, was growing increasingly heavy. Had we known that the village would eventually be bypassed, it is quite likely that we would have sought to buy the house. Instead, because we heard that, at some moment in the future, St. John's College might be making building land available in Cambridge, we decided to wait until it was, which proved to be a long time.

My wife always managed, somehow, to provide for holidays in the summer. In 1945, it had been Achill Island. In 1946, it was a camping holiday in which the Singer was given its head. We worked our way into the Lake District, where the cable brakes of the car gave out on the long steep hill that goes down from Little Langdale to Elterwater. There were several parties coming up the hill, walking with bicycles and talking. The Singer being open, my wife simply stood up and shouted down at them, waving her arms, and so disaster was avoided as we slithered uneasily into Elterwater. Then there were the pigs that got into our tent at a camping site in a field at Wall End Farm in upper Great Langdale. The pigs nipped the cork off a large bottle of methylated spirits that we had brought along to fuel our primus stove. Three-quarters of the methylated spirits was gone, so we decided to quit that site immediately, before the farmer took a good look at his pigs.

The same summer, the biochemist John Kendrew was driving to visit his mother in Florence. My wife insisted I should go along on what would prove to be a trip of five weeks. We did some walking, first in the Kandersteg–Lötschental region. Then we crossed Switzerland by car, with John going down into Italy, leaving me to walk the alpine passes, and finally up to mountain huts above Zermatt. John didn't like my driving, except when he went down with an attack of malaria, and then he had to put up with it. It was the first time I had seen malaria, and I decided to do what I could to avoid it.

John Kendrew sometimes brought his friends to Ivy Lodge. I remember Jim Watson, of double-helix fame, at the big kitchen table drawing something on the table as he engaged in an avid discussion with other guests. Watson was as thin as a rail, so it seemed natural to me that he was ravenous at tea. Afterwards, my wife exclaimed, in a steeply rising voice, "That man has eaten our week's butter ration." I mentioned the matter casually to Kendrew, and the remark went boisterously around microbiological circles. It explains why, in a film reconstructing the discovery of the structure of DNA, Watson was shown as perpetually eating.

In 1947, there was a meeting of the Royal Astronomical Society in Dublin. I received a grant from the RAS, probably because I was to read a paper at the meeting. At all events, we were able to stay at the Gresham Hotel in O'Connell Street. Arriving a week ahead of the meeting, we hired a car, drove west to Kerry, and made a first tour of country I was later to see in detail with Paddy Browne. It was at this meeting that I first met Paddy, thus setting the ground for my second meeting with Erwin Schrödinger, when he was chasing after Miss Y. Unfortunately, it was also at this meeting that E. A. Milne died of a longstanding illness.

In 1951, there was a meeting in Paris between two international organizations, the International Astronomical Union and a similar organization for those concerned with mechanics. This was at a time when turbulence was supposed to solve every problem in astronomy. Heisenberg was there as a turbulence man. Jan Oort, from Holland, was there as an astronomical man. Lyttleton and I ate at the same café every night, and Ray Lyttleton ordered the same meal every night, a phenomenon that eventually attracted the attention of the café proprietress. She was a powerful woman, well suited, you might think, to bringing down the Bastille. She tried all she knew to get Lyttleton to change his diet, but it was a case of the irresistible force meeting the immovable object. I have forgotten the name of the place, but it would be identifiable because there was a big bronze statue of a lion at its center. The sculptor was probably Rodin.

The bigger quadrennial meeting of the International Astronomical Union followed in Zürich, the meetings being held at the Technische Hochschule there. I remember the meeting on several counts. I caught an abominable cold, which is a strange thing to do in August. I lurched from one commission to another, speaking volubly in a thick voice on every possible occasion. It was at the Commission for Extragalactic Nebulae that I first met Edwin Hubble. There was an invitation to tea and cakes, with traditional musical instruments, at a restaurant high on a hill above the Lake of Zürich, the sort of place where they always fly the Swiss flag—not so much for patriotic reasons, I decided, but to attract custom from afar. I remember that I walked back down to Zürich with Lyman Spitzer from Princeton, our memories going back to student days in Cambridge in the late 1930s. I also remember the first of several insults from the British astronomical establishment. I suspect it was from an inner feeling of inadequacy that the insults came, the establishment having been seriously diminished in quality by the untimely deaths of Jeans, Eddington, and Milne.

It was the practice of the American contingent to give a dinner for the rest of the union. But, because numbers had become too large for everybody to be invited, quotas were issued to the official International Astronomical Union representatives of the various nations, who were then left to decide who should go and who should not—the sheep from the goats as it were. By now I had done quite enough to have been included among the goats, but I was not. The same petty malice was repeated four years later in Rome—which, in that case, was absurd, because Lyttleton and I were the only British astronomers invited to speak at the plenary session.

My fellowship at St. John's was within an ace of running out when the fates would have it that Leslie Howarth should be offered the Professorship of Applied Mathematics at Bristol University. He accepted, and I inherited his position as College Lecturer, now with a fellowship to my retiring age in 1982. Suddenly, I was home and dry, and in a position again to acquire a car.

Lawrence Balls, a trim man in his sixties, had advised the Egyptian government on the growing of cotton. After retirement, he returned to Cambridge and had been elected, on the recommendation of his lifelong friend George Briggs, the Professor of Botany, to a nonstipendary fellowship at St. John's. Well, Lawrence Balls advised me to "go for" a device of a car built before the war by DKW in Germany. It was a two-cylinder, two-stroke, front-wheel-drive machine, "with absolutely nothing to go wrong," Balls pointed out with cogency, for it was clear that few, if any, garages in the United Kingdom would be able to cope with such a job, except by trial and error. Anyway, I acquired a DKW from a garage near Paddington Station in London for £100. Except that the device that protects the dynamo or alternator when a car is started worked only sporadically, the so-called "cutout," the car ran well. The bodywork was of papier-mâché, which meant there was no rust to worry about. During a visit to Switzerland in the middle of an incredible heat wave, I found new parts could still be bought there. I snapped up a new cutout and one or two other suspect components, and thereafter the DKW ran 60,000 miles without trouble at 40 miles per gallon, just as Lawrence Balls had said it would. When convalescing from an operation, the historian Michael Oakshott came to stay with us for a week and was much taken with it, and, in the fullness of time, I sold the DKW to him. I never heard what happened thereafter, which might indicate that it came to a bad end in London traffic, for Michael moved just then from Cambridge to the London School of Economics.

In mid-January of 1950, Peter Laslett, a research fellow of St. John's, approached me with a request. St. John's did not require its research fellows to be in residence. They could have a job elsewhere, and Peter had a job with the British Broadcasting Corporation's Third Programme. Unfortunately for him, he was let down by a fellow historian, and the story was this. In what was for him a fairly typical brainstorm, he conceived of running a series of university-style lectures on the Third Programme. Peter had managed to secure prime listening time at 8:00 P.M. on Saturdays for five such lectures to be delivered, I understood, by Herbert Butterfield, later the Master of Peterhouse College. But, at a late stage, with the program slots already penciled in, the proposed lecturer had decided that time was too short and had asked to be excused. One evening, Peter told me of his dilemma, and he asked if I would care to stand in with five lecture-type talks on astronomy. Afterwards, Peter told me he had looked up my file at the BBC and had found an entry from a science producer that read: "*Do not use this man.*" Old-college-tie considerations caused him to ignore the file, however, with eventually unexpected results, for which Peter must take much of the credit. What I was to receive was £50

per lecture, plus a likely £10 for publishing the lecture in the magazine
The Listener, giving a total of £300, a sizable fraction of my annual salary
as a lecturer in the university.

By now, I had managed to write an average of about two scientific papers per vacation, some of which dealt with topics other than astronomy,
such as the formation of the E layer in the Earth's atmosphere. A paper
on that subject that I wrote with David Bates of University College, London, is still, I believe, the favored explanation for the origin of the E layer.
So, when I acceded to Peter's request, my head was crammed with possible material. All I had to do was sieve it. I started with a lecture in hand;
I had the second ready by the time I had given the first. But, owing to my
university lectures and college supervisions, the margin of advantage
shortened bit by bit, until things came right down to the wire on the fifth
lecture. After giving the fourth lecture, I drove home from London in the
DKW to Ivy Lodge, getting there about midnight on the Saturday. All
Sunday I worked on the fifth lecture, and I worked on it also in spare moments on the Monday. I gave Peter my handwritten manuscript on Tuesday, and, as usual, he tore it to shreds—or, more accurately, he made a
mountain of suggestions, which I attended to on the Wednesday. Peter's
wife, Jan, typed the manuscript on Thursday, and, on Friday, Peter took
the eventual product to be censored at Broadcasting House, the headquarters of the BBC. Every word that was said over radio had to be censored in those days. Then, on Saturday, I drove to London, and, in the
final hours, Peter and I endeavored to accommodate the censor's remarks—or, in some cases, we had arguments with the censor, as I recall—
right up to the beginning of the lecture at 8:00 P.M.

I had expected £300 for the project, but, in the event, it turned out to
be much more. There were repeats, to start with. Then the BBC foreign
services wanted the same material chopped into smaller blocks. Then
Basil Blackwell wanted to print the lectures in a little book called *The
Nature of the Universe*. Then Harper Brothers in New York wanted to
produce an American edition. So, in the end, it was more like £3000. Suddenly, we had the bit of money that had eluded us for ten years. Both my
wife and I came from families where money wasn't too plentiful, and,
through the first ten years of our marriage, we had been obliged to view
our expenditures with an ever-watchful eye. Neither of us now had any
rush of blood to the head, no impulse to plunge into a different lifestyle,
except that we plunged to the extent of buying our first refrigerator. At
last, we had got ahead, and we determined to stay that way. If we had not
done so, it would, I think, have been impossible to break out of the academic system when, in 1972, it became necessary to do so. If I have been

wearisome in mentioning my monetary affairs, it is their relevance to the events of 1972 that I really have in mind.

Those broadcast talks hit the top of the annual national ratings, much to Peter's delight, thereby vindicating the old-college-tie approach. The lectures earned me a lot of opprobrium, however, some of it justified. Partly because of the need for clarity, and partly because of my haste in the construction of the talks, there was a temptation to convert the possible into the certain. The lectures were too apodictic, as one famous scientist put it, but they won approval in other directions, which I valued. They led to a lifelong friendship with the novelist Jack Priestley and to a correspondence with the biologist J. B. S. Haldane that continued up to his death. And, in the press of that last lecture, I coined the term "big bang," which became a permanent addition to the language of cosmology.

Chapter 18

An Unknown Level in Carbon-12

THE PREDICTION in early 1953 of the existence of a hitherto unknown level at an excitation of about 7.65 million electron volts (7.65 MeV) in the nucleus of the common isotope of carbon, carbon-12 (commonly written ^{12}C), has achieved some notice in recent years on account of its being an early application of what is known nowadays as the anthropic principle. Indeed, it is sometimes referred to as the only, predictive application yet made of the anthropic principle. In this chapter, I will explain how, as an outcome of another lucky chain of accidents, the prediction came about.

The anthropic principle is a curious notion that appears to be clear to some people but not to others, as often is the case with concepts to which the word principle is applied. No one can quarrel with the statement that the physical properties of the universe must be consistent with our own existence, a statement so obvious as hardly to justify the term principle, one might think. Nevertheless, this trite observation can be made to do some work, as the ^{12}C story showed. Application of it to cosmology has some predictive possibilities. Thus, in evolving cosmologies, human life could not have existed in earlier times, when the temperature everywhere was too high, nor will it exist in later times, when all dwarf stars, such as the Sun, will have burned out. These two requirements set the epoch of our existence as between about 10 million years and 30 billion years after the Universe started in a big bang. If one believes in the big bang, this is not a negligible deduction. In other forms of cosmology not of the big-bang type, galaxies can be of variable ages—some old, some young. The anthropic principle requires us to live in a galaxy that is old enough for the evolution of life to have led to our existence. Should this require us to live in an exceptionally old galaxy, then that is where we must be, regardless of the comparative rarity of exceptionally old galaxies. Probabilities do not arise in such a case, as they would for a randomly chosen observer.

More subtle issues appear when the existence of life is seen to depend crucially on a fine tuning of the laws of physics themselves, in the sense that it is not hard to imagine minor changes in the laws that would have made life impossible. Thus, we are faced with several alternatives. Either our existence is a freakish accident, or the laws of physics are not invariant. Either the Universe is more complex than we think, in that variations of the laws are realized (in which case we happen to pick out the particular choices that are suited to our existence), or the Universe is teleological, with the laws deliberately arranged by some agent to permit our existence. The latter view is, of course, common to most religions, but it were better for a scientist to have a millstone hung around his neck than that he should admit to such a belief—yea, verily. If he does so, his papers will be rejected, he will receive no financial assistance in his work, the publishers of his books will receive threatening letters, and his children will be waylaid on their way home from school. As well might he seek to pass through the eye of a needle, for to hold such a view is the greatest possible scientific heresy. On the other hand, it is possible to hold the inverse view (and even to win plaudits by doing so)—namely, that it is our existence that requires the laws to be the way they are. Stated this way round, you have what is called the strong anthropic principle. Our existence dictates how the Universe shall be, a fine ego-boosting point of view on which you may travel, fare paid, to conferences all over the world.

A balanced person will be surprised to hear that a respectable argument can be given in support of this seemingly outrageous point of view. It concerns the phenomenon of the condensation of the wave function in quantum mechanics. It is possible to set up a physical either/or kind of experiment without difficulty. A radioactive atom either decays in a specified time or it does not. If it does, the decay products are used to trigger a camera, and a picture is taken of the House of Commons in session. If it does not, the absence of decay products triggers a different camera, and a picture is taken of St. Paul's Cathedral on Easter Sunday. So, inevitably, some picture is taken, and the question is, How do we discover if it is the House of Commons or St. Paul's? Not by calculation—absolutely not. Any attempt to fiddle the calculations to yield a definite prediction results in abysmal contradictions. Calculation can only assign relative probabilities to the two possible results of the experiment. The widely accepted answer to the problem, given in my youth by what was called the Copenhagen school of quantum mechanics, associated particularly with Niels Bohr and Werner Heisenberg, was that decision is made in the matter by the experimental equipment, the circumstances of the triggers, the electronic storage devices, and the cameras. Schrödinger differed, however, arguing (in what came to be called the "Schrödinger cat experiment") that the ultimate decision rests with the human who eventually takes a look at what the cameras have done. This was very much a minority view, as I discovered in 1938, when, in ignorance of the existence of Schrödinger's now famous cat, I arrived at a similar conclusion. The thing hit me as I sat near the village of Grantchester on the banks of the river Cam, after swimming in its polluted water, which gave me a sore throat for life. The respectable position in physics today is different. The Copenhagen school is in eclipse, with most of the younger generation tending to agree with Schrödinger. So it is human consciousness that makes an otherwise indeterminate world determinate, the process known as the condensation of the wave function. After human consciousness resolves a quantum uncertainty, the future behavior of the world is calculated for new, more specific conditions.

All this rather trumps the trick of the religious person who sees the Universe as a sort of factory set up by God. It makes more sense to suppose that a bit of God is operating in all of us, and not mainly over high-flown moral issues but even over such issues as which pictures are produced by our two cameras, when it is God who decides whether the picture is to be the House of Commons or St. Paul's. It is almost as if, without such interventions, God doesn't know what is going on in the Universe, as if this is the way that a record of happenings is kept. I pointed out earlier that one has no affection for a pulled tooth, despite the tooth having served one well by chewing up many thousands of meals. It is as if the bit of God that is in us doesn't give a hoot about the hardware by which relevant observations are made. The hardware matters only so long as it continues to work. When it doesn't, it becomes no more important than any other collection of atoms of hydrogen, carbon, oxygen, nitrogen, and so on. This is the kind of speculation, however, that is only too likely to get one's papers put into the shredder and to cause one's publishers to be hounded to extinction, as well as to ensure that one's grant applications to government agencies are thrown instantly into the deep freeze. But, in a common-sense view, it is the way things seem to be.

Yet again, I owed the prediction of the 7.65 MeV level in ^{12}C to the strange structure and history of the University of Cambridge. To understand nineteenth-century Cambridge, one should compare the research student of today with the undergraduate of 1850. The equivalent then of the modern undergraduate course was the excellent instruction given in upper forms of independent schools and grammar schools, and the equivalent of the modern B.A. degree was university matriculation, which was dependent, in Cambridge, on an examination referred to generally as the Little Go. Just as, in my day, some research students engaged in curious pursuits, like the one who spent three hours a day brushing his hair, or the one who enjoyed consuming immense bundles of bananas, so, in the nineteenth century, there were many wealthy undergraduates who did not take university life any too seriously. We did not think the chap who spent three hours a day brushing his hair to be narcissistic, by the way. It was just that his mental processes were very slow, like those of the elephant.

And, just as modern research students have a supervisor, so undergraduates then had a coach, who was paid by the student according to his skill and reputation, with the best students wanting the best coaches and the best coaches wanting the best students. Many stories are told from the period. James Clerk Maxwell, as his later brilliant career might indicate, was an unruly student given to following his own bent. On the first day of his final examination, Maxwell's coach observed to a colleague that he had never sent a student "in" so ill-prepared: "Yet," added the coach, "I have some confidence in him, for he is incapable of thinking wrongly physically."

Over the half century from 1860, the coaching system developed into the college lecturing system, and college lecturing then developed into university lectures. The best coaches probably had little need of college patronage. They became well off financially from their own efforts, likely enough nurturing some envy in less fortunate colleagues, who compensated by obtaining a supply of students through individual colleges, of which they became fellows. I suspect there was a clash of interest, because a measure of contradiction developed that persisted into my own day. On the one hand, there was the one-to-one relationship of coach to individual student, in which the student took along his difficulties to be ironed out by the coach, and, on the other hand, there was the situation of a lecturer talking at students without much reference to their individual difficulties. It was the difference between a hand-made article and a mass-produced one.

There was still a one-to-one mystique during my undergraduate days in the 1930s, but, in practice, it turned out to be one-to-eight. When I became a college lecturer myself in the late 1940s, the normal situation was one-to-six for science students and one-to-three for mathematics students. Either way, I never felt the arrangement to be satisfactory. Students in a group are never sufficiently uniform to be treated as if they were an individual. If questions are asked, the tendency is for the best of a group, or the most extroverted, to do all the asking, with the rest probably getting less out of the proceedings than with the supervisor in the driving seat. Tutorials of the kind practiced in some universities are different and probably better. In a tutorial, the students are set questions to work by the supervisor, who circulates from one to another, helping when a student gets into difficulties. The problem with tutorials at Cambridge was that our examination questions were just too hard for the method to work. Nevertheless, an attempt at it would have been salutary, for a pretense developed in which everyone was at fault, a pretense in which students appeared to be much better than they were—actual examination results were mostly a shock to supervisors. If we had seen the situation clearly, we would have realized that it was essential to begin with questions that were much less difficult.

The most satisfactory memories I have of undergraduate supervision are of exceptional circumstances that led to the old style of coaching. One week, five out of six science students in a group were absent with flu. The remaining chap was a scholar of the college, one of our best students. He asked about a relatively simple problem in elasticity, which I managed to solve. In the ordinary way of things, this would have ended it. The group of six-plus-supervisor would have moved on to something else. But now I was able to see that the lad had not understood. So I worked the problem again from a different point of view, and again I found the student was not understanding. At that stage, I revealed my limitations as a teacher, for I began to get exasperated. How could a chap of such ability not see something so simple? Our somewhat fast and furious conversation stopped suddenly. The student's face went red, and he began to laugh. It was a fine example of the penny dropping suddenly. Evidently, the lad had been suffering from the sort of logjam in the brain that sometimes happens with all of us. Once the jam was gone, the situation was instantly clear to him. Only in the old system could this essential kind of correction happen. In the way we had it from my own student days onwards, there was normally no way to get things right once they had gone wrong, a situation that every student comes to fear.

One post-Easter term, I was fortunate to have an ideal student on a one-to-one basis. He never asked me to look at a problem without having made a serious shot at it himself, and he never came without bringing his attempt clearly set out on paper. Usually, it was only a detail he wanted to discuss. In the few cases where he had become stuck, as soon as I suggested a move he had overlooked, he had no wish for me to continue; he wanted to go away and finish the job by himself. It was no surprise when he became the top student of the university in the final examinations.

Coaching in the nineteenth century did not extend to research. For research, you were on your own, with neither college nor university to guide you. Harold Jeffreys, the St. John's mathematician, told the story of a student who did very well in the mathematical tripos (judging from the details, I think the student must have been Jeffreys himself). After the examination results were out, the student went to see one of the college lecturers in St. John's, the expert in elasticity, A. E. H. Love. After congratulating the student, Love then kept his mouth shut, and the student also kept his mouth shut. When an embarrassingly long time had elapsed, Love gave way by congratulating the student again, when, once again, Love kept his mouth shut. Eventually, the student himself broke the second embarrassing interval by asking Love if he had any suggestion as to what might be a suitable research topic. Love thought for a moment and then said: "Young man, do you think that, if I were fortunate enough to have an idea, I would give it to you?"

The key word here is "give." In the old system, a student had no right to expect to be helped in research, but, once he heard of an idea, it was his to fasten on to. It could not be taken away or even shared by someone else. This explains why there were very few shared pieces of research of a mathematical nature in the nineteenth century. The system came through into my time, but with a crucial addendum that greatly loaded the scales in the student's favor. The degree of Ph.D. was introduced in the 1920s for the benefit of research students who came from outside the university, particularly those from the United States. It was an innovation that Maurice Pryce deplored as a debasing of the coinage. Maurice showed his contempt for the thing by satisfying all the requirements for the Ph.D. but then not bothering to join a congregation in the Senate House to have the degree officially awarded. It will be recalled how he persuaded me to do likewise—not for high-flown moral reasons, I fear, but because, in that way, I avoided having to pay income tax on a generous studentship that I had unexpectedly been awarded. Anyway, after introducing the Ph.D., the university required all registered research students to have an official supervisor, and, from then on, A. E. H. Love's option of keeping his mouth shut was taken away. The research student could now expect to receive ideas from a supervisor, and, once received, they became his or hers. I could not object myself when it came to my turn to provide ideas, because I had profited myself from the system when, in 1936, I had become a student of Rudolf Peierls. It was due to the good ideas that Peierls had given me that I had been able to get ahead so fast during the following two years. There were other occasions in which the situation was far from being fair, however. At the end of 1930, E. C. Stoner, later Professor of Physics at Leeds University, published a paper in *Philosophical Magazine* in which he showed that he had obtained the critical stellar mass known nowadays as the Chandrasekhar limiting mass. Stoner used a mean-value theorem for the astronomical part of his paper rather than a full stellar integration, which led to his numerical answer being slightly different from the eventually accepted value. But it was an exceedingly important result, and it can hardly be doubted that it was Stoner's supervisor, Ralph Fowler, who lay behind it, the Fowler whose lectures on statistical mechanics I had taken over in 1945. If this is correct, it seems unfair that Fowler should be accorded no part in an important and widely acclaimed result.

The satisfaction for a supervisor lay in the student making a good job of what he was given. I never had any complaint on this score—except once. The exceptional case happened in the period 1949–1951. I gave what eventually turned out to be a good idea to a student who decided to ditch the problem about two-thirds of the way through. But, since the student did not cancel his Ph.D. registration, I had no option but to wait until the registration lapsed, or until the student completed the Ph.D. requirements in some other topic. The problem in question had already been foreshadowed in Hans Bethe's 1939 paper on the conversion of hydrogen to helium. It was to make the step from helium to carbon at higher temperatures than in main-sequence stars, but not so hot as in the statistical calculations I had done in 1945–1946. The first maneuver was to calculate the statistical abundance of beryllium-8 in the reversible reactions $^4He + {}^4He \leftrightarrows {}^8Be$. Then one had

$$^8Be + {}^4He \rightarrow {}^{12}C + \text{gamma ray,}$$

but also

$$^{12}C + {}^4He \rightarrow {}^{16}O + \text{gamma ray,}$$

and so on,

$$^{16}O + {}^4He \rightarrow {}^{20}Ne + \text{gamma ray.}$$

It was this chain that I set the student to calculate and with which he became fed up. Then, in 1952, Ed Salpeter of Cornell University published a paper on the synthesis of carbon from helium, and I became thoroughly frustrated by the system, my frustration serving as a critical link of the chain that led to the discovery of the 7.65 MeV level in carbon-12.

The second link came with the meeting of the International Union of Astronomy in Rome in the summer of 1952. Increasing prosperity in Italy could be measured by the nature of the traffic in Rome. In 1952, it was by motor scooter, when it was not by train or bus. By 1958, it was by the little Fiat 500 c.c., and so on up and up the scale. In 1952, then, the piazzas were a bedlam of scooters. Those were the days in which Italian women were not supposed to appear in public with bare arms. Bernard Miles told me a story belonging also to those times, but not in Rome, in North Africa. He was doing a film with a Scandinavian blonde of classic features and statuesque proportions. They walked one day in a town, she with bare legs and arms. Arab males began to mutter, and a crowd around them began to advance in what looked like a tight situation. The Scandinavian blonde waited until the nearest male was within two yards. Then she closed the gap, swung an arm, and lifted the fellow clean off his feet. Bernard said he expected knives to be out at a trice. But this didn't happen. The crowd just suddenly became quite silent, the men feeling the way male spiders must feel, I suppose.

Anyway, Walter Baade was there in Rome at the 1952 meeting of the International Astronomical Union. Walter was then the Chairman of the Commission on Extragalactic Nebulae (galaxies). It was typical of his disregard for bureaucratic detail that he had overlooked the need for a secretary to take the Commission's minutes. So what does Walter do but ask me, on the spot, to become secretary, without bothering to verify that I was a member of the Commission, which I wasn't.

My minutes were eventually to have importance in establishing Walter's priority for breaking a fifteen-year-old logjam concerning the distance scale of the galaxies, which affects the so-called age of the Universe. Edwin Hubble's estimate for the age had been 2 billion years, and Baade's was 3.6 billion. Eventually, the age was to rise higher and higher, attaining a maximum of about 15 billion. But nothing in the later increases caused quite the sensation of Baade's first jump from 2 to 3.6. My own increase to above 10 billion passed by quickly enough in 1958–1959, although the increase to the range 12–15 billion by Willy Fowler and me in 1960 attracted a little more notice.

Baade's announcement was made orally. When one makes a clear-cut statement to a considerable body of people, it seems impossible that anybody could be in doubt about the precise circumstances of the announcement. Yet experience shows that very few members of an audience ever do remember the exact details of such situations. Within a year, it becomes impossible to recover what was said, unless explicit details have been written down. This failure of the human memory opens the floodgates of plagiarism, permitting a subculture to flourish in which certain individuals make the round of international meetings picking up bits and pieces that they then proceed to represent as their own. It is a maxim of biology that, wherever a niche of survival can exist, it does. It was a maxim that the unworldly Baade had not learned. Surprising as it may seem, attempts were made to rob him of his priority over the age of the Universe, and, at the pinch, it was my minute that saved the situation.

This was another link in the chain leading to the 7.65 MeV level in carbon-12, for Baade sat on the combined astronomical steering committee of the Mount Wilson Observatory and the California Institute of Technology. It was to his good offices there, and to our meeting again in Rome, that I ascribe the invitation I received, in the autumn of 1952, to spend the first three months of 1953 at Caltech, to be followed by two months at Princeton University. Because living conditions in the United States were then so much superior to those in Europe, such invitations were greatly prized.

When the autumn term of 1952 ended, I flew to New York, where I made my first contact with George Jones at Harper Brothers, then on East Thirty-third Street, George being the American publisher of *Nature of the Universe*. I went then to Princeton, where I made arrangements for the lectures I was to give there a few months later. In Princeton, under Martin Schwarzschild's tutelage, I bought a car. My intention was to drive to California more or less by the route Frank Westwater and I had flown over eight years before: First, the Pennsylvania Turnpike to Washington, D.C.; then the Blue Ridge as far as Asheville; through Tennessee and over the Mississippi River at Memphis; then the long drive across the prairies to Meteor Crater (near Winslow, Arizona), the Painted Desert, and the Grand Canyon. I spent Christmas Day at Phantom Ranch at the bottom of the Grand Canyon, winning a few dollars from a party of mule drivers in a card game I found difficult to understand. Then I drove across the Mojave Desert and over the Cajon Pass to San Bernardino, and thence mostly through orange groves to Pasadena, halfway between Christmas and the New Year.

In preparation for the lectures I was to give at Caltech, I began once again to work through the details of the production of carbon from helium. But the details wouldn't go properly, not unless the carbon atom had a state at an energy level where, according to experiment, there did not seem to be a state. The trouble was that, otherwise, carbon was scoured out as it was slowly produced, scoured out to oxygen by this reaction:

$$^{12}C + {}^4He \rightarrow {}^{16}O + \text{gamma ray.}$$

Eventually, I felt so convinced about the point that I went to see Willy Fowler. I explained the situation and asked if there was any possibility of an experimental oversight. Fowler has said, in later years, that his first impression was that I had somehow gone a long way off my mental compass bearings, but I can't remember any such suggestion being made overtly. Instead, I remember him calling a small group of experimentalists into his office, the argument being repeated for their benefit, and there being a long technical discussion of whether the experimental methods used thus far might have missed the state I was looking for. The outcome was indeed that an even state of zero spin could possibly have been missed. With this information, together with my calculation of the actual energy of the state, about 7.65 MeV above the lowest energy state (known as the ground level), the group was able to design a new experiment to look for it. As I recall, it took about ten days to verify the prediction. There was indeed a state in carbon-12 and at about the energy level I had predicted. The day I heard the result, the scent of the orange trees smelled even sweeter.

I suppose the nearest one can come, in the ordinary way of things, to reproducing the way a scientist feels, when a prediction of his is being tested, is to be in court, in dock, with the jury out. Except, of course, the prisoner in dock knows already whether he is really guilty or not. In court, the prisoner hopes the jury gets it right if he knows he's innocent, and he hopes the jury gets it wrong if he knows he is guilty. In physics, on the other hand, the jury of experimentalists can be taken always to be right. The problem is that you don't know whether you're innocent or guilty, which is what you stand there waiting to learn, as the foreman of the jury gets up to speak.

After some struggle in classically minded quarters, the world came to believe in the general theory of relativity on the basis of only a single predictive test, the deflection of the direction of starlight passing close to the surface of the Sun. Einstein is reported to have said, after the test had been made successfully, that he never felt much strain while it was going on. Perhaps the five years or more needed to make the test helped to lighten the tension, but, even so, if indeed the report is true, Einstein must have had nerves of steel. Even in my small case, I felt the hot wind on my neck as I crept each day into the laboratory, escaping into the open air with relief as, for a period of two weeks, each day passed by without a result.

When the modest euphoria of success had faded, I was left in some awe at the broad picture that had emerged. It involved beryllium-8 and oxygen-16 as well as carbon-12. Chance coincidences appeared to be involved in all three of these nuclei, if the twenty or so indispensable elements (that is, the elements apparently required for the existence of life) were to be synthesized successfully in nature. Calculations of stellar structure led to the conclusion that, were the beryllium-8 nucleus stable, the production of carbon from helium would always proceed so explosively that stars in which "helium-burning" occurs would be blown violently apart, thereby making the synthesis of such elements as magnesium, sulfur, calcium, and iron impossible. Because the beryllium-8 nucleus is actually unstable, it splits quickly apart into two helium nuclei, making helium-burning a safe, slow process. What happens is that a beryllium-8 nucleus forms and then splits apart, with statistical balancing set up in the reversible reaction $^8\text{Be} \leftrightarrows {}^4\text{He} + {}^4\text{He}$, the concentration of beryllium-8 in the statistical balance being very low, making carbon production proceed only slowly by the reaction $^8\text{Be} + {}^4\text{He} \to {}^{12}\text{C}$. Too slowly, I found, unless some special property was added in order to favor the latter reaction. The special property was the existence of the state in carbon-12, which was required to be present at such a level that the reaction would be in resonance at an energy nearly equal to the sum of the energies of beryllium-8 and helium-4. This was how the explicit value of 7.65 MeV for the level was calculated. But this was not all. It would be no use having carbon produced at a satisfactory rate if all of it were lost in the formation of oxygen in the reaction $^{12}C + {}^4He \rightarrow {}^{16}O$. On the other hand, it would be no use either if *no* oxygen were formed—at any rate, at least as far as life is concerned. Not only had carbon production by the reaction $^8Be + {}^4He \rightarrow {}^{12}C$ to proceed in a suitably controlled way, but so had the reaction $^{12}C + {}^4He \rightarrow {}^{16}O$. This, I was able to show, was dependent on the oxygen nucleus *not* having a level in which oxygen production could proceed in a resonant way, the position for oxygen being the opposite of that for carbon. The experimental evidence for the nuclear structure of oxygen showed there was actually a state, very close to the danger point, given by the sum of the energies of carbon-12 and helium-4, at an energy excitation of 7.19 MeV, just a little above the experimental state at 7.12 MeV. Technically, this meant that the situation was safe—but just barely so. With 7.12 MeV below 7.19 MeV, there could be no resonance. But, if the energy of the state had been above 7.19 MeV, all would have been lost. Essentially, all carbon would have gone to oxygen in short order. So the situation turned crucially on two numbers very close to each other being just the right way around. Were they the other way, there would be no life.

All of this suggested to me what I suppose might be called profound questions. Was the existence of life a result of a set of freakish coincidences in nuclear physics? Could it be that the laws of physics are not the strictly invariant mathematical forms we take them to be? Could there be variations in the forms, with the Universe being a far more complex structure than we take it to be in all our cosmological theories? If so, life would perforce exist only where the nuclear adjustments happened to be favorable, removing the need for arbitrary coincidences, just as one finds in the modern formulation of the weak anthropic principle. Or is the Universe teleological, with the laws deliberately designed to permit the existence of life, the common religious position? A further possibility, suggested by the modern strong anthropic principle, did not occur to me in 1953— namely, that it is our existence that forces the nuclear details to be the way they are, which is essentially the common religious position taken backwards. Before ridiculing this last possibility, as quite a few scientists tend to do, it is necessary, as I pointed out before, to explain the condensation of the universal wave function through the intervention of human consciousness. While this could be seen as a matter for philosophical discussion, I suspect its resolution will eventually come from exact science.

I set to work immediately to broaden the base so favorably revealed. Before I left Caltech in March of 1953, I had found the additional processes of synthesis that are referred to nowadays as carbon-burning and oxygen-burning, which, in the 1957 paper of Burbidge, Burbidge, Fowler, and Hoyle, were referred to as the alpha-process. Before I left Caltech, I had also made a preliminary draft of a paper that was to appear the following year in *Astrophysical Journal, Supplement Series*, under the title "I. The Synthesis of the Elements from Carbon to Nickel." The "I" in the title meant that all neutron processes were left over for a second paper, which was to be part II. In the event, part II never appeared. With many improvements from others, it passed through into the 1957 paper of Burbidge, Burbidge, Fowler, and Hoyle.

Chapter 19

Steps to the Watershed

THE LUCKY CHAIN of accidents referred to in Chapter 18 was offset in the same years by an unlucky chain. Back on Edward Appleton's propagation committee in 1944, I had learned of the existence of cosmic radio sources from J. S. Hey, who discovered the first of them, the source known as Cygnus A. I had discussed with Hey the physical processes that might explain such an unexpected discovery, a discovery that appeared to the observer as an island of high radio intensity on the sky. Hermann Bondi, Tommy Gold, and I had even discussed the possibility of building equipment suited to what later became the new science of radioastronomy. By 1945, Gold had already come up with the concept of using a natural concavity in the ground as a place in which to build a stationary paraboloid with a very large collecting area, and it is a curiosity of later personal history that Gold eventually became the director of the very large radiotelescope at Arecibo, in Puerto Rico, which is of this very design.

After his return from Admiralty Signal Establishment to the Cavendish Laboratory in Cambridge, Gold sought to move towards radioastronomy, but J. A. Ratcliffe, who had become the laboratory's administrative head, favored the appointment of Martin Ryle from Oxford. Tension developed between Ryle and Gold, so that Gold was soon forbidden by Ratcliffe to visit the part of the Cavendish given over to radioastronomy, despite his official university appointment as a demonstrator at the Cavendish.

But it is an ill wind that blows nobody any good. I owed the start of a lifelong friendship with the Australian radioastronomer John Bolton to Ryle's acute dislike of anybody from outside visiting his own tightly controlled group. Bolton had been a wartime undergraduate at the Cavendish Laboratory. After the war, he emigrated to Australia, where he obtained a post in Taffy (E. G.) Bowen's radioastronomy section of the Commonwealth Scientific and Industrial Research Organisation (CSIRO). By the time of his visit to Cambridge circa 1950, Bolton had achieved worldwide notice by identifying two important radio sources, the Crab Nebula and the giant external galaxy NGC 5128. Yet, when he called at the Cavendish, his old laboratory, he was directed to Ratcliffe's office, where he was told that his presence was not welcome. So it was a somewhat distraught Bolton who knocked an hour or so later on the door of my rooms in St. John's College. Thereafter, we spent the day together, the beginning of many.

By 1950, enough radio sources were known for it to be apparent that their distribution was approximately isotropic—that is, similar in all directions. This meant that either they were local to the neighborhood of the Sun or they were very distant, outside our own galaxy. The local explanation was generally preferred, even to the extent of the sources being named "radio stars," the idea being that they were simply local stars that happened to emit radio waves more intensely than the Sun was known to do. If the positions in the sky of the radio sources had then been known with sufficient accuracy, it would of course have been possible to test this view directly by looking for visible evidence of the supposed stars. But the positions were not known with anything approaching the accuracy necessary for such a task.

Here is an example of what seems to be general practice in astronomy: When two alternatives are available, choose the more trivial. It was so with the discovery of pulsars—white dwarfs, everybody said they were, until confrontations with fact showed otherwise. And it is so today throughout cosmology. For a long time, I was puzzled by the fervor with which the principle of maximum trivialization is held, until an interesting explanation was recently offered by the Oxford astronomer Victor Clube and Bill Napier, from Edinburgh, an explanation that will become an important feature of the final chapters of this book. When one considers all that is written of the wonders of astronomy, all that is spent by nations worldwide in exploring those wonders, is it surprising that, with the exception of the Sun's gravity, which holds the Earth beneficially in its orbit, nothing from outside is supposed to affect the conduct of life on the Earth? This is the fiction that the trivialization of astronomy is designed to support.

The ill feeling between Martin Ryle and me, which became public in the 1960s, had its source in 1951, a few hours before Tommy Gold and I ate chips-with-something at midnight at Jack's Cafe, situated on a hilltop on the ancient A1 road from London to Edinburgh, about halfway between the towns of Stevenage and Baldock. We were returning in Tommy's £50 Hillman from a meeting that had been held at University College, London, a meeting that subsequently became known as the Massey

Conference, named for Harrie Massey, a pioneer in the field of quantum atomic collisions, although it had actually been organized by Massey's student Bob Boyd (later Sir Robert).

Controversy had been far from my mind as I gave my paper on the technical problem of the conductivity of an ionized gas in a general electromagnetic field. Nor should there have been serious controversy over Gold's paper. Everything he said was true. By 1951, about half a dozen radio sources had been definitively related to astronomical objects, not one of which had turned out to be a star. There had been identifications at Jodrell Bank with weakly emitting nearby galaxies, and there had been the two cases identified by John Bolton, the Crab Nebula and the galaxy NGC 5128. So Gold said that perhaps the other possibility for explaining the isotropic distribution of radio sources—namely, that most radio sources were very distant—should be taken seriously. He said nothing more than that, and he expressed it temperately. It was in these circumstances that Ryle began an attack that was to persist for almost two decades.

If my mouth was open when Ryle began speaking, it must have closed immediately with an audible snap. There is all the difference in the world between a critic saying "I don't agree with you" or "I get a different answer" and Ryle's habit of flat denunciation. On this occasion, he began, "What the *theoreticians* have failed to understand . . ." (with the word theoretician implying some inferior and detestable species). Actually, as it turned out, the theoreticians hadn't failed to understand at all. A sufficiently accurate position on the sky was eventually obtained for the source Cygnus A to be definitively identified by Walter Baade at the Mount Wilson Observatory. It turned out to be a galaxy at a distance of 500 million light-years or more from the Earth. Baade made this announcement at the 1952 meeting of the International Astronomical Union in Rome, the same meeting as the one at which he revised the age of the Universe from Hubble's 2 billion years to 3.6 billion years. The irony of the situation was that it had been Ryle's brother-in-law, Graham Smith, who obtained the accurate position for Cygnus A. So, as it turned out, the supposed radio stars really were external galaxies at large distances, just as Gold had said.

I do not think it unreasonable to say that Ryle's motivation in developing a program of counting radio sources, a program that was to occupy a major fraction of his group over the next ten years, was to exact revenge for his humiliation over the radio-star affair. This was to be done by knocking out the new form of cosmology with which Gold, Bondi, and I were associated. Nothing really new, beyond what was known already in 1930, has happened theoretically, except for what the three of us did in

1948 and what the American cosmologist Alan Guth did in 1981. The form of cosmology I invented had mathematical points in common with Guth's work, and it might be said to have been an early example of what is known nowadays as inflationary cosmology, although my interpretation of it in 1948—in conformity with the views of Bondi and Gold—may well have been wrong or, at any rate, partially wrong.

Establishments in all realms of life do best when nothing is happening. Great reputations are made over the supposed virtue of doing nothing, which is just as true in cosmology as elsewhere. It is a situation well summed up by a remark of Johann Sebastian Bach. Following a visit from a supposedly famous violinist, Bach wrote to a friend: "He played well enough, but only from known music." Throughout my troubles with Ryle, it always seemed to me that the cosmological work of his group was rather like the playing of Bach's visitor.

Walter Baade's demonstration that Cygnus A is a galaxy at a distance of at least 500 million light-years raised the important problem of the nature of the physical process that generated the radio radiation. The favored idea of the day was plasma oscillations, which I doubted, perhaps because I didn't understand it very well, but perhaps also because there was an idea I understood better. In 1948, Julian Schwinger, then of Harvard University, had investigated the unexpectedly large energy loss from a type of particle accelerator known as the synchrotron. He found that, when highly relativistic electrons precess around a magnetic field, they emit energy not just at the precessional frequency but, more importantly, at high harmonics of that frequency. In 1950, this process of synchrotron radiation was proposed by Hannes Alfvén and N. Herlofson as the origin of radio emission from stars, the elusive radio stars. In 1952, after Baade's identification of Cygnus A, I wondered if the same process might not apply to radio galaxies. Now, in those days, one did not rush precipitately into print the way we all do nowadays. So I published nothing immediately, preferring, after I had received the invitation to the California Institute of Technology in 1952, to wait until I had had the opportunity to discuss the matter with Baade and with his friend Rudolf Minkowski, the nephew of a famous mathematician of the early twentieth century, Hermann Minkowski. Then came the affair of the energy level in carbon-12 and the writing of my paper "I. The Synthesis of the Elements from Carbon to Nickel," which, with my teaching duties back at Cambridge, took up a full year—that is, to the autumn of 1953. So it was not until then that I wrote up the radio-emission idea, which I eventually presented at a conference held in Washington, D.C., in December 1953, with the work appearing eventually in the international journal *Nature* in January 1954.

There was a complaint by Alfvén from the audience at the Washington meeting to the effect that I was infringing the priority of his 1950 paper with Herlofson. Forty years on, I still rather doubt it. The priority for the synchrotron process itself clearly belonged to Schwinger. All that remained after Schwinger was to make the appropriate application of a known process to astronomy, and this Alfvén and Herlofson had unfortunately not done. But as it was soon to turn out, during the year I had just let pass, the same step had been made in the Soviet Union and had been published in journals in the Russian language, of which I was ignorant. Once I learned that, I abandoned the matter, for it has always been my view that whoever publishes first, regardless of all other considerations, should take the priority.

But this is by no means the way the world always sees it. The world behaves in strange ways where priorities are concerned. A noteworthy example is Darwin's being accorded priority in 1859 on the theory that bears his name, despite the same theory being stated by Patrick Matthew in 1831, by Charles Naudin in France in 1852, and by Alfred Russel Wallace from the East Indies in 1855 and 1858. This was done on Darwin's assurance that the theory had lain unpublished for many years in his escritoire, despite there having been no independent witness to that effect. Another interesting case occurred over the years 1842–1846. A young man at the top of the mathematical tripos in St. John's College, John Couch Adams, had the idea that deviations of the recently discovered planet Uranus from its expected orbit might be due to the existence of an undiscovered planet still more distant from the Sun, a planet that eventually came to be known as Neptune. Adams set himself to use the deviations of Uranus to calculate the position in the sky where Neptune would need to be at that date, a calculation he completed by 1844 or thereabouts. Adams then communicated his results to the two most senior members of the British astronomical establishment, but he did not actually publish it. Owing to the ineptness of the establishment, no different then than in the 1950s, Neptune was not found by a British astronomer.

Meanwhile, the French astronomer Urbain Leverrier took up the same problems, sending his result to the German observer Johann Galle in Berlin. Almost immediately, Galle detected the new planet, and there was much rejoicing in France and Germany, the detection being considered a big publicity affair.* But now the big guns of the British establishment

---

*Present-day German astronomers tell me, with a certain cynicism, that the way this story is usually told obscures the essential point that Galle, at that time, was in a very junior position at the Berlin Observatory. It was because his superiors happened to be away from the observatory that he was able to make the discovery. Otherwise, it would very likely have been the same as in Britain.

raised themselves, and, in the ensuing display of political muscle, a half-share of the priority for the discovery of Neptune was salvaged for the fortunate John Couch Adams. The point being that, where international priorities are concerned, a great deal depends on the strength of the national establishment standing behind the scientist, with the present-day situation greatly favoring scientists in the United States. A major reason why the smaller European nations are joined into wider scientific organizations, such as CERN and ESO (European Southern Observatory), is to ensure that their scientists get a fair crack of the whip, the situation for Third World countries remaining dismal in this respect. Had I been American in 1953, the situation for me would perhaps have been different. The Washington meeting would, I suspect, have secured some share of the radio-source priority.

The first I heard of the synchrotron explanation of radio-source emission was from George Hutchinson, who, I think, held an industrial fellowship in the Cavendish Laboratory. Hutchinson, who later became Professor of Physics at Southampton University, used to lunch in St. John's, and it was over lunch that he used to talk to me about such things. Since, to my knowledge, Hutchinson did not publish (although there may be a discussion in his Ph.D. thesis), I cannot recall whether his work applied to galaxies or whether it was restricted to stars, as with Alfvén and Herlofson. If it applied to galaxies, then Hutchinson was close to having priority—oral priority, at any rate.

I never learned exactly what happened in the Soviet Union. Dates attached to papers often do not tell the full story. There was a moment when I came close to learning the story—or at least to learning one side of it. In the 1960s, the Russian physicist Vitaly Ginzburg gave three lectures in Cambridge on cosmic rays. In one of them, he came around to the issue. After beginning, he stopped, shook his head, and said he wouldn't continue, "because," as he correctly observed, "priorities is dirty business."

The story of the Washington meeting in December 1953 has been worth telling because of a remarkable paper given there on a subject quite different from radio sources. The detection of 21-centimeter radio radiation from atomic hydrogen was big news in those days, and a major section of the conference was given over to it. The remarkable paper was not on hydrogen, however, which was probably why it did not arouse as much notice as it should have.

The remarkable paper was by Charles Townes, and it was a discussion of what further radio spectrum lines might be expected from other atoms and molecules, a forerunner of the millimeter-wave astronomy of modern times. I remember that Townes had condensed his results into a big table. I also remember a talk with him, especially about detection possibilities. Then, faintly on the wings of memory, it seemed to me that I had met

Townes before, perhaps on the occasion of my first visit to the United States in the autumn of 1944. I remember what must have been a visit to the Bell Telephone Labs in New Jersey, and our meeting was either there or in Washington. The disturbing thing is that I have lost the connecting links. What I remember without question is a train from New York stopping at every station, a crowded train, even sordid, compared with the fast trains made up of Pullman cars. As Marcel Proust said, it is the irrelevant details that show a memory to be correct, the irrelevant details in this case being snow at the side of the railway line and of stepping into slush with inadequate shoes. This means, certainly, that I didn't visit Townes on the occasion when Frank Westwater and I had called on Henry Norris Russell in Princeton because, over those days, New Jersey was enjoying the late stages of an Indian summer. Exactly where and when it was that I met Townes on that trip in 1944 I can't dig out, but, because of my memory of the snow and slush and of the repeated stopping of the train, I am convinced that it must surely have happened.

Now to a different sequence of events, a sequence that began in the spring of 1953, on the day I left Caltech to drive back east to fulfil my commitment to spend two months in Princeton. The journey was to prove epoch-making, with the elderly Dutch astronomer G. von Biesbroeck practically saving my life in circumstances shortly to be related. It is remarkable how the journeys one makes in life mostly pass by without incident, conspiring, it seems, to pack all exceptional situations into a single episode, the way this one did.

I started off from Pasadena north across the San Gabriel mountains, because Otto Struve had asked me to give a lecture in Berkeley on the morrow. It was pretty well solid traffic lights for the first thirty miles, in those days. The road north along the San Joaquin Valley was a seemingly endless string of towns and cities. By the time I reached Fresno, with Merced and Modesto still ahead, I was reduced to shouting incoherently at every red light. Never thinking I would have reason to be thankful to Fresno, I continued grimly, reaching the Faculty Club at Berkeley at about 10:30 P.M. It was just about a decade later to the month, as I was attempting to order breakfast at a hotel in Yerevan, Armenia, from a menu written in Armenian, when an American voice said, "Perhaps I can help translate. I'm from Fresno." Then I learned that Fresno, city of traffic lights trailing away to infinity, was also a center of Armenian culture, in which translating menus was a small matter.

I had a hard time giving that lecture in Berkeley the following morning. Struve introduced me by saying that I was the author of a dubious theory, for which he intended to reward me appropriately, at which he proceeded to wave a check at the audience. First putting the check in a long white envelope, he handed it to me with a flourish, sitting thereafter and watching me closely with his usual cross-eyed stare. If Struve had been intending to invite me to dinner after the lecture, he didn't succeed, although he did try to detain me with information that a blizzard was raging in the Donner Pass. Undeterred by this intelligence, I started off, at about 2:30 P.M., on my way east to cross the Sierra Nevada, aiming to reach Reno that night.

The snow had appeared already on the upgrade east of Sacramento. I was encouraged by the road having been plowed, but the drifts at the sides of the road became higher and higher as I gained altitude, eventually reaching a level above the roof of the car, which was, most fortunately, a reliable old Chevrolet. I saw nothing to worry about, however, because, in Canada in 1944, I had observed how drivers corrected even bad skids by bouncing their vehicles off big snow banks just like these. A skid at 60 miles per hour prompted me to try this maneuver, with a result less favorable than I had expected. To be frank, I ended decisively off the road, turned by 180 degrees. Considered over my entire life, good luck and bad luck both come out about average. But I tend to accumulate bad luck in many small pieces and then to return to the average in a single large slice of the good. So it was on that occasion. The United States is a big country, and, the way it normally is, I might have walked for miles before reaching help. As it turned out, I came on a chap with a tractor in less than a quarter mile. He pulled me out with little harm to the Chevrolet but with inevitable inroads into Otto Struve's check for $50. The descent from the summit of Donner Pass down to Truckee was the second hairiest descent of my experience. Indeed, I hadn't been long on the descent when I met the cops coming up to close the pass. A couple of months later, Struve sent me the true story of a party of easterners immigrating to California who, in the winter of 1846–1847, had been overwhelmed by snow on what is now known as Donner Pass (subsequently named after the leader of the party) and who had eventually been reduced to eating the corpses of their less hardy comrades in order to survive the winter. This, I suppose, was another example of Otto's curious sense of humor.

I started off east the following morning through Nevada towns with the contrasting names of Wadsworth and Fallon, where I turned south for Tonopah, Nevada. Every time I have been in Tonopah, I have been escaping from snow. For instance, there was a surprising escape from a big storm a few years later on the road from Ely, Nevada, made possible by the car I was then driving having a differential lock on its rear wheels. So Tonopah is associated in my mind with heavy snow, which I'll just bet isn't in the minds of its inhabitants, accustomed as they are to their place not far from Dante's View into Death Valley. And so I continued through Stovepipe Wells to Las Vegas, where I intended to dissipate the last of Otto Struve's check.

I have no patience with fruit machines or with their high-class equivalent, the roulette table. For me, these devices represent the lowest form of gambling. Because he thought the female card-dealers were "something," Willy Fowler always had a good word for the game of twenty-one, or blackjack. And, of course, the discovery that knowing the odds, combined with the ability to remember exposed cards, permits a player to beat the bank makes the game attractive to many. But winning that way has always seemed to me to be more business than pleasure. Better H. P. Robertson's method of playing poker. Robertson, known to his close friends as Hot Dog, was the author of a famous prewar article on cosmology. Robertson and John von Neumann, his colleague at Princeton, had worked out accurate probabilities for the various combinations in poker; and von Neumann had also demonstrated mathematically that the optimum strategy in poker is simply to place your bets in proportion to the odds. This is the kind of mathematics that has had sociological importance for politics and for the strategy of war. In this respect, some credit belongs to mathematics for the fortunate circumstance that a Third World War did not break out in the years from 1945 to 1990.

At the top of the gambling world is proposition betting. A proposition can be formulated over an essentially infinite range of possibilities. For example, assume that there are thirty people, chosen at random, in a room and that I bet you that at least two of them will turn out to have birthdays on the same day of the year. To the uninitiated, this looks like a good bet to take, but the odds are actually a healthy 2.3 to 1 in my favor. The aim of proposition gambling is to deceive the other chap over the true odds in some unusual situation. Richard Feynman is reputed to have been exceedingly good at it, to the extent that the real professional gamblers in Las Vegas had agreed among themselves to keep away from Feynman. How could it be otherwise, when you consider Feynman's resounding victory in arithmetic over an abacus salesman in a backstreet café of Rio de Janeiro? I can't conceive of Einstein scoring such a victory in a backstreet café of Rio de Janeiro. Or Dirac, or Heisenberg. Wolfgang Pauli had the speed to have won such a competition, if he could have conceived of it.

The knowing ones, the cognoscenti, will have wondered why, in driving from San Francisco to Princeton, I had routed myself through Tonopah and Stovepipe Wells. The answer to this conundrum is that I had an appointment at the Lowell Observatory, to discuss color-magnitude diagrams with Harold Johnson, this being the real point of the whole story, to which I shall come eventually. Harold Johnson said that, if I could stay an extra day, he thought he could arrange for me to have lunch with V. M. Slipher the following day, to which I happily assented, since it was Slipher who discovered the expansion of the Universe, which forced Einstein to abandon his affection for a static Universe in favor of the dynamical models of Aleksandr Friedmann, which Einstein had at first opposed.

Slipher did not have red shifts of sufficiently distant galaxies to establish the linear relation between red shift and distance discovered by Edwin Hubble. On most Sundays while at Caltech, I had had the pleasure of walking with Hubble and his wife Grace, so such motivation as I have on priority issues would favor Hubble over Slipher. Attention to the historical facts, however, suggests the position as I have just stated it. Indeed, I feel Hubble's first statement of the linear law in the late 1920s was really somewhat premature. If one looks at the sample then available of galaxies with known red shifts, a quadratic dependence on distances appears to be just as likely as a linear dependence. It was only with red shifts for galaxies at much greater distances becoming available in the 1930s that the linear law could be seen to be correct. Slipher, working at the Lowell Observatory, had a weaker organization behind him than Hubble had at Mount Wilson, and it is regrettable that such inconsequential issues can have an effect on how priorities are accorded in science, just as the political importance of whole nations can. The remarkable thing about Slipher's career was that he started as a young assistant to Percival Lowell, with instructions to find canals on Mars, and he ended by discovering the expansion of the Universe. I remember him as a slender, gray-haired man (in his middle sixties, I would think), a little embittered, perhaps—and justly so, one might say. However, being embittered butters no parsnips.

Staying to meet Slipher delayed my schedule, leaving me a day less than I had planned in which to drive to the Yerkes Observatory at Williams Bay, Wisconsin, eighty miles north of Chicago. Therefore, I started out very early the following morning. At a red light in the center of Flagstaff, two lads in U.S. Navy uniform asked for a lift—"a ride," as they put it. They were just back from the Korean War and head-over-heels to get back to their homes in Oklahoma City. So, by sharing the driving, we made the best part of a thousand miles from Flagstaff to Oklahoma City, putting me right back on schedule. One of the lads said: "It was so cold in Korea that I'll never bitch again about the heat." I wonder if he has kept to his vow.

At Yerkes, I stayed two days with Subrahmanyan Chandrasekhar and his wife, Lalitha. I think I gave a talk on the spare day, but, more notably, I bought toothpaste and razor blades at a village shop in Williams Bay, putting my purchases into the glove compartment of the Chevrolet. Chandra was due to lecture early in Chicago on the morning of the third day, which required that he and his wife leave home at about 6:00 A.M. Rather than ask me also to leave as early as that, Chandra proposed that I follow in my car at a later, more civilized hour, to be in time for lunch at the University Faculty Club with Enrico Fermi, and then to give a talk at the Physics Department of the University of Chicago in the afternoon. Agreeing readily to this plan, I rolled out of bed at a little after 8:00 A.M. on the morning of the third day, making my way to the bathroom, where I stripped off my pajama top and prepared to shave. My razor being blunt, the first stroke of it reminded me of the blades in the glove compartment of my car. So, back in the bedroom, I fished the car keys out of a pocket and decided, because of the snow outside, to slip on my shoes. Ever one to minimize the number of physical movements I have to make, I didn't bother to lace up the shoes. The house had a verandah and steps down to the pathway outside. I had just about maneuvered down the steps in my unfastened shoes when I heard an ominously loud click from the house door behind me.

A biography of Chandra has recently appeared, and, frankly, I must say that there is one thing the biographer doesn't know, and that is, when Chandra locks up a house, he locks it good. No downstairs window was openable—hardly so even if I had been prepared to break the glass, for they were all double windows with big outer screens. Then I had an idea. Nearly all American houses have cellars. Maybe householders were not so careful in preventing you from breaking down into the cellar as they were in preventing you from breaking up into the house. So it proved. With a measure of ingenuity, I managed to break down into the cellar, thinking that now my problem was solved. But a flight of wooden stairs led up into the house with a solid looking door at the top—locked, as it turned out, much to my dismay.

In my pajama bottoms and unlaced shoes, I shuffled my way out of the cellar again and back to my car. The saving grace was that at least I had the keys and could start the engine, eventually thawing out from the morning's icy cold in the warm blast of the heater. All kinds of mad ideas raced in my mind. I had no money, my wallet being inside Chandra's hermetically sealed house. But there was enough fuel in the car to allow me to reach Chicago. How about turning up at the Faculty Club for lunch with Fermi attired as I was? Could I also get away with giving my lecture to the physics department "come as you are"? That was the question.

At last, it broke through into my consciousness that Chandra's house was constructed from wood. It had taken a long time for this significant fact to present itself to my dulled mind because, in my boyhood, houses had all been made of stone or brick. Houses of wood were thought unsafe because of risk from fire. Of course, I had no matches with which to start one, but, surely, wouldn't the risk of fire require there to be a spare key somewhere around the observatory? A mile's drive brought me to the front of the observatory, where a fair fraction of the young women of Williams Bay were now arriving to begin their day's work. It had been a tradition, dating from the days of Pickering at Harvard, for all American observatories, from Chicago to the east, to employ an immense quantity of female labor, and here it all was now, all around me as I shuffled in pajama bottoms and untied shoes up the great steps of the Yerkes Observatory.

The builders of prestigious institutions in olden times went in for a lot of polished stone and marble, as they had done on the great steps of Yerkes. Each year, the entire observatory staff was photographed sitting on the top step. Immediately inside the massive doors, you could study pictures of the staff taken over thirty or forty years, learning therefrom as much of human aging as you cared to absorb. Interesting too was a year-by-year comparison of the photographic images of Otto Struve. To one side of a critical dividing point was a glum, scowling Otto, becoming younger and younger with increasing distance from the dividing point. On the other side was a widely smiling Otto, becoming older and older with each passing year. The sharp dividing point corresponds, believe it or not, with the year in which Otto became the Director of the Yerkes Observatory.

Anyway, I shuffled along the corridors of Yerkes until I found a staff member who had arrived early. It was von Biesbroeck, neatly dressed and with a neatly cut beard. I swear he never flickered an eyelid when I appeared, a grotesque figure, still unshaven, in the doorway to his office. As if he had such situations on his hands every day of the week, he soon produced a key, which is the story behind the story of how I contrived to have lunch with Enrico Fermi.

Like Eddington, Fermi resembled his photographs—it is amazing how many people don't. Because I had begun research in 1936 with Fermi's theory of beta decay, I had a disposition to be serious, but Fermi would have none of it. To Chandra's annoyance, Fermi was not in a serious mood, and the best we could do was to take part in a complex discussion, promoted by Fermi, on the trunks of elephants. In that, I missed a great opportunity. If I had thought of elephant ears as cooling devices for the brain, I would have risen high in Fermi's esteem. I was always puzzled by why there should be so much blood in the human ear, until I heard about the situation for elephants.

In retrospect, it seems surprising that, in all the talks I gave in April and May of 1953, I never spoke about the discovery of the 7.65 MeV level in carbon-12. The explanation is that astronomy is a subject of fads and fetishes. In 1953, the fad was plasma oscillations. In 1936–1939, when I had begun research, it had been radiation pressure. Then it was turbulence, then magnetic fields, and now plasma oscillations. In the future, it would be black holes, accretion disks, and gravitational lenses. There is a close psychological connection between these fads and children's games. A game is "in" one moment and totally "out" sometime later. Astronomical fads have always involved miracle working to some degree, and their discussion in so-called workshops and in the streams of papers that pour into the journals have affinities to the incantations of Macbeth's witches on the blasted heath. So it was on the chimera of plasma oscillations that I spoke in Chicago, not on the collapse of the central cores of supernovae, which, in those days, was considered outré.

An Italian associate of Fermi's had his own views on plasma oscillations, which he invited me to discuss at dinner that night. When I arrived at the restaurant he had named, it was to find him in company with two young Japanese who, it seems, had come temporarily to the United States on the Italian professor's contract. One of the Japanese spoke English, the other did not. So this is the picture: The restaurant is dark—not so dark that I can't see the other side of the table, but everything in the intermediate distance is in shadows. The Italian's ideas are crazy, in my judgment. One of the Japanese spends much of his time in translating what we say for the benefit of his non-English-speaking companion, and, at no great distance away, music with an insistent, crude rhythm is thumping relentlessly. This goes on for a long while, and I am thinking about escaping into the open air, when the Japanese who doesn't speak English at last begins to speak. Not a syllable do I understand until his companion begins to translate his remarks into English, and then, at last, the situation becomes interesting. What the dark figure in the corner has said in Japanese, with all the politeness stripped away from it, is that, mathematically speaking, the Italian professor is crazy, and for exactly the same reason I think he is crazy.

So far, I have spoken of mathematics as if it were a known entity, like a pound of butter. But what is this mathematics? At the turn of the century, Alfred North Whitehead and Bertrand Russell sought to define mathematics by purely symbolical processes. Henri Poincaré, the French mathematician, viewed the question otherwise. In a lecture given in St. Louis, Missouri, in 1904 (in the same series of lectures as the one in which he stated the principle of relativity—all observers find light to have the same speed—which Einstein showed, in 1905, would lead to no inconsistency), Poincaré pointed out that the symbols of mathematics are always defined in words. Symbols are only a shorthand that save using an enormous number of words. In modern computer language, we would say that symbols are subroutines written in words and that mathematics is a macroprogram using symbols as subroutines. I always thought Poincaré had the better of this argument, and, like him, I thought of mathematics as a shorthand for an argument that, if spoken out in full, would be longer than any political speech or any sermon from the pulpit.

This was my position until the affair in the Chicago restaurant, where I saw that there must surely be something still more basic behind words, for how could languages so different as English and Japanese throw up just the same set of mathematical ideas? To take a parallel situation, is a novel really a collection only of words? Evidently not, for the reader is required to possess an emotional structure, without which the novel would be nearly meaningless, and indeed also a cultural structure, which often prevents a highly successful novel in one culture from being transferred to another—although there are rare authors, of which Shakespeare is perhaps the best example, who manage to cross cultural barriers without too much difficulty. Yet a child cannot read Shakespeare as an adult can, and we read Shakespeare differently as we grow older.

A detail in Homer's *Iliad* makes the situation clear. Pegasus is an ordinary horse who is included as a makeweight beside the horses of Zeus in the team that powers the chariot of Achilles. Yet, throughout the strain of battle, the determined Pegasus yields not a step to the swift immortal horses at his side. Towards the end of the day, Pegasus is hit by a miscast spear. Being only a mortal horse, he dies. He is cut loose from the chariot, and, as the Sun goes down, his body is left alone on the battlefield. So, in but a few strokes, Homer shows us the frailty of all mortal beings, even if, for a while, we are able to keep pace with the gods.

But we have to know this already, before Homer's story has meaning, which shows that something more fundamental than words must lie behind the words, just as it does behind the symbols of mathematics, something that it is clearly useless to attempt to express in words. Mathematics differs crucially from literature, however, in that its more fundamental concepts cross over all races, languages, and cultures. And the concepts of mathematics can have the unique property of representing the world exactly in the most revealing cases, as in the seemingly limitless accuracy of quantum electrodynamics. The writer, on the other hand, can permute and combine what lies behind words in ways that are very far from unique. In nine cases out of ten, Othello will spot Iago's villainy without too much difficulty, and he will respond by running a well-deserved sword through him. It is the tenth case, the unusual case, that makes Shakespeare's play possible. Unlike mathematics, which is universal, the storyteller, working with our underlying feelings as his material, deliberately seeks out the abnormal. But both the storyteller and the mathematician trade on what it is that lies behind mere words.

Animals lack words, but do they also lack what lies behind words? From my own observations, I would think not. In the Lake District of northern England in later years, we had a couple of acres of rough ground with many trees and bushes. One day, I came around a bush to find a female pheasant that hardly moved as I approached. If I had swooped, the way John Gillams did on the squirrel, I could easily have snapped her up. I thought for a moment that the creature was seriously ill, until, around the next bush, I came on a mass of gaudy feathers, making it clear that the mate had been taken recently by a fox. Birds, when they are frightened, fly desperately away from danger, so I knew the female pheasant wasn't simply frightened. She was shattered by grief.

And how about what lies behind mathematics? Again, in the Lake District, there was a spotted flycatcher that came back from central Africa each year to make its nest no more than five yards from our front door. The ability to navigate four thousand miles to this degree of precision would seem to me impossible without employing concepts similar to those we use in mathematics. The accuracy of the flycatcher and the accuracy of our calculations in quantum electrodynamics seem to me to spring from the same source, a source behind and independent of words, a source that I referred to previously as the godlike bit in all of us.

In the spring of 1953, during my drive from San Francisco to Princeton, an old problem was much in my head—or, at any rate, it was in my mind from Stovepipe Wells in Death Valley onwards. In my argument with Eddington on the day, in 1940, I had walked from St. John's to the Observatories, the question had been to explain the large radii of giant stars. Lyttleton and I had integrated a stellar model with a region of higher molecular weight (more helium) inside an outer region of lower molecular weight (less helium), and we had found that the radius extended beyond what was to be expected for a main-sequence star, sufficiently extended to explain a case like Capella.

In our example, the inner region had a considerably higher mass than the outer region, and the material in both regions was a normal nondegenerate gas. Just at that time in 1952–1953, Allan Sandage and Martin Schwarzschild had obtained a similar but more dramatic result that greatly puzzled me. They had an inner region of smaller mass in which hydrogen had been entirely converted to helium and from which—temporarily, at any rate—the supply of nuclear energy had been lost. The energy of the star in their model came from a thin skin at the surface of the inner region where hydrogen was still available. With this kind of structure, they found that the radius of the star became larger and larger as more material was added to the inner region through steady helium deposition from the skin. Without energy sources, the material of the inner region became a degenerate gas, as in a white dwarf, and, eventually, the outer radius went off to infinity.

This could not be a physical result because, of course, no physical structure can have an infinite radius. While I continued to be puzzled about the relation of the inner degenerate core to the grossly distended outer envelope, I could see that the wrong infinite radius was probably due to the use of mathematical conditions at the outer boundary of the model that were not correctly related to physics. Sandage and Schwarzschild had used the same boundary conditions that Eddington had used in his famous work of the 1920s, which again took me back to a controversy of the 1920s between Eddington and E. A. Milne. Milne had maintained that Eddington's models were mathematically wrong because of unphysical conditions at the boundary, but Lyttleton and I had long satisfied ourselves that the difference between the true conditions and Eddington's unphysical ones made little difference to the results he had derived. But this, again, was for main-sequence stars. Could it be the case, I now wondered, as I thought about the models obtained by Sandage and Schwarzschild, very far removed from main-sequence stars, that for these Milne was right? The more I thought about it, as I drove along, the more I thought that the key to getting reliable radii must lie in using correct conditions at the outer boundaries of the stars. I therefore resolved, on reaching Princeton, to discuss the matter with Martin Schwarzschild and to suggest, if he were agreeable, that we investigate the matter together.

Martin Schwarzschild and his wife Barbara had obtained a result, at about that time, that appeared to me to be of profound significance. From the spectrum lines of comparatively nearby stars, they had found that the ratio of abundance of iron to hydrogen was variable from one star to another. The Sun's ratio was towards the top end, with those of most other stars being lower—in some cases, by a factor of about 3. The result was profound, because it was the first clear-cut demonstration that such elements as iron had not been produced in the first moments of an exploding universe. If they had been so produced, the distribution of iron should have been uniformly mixed with respect to hydrogen, and no such variations should have been found. Pretty soon, variations by as much as a factor of 100 were discovered for more distant stars, emphasizing the lack of uniformity. Plainly, then, the bulk of the elements in the Universe had to have been produced in the stars.

There was just one case for main-sequence stars in which Eddington's methods hadn't worked any too well—for those stars with masses appreciably less than the mass of the Sun. Faced with discrepancies, Eddington, in his book *The Internal Constitution of the Stars*, took the view that perhaps it was the observations, rather than the calculations, that were wrong in this case. I arrived in Princeton to find that the issue was in the process of being settled—by Don Osterbrock, then a student of Martin Schwarzschild. The resolution again lay in the outer boundary condition. So, when I proposed to Schwarzschild that calculations of the radii of giant stars had been going wrong because of the outer boundary condition, he was very ready to agree, and pretty soon we had decided to investigate the matter together.

We were greatly assisted in our investigation by R. Härm, who was one of those immensely valuable individuals who are happy to be calculating perpetually to the greatest possible accuracy. In a way, it was a restful life, free from committees, free from worries about which theory was right, about which path to take. Härm always knew he was right. If anything was adrift then, it was others who were to blame—in this case, Schwarzschild and Hoyle. Härm was perhaps the last of the human computers to be of real importance in astronomy. In a year or two more, we would all be trusting our fate to digital computers, which have the disadvantage that they don't smile at you in the morning.

Walter Baade had startled the astronomical world by dividing stars into two types—populations I and II, as he called them. They were distinguished by different color-magnitude diagrams. This was why I had driven south to Flagstaff to visit Harold Johnson. In essence, a color-magnitude diagram is a plot of a star's luminosity against its radius, with each individual star represented by a point in the diagram. A cluster of stars gives a distribution of points in the diagram, and, because the stars of a cluster could be taken all to have the same age and the same composition, their differences in the diagram could only be caused by differences in their masses. Now differences caused by masses in a cluster of stars observed at the same time could readily be converted into differences caused by ages at the same mass. Color-magnitude diagrams, therefore, showed how a star of fixed mass changed as it aged—in other words, the evolution of the star, which was the essential point of interest in the whole business.

What Baade's populations I and II told us was that stars of a particular mass could evolve in two distinct ways that were characterized by distinct color-magnitude diagrams, and what Schwarzschild and I set out to prove was that the difference was caused by the ratio of iron and other iron-group metals to hydrogen. Stars with ratios like that of the Sun were designated population I, and stars with much lower ratios were designated population II. The investigation was to take about a year to complete, and it was not to appear in print until 1955. We never encountered any controversy over the result, which was quickly and widely accepted. The result, together with Baade's populations I and II and their different distribution in our galaxy, was to form the basis of a conference, involving a fair fraction of the world's leading astronomers, to be held at the Vatican in May 1957.

Only once do I recall Martin Schwarzschild and me having even a moderately sharp word. This was when I was injudicious enough to say that I thought it curious that we had hit on a situation in which E. A. Milne was right in his argument with Eddington. No, Martin said, Eddington had been concerned with calculating stellar luminosities, and these were quite unaffected by the surface boundary condition. It was the radii of stars that were affected, and these had not really been Eddington's concern. Martin's quickly springing to Eddington's defense was, I thought, a reflection of how, in the early years after Einstein's publication, in 1915, of the general theory of relativity, Martin's father, Karl Schwarzschild, had been Einstein's first major supporter, followed by Eddington in England and by de Sitter in Holland. Between them, the three had pretty well swayed the entire scientific world into acceptance of Einstein's theory.

My visit to Princeton in the spring of 1953 was the time, if ever, for me to try to meet Einstein. Since he was living there, the geography would have been easy; but I had heard that Einstein was not too well and that he was exposed to a never-ending stream of visitors. I asked Schwarzschild for advice on what I should do, aware that, because of his father, Martin would know Einstein well. He asked for a day or two to think about the matter. Eventually, he said it would be a kindness to desist. So I did, which was why I never met Einstein.

Not long before my visit to Princeton, an eruption over a seemingly trivial affair had occurred among American astronomers. Immanuel Velikovsky published a book with the title *Worlds in Collision* in which he claimed that a sequence of dynamically impossible events had shaped society circa 3000 B.C., giving rise to well-known myths and legends. The book cared naught for science but a great deal for Velikovsky's reading of ancient documents, his expertise being in the deciphering of such papers. The book caused a sensation both with the public and among astronomers, the latter becoming stirred to near-violent displays of outrage. Such eminent figures as Harlow Shapley were heavily involved. It could be said that Shapley became angry even to the point of incoherence.

During my stay in Princeton, Velikovsky attended the weekly astronomy seminars, whether to learn something of astronomy itself or to learn what made astronomers tick, I never really discovered. Perhaps because I couldn't see what the fuss was about, and so was calm about it, Velikovsky used to come to talk at the tea intervals that preceded the seminars. I managed to convey to him that our ground rules were different from his. He believed in the primacy of documentary evidence, whereas we believed in the primacy of mathematical rules, rules that enabled us to predict, with a high degree of accuracy, where and when the next total eclipse of the Sun was going to occur. This made Velikovsky look sad, which is how we parted.

It must have occurred to me that the intensity of the fuss was peculiar. After all, the world is full of crazy notions, notions the public keep on buying, as Barnum's remark about a sucker being born every minute makes clear. I suppose I passed the situation off with the thought that Americans are highly emotional people, which, in some degree, is true. But is it sufficiently true to explain why Harlow Shapley, a most distinguished astronomer, was prepared to parade himself on the national stage over what was claimed to be nothing but claptrap? Or could it be that Velikovsky had revealed, admittedly in a form that was scientifically unacceptable, a situation that astronomers are under a cultural imperative to hide? Could it be that, somewhere in the shadows, there is a past history that it is inadmissible to discuss? This will be a topic for a later chapter.

Chapter 20

The Watershed

NOT LONG AFTER returning from the United States, I received a package from Chandrasekhar. It was a paper that had been submitted for publication in the *Astrophysical Journal*, of which Chandra was then the editor. The paper, which had been written by a young Canadian, A. G. W. Cameron, had been submitted from Chalk River, Ontario, near Ottawa and Montreal. The paper had been seen by two referees, both of whom had recommended its rejection. Before taking such a step, Chandra asked me for a third opinion, a procedure that might be considered unusual and that indicated to me that Chandra was uneasy about the content of the paper, or about alienating the Canadian astronomical establishment, or perhaps even both. After casting an eye over the two very hostile reports, I turned to the paper itself, finding it to contain a marvelous idea, one that blew some of my own ideas sky high.

The reason my paper on the work I had done at Caltech in 1953 ended with nickel was that I had decided nickel was the heaviest element that could be synthesized by charged-particle nuclear reactions alone. The synthesis of the heavier elements (those in the lower three-quarters of the chemist's periodic table) would need a supply of free neutrons inside stars as well as charged-particle reactions. I had found free neutrons to be generated in the processes of carbon-burning and oxygen-burning, and it was these that I intended to invoke in a second paper on the genesis of the elements (the paper on the heavier elements, which never appeared). What Cameron's paper did was to suggest a simpler source of free neutrons that I and everybody else had overlooked.

Bethe's process for converting hydrogen to helium had made use of trace quantities of carbon as a catalyst. During the catalytic reactions, the uncommon heavier isotope of carbon, carbon-13, was produced in low abundance. Should some of this carbon-13 survive to a higher temperature, when helium was synthesized by

$$^8Be + {}^4He \rightarrow {}^{12}C + \text{gamma ray}$$

and lost by

$$^{12}C + {}^4He \rightarrow {}^{16}O + \text{gamma ray},$$

there would also be

$$^{13}C + {}^4He \rightarrow {}^{16}O + \text{neutron}.$$

Nothing much could be made of Cameron's idea in this form, however, because the neutron yield would be too small. So what one had to do, in a manner of speaking, was to eat one's cake and have it too. Start by exhausting all the hydrogen. Let the star evolve to higher temperatures, bringing the first and second of the preceding reactions into play. Then introduce a relatively small quantity of hydrogen from out of a hat. The higher temperature would cause the hydrogen input to be quickly destroyed by

$$^{12}C + {}^1H \rightarrow {}^{13}N + \text{gamma ray},$$

after which nitrogen-13 would quickly decay to carbon-13 by emitting a positron and a neutrino pair. Then, the reaction $^{13}C + {}^{14}He \rightarrow {}^{16}O + \text{neutron}$ would again take place, providing a supply of neutrons to equal the hydrogen input. One could object, of course, to this sleight of hand, and the first two referees of Cameron's paper had done so. But Martin Schwarzschild and I had already seen how complicated things could become when a star with a dense core evolves towards a state with more than one source of nuclear energy, different sources operating in different parts of a star. So I felt it was impossible to be sure that carbon-13 would not be produced in appreciable quantity, either in this or in some other way.

Since, moreover, Cameron's nuclear physics was correct, I saw no reason not to recommend publication of his paper. Chandrasekhar accepted my recommendation, and the paper was subsequently published—to the chagrin, I suppose, of the first two referees.

From my first contacts with astronomers in 1939–1940, I have always maintained an antiestablishment position with respect to the refereeing of papers. Refereeing had its origin in the nineteenth century. It was brought in to ensure that authors did not make unacceptable personal remarks about each other, especially about dishonesty practiced in reporting observational and experimental results. Since dishonesty remains a perennial issue, with the near certainty that reported cases are only a minor fraction of the total, this early application of the refereeing system has not been an unqualified success, even though one might think it to have been otherwise generally desirable.

Editors of journals receive a trickle of papers written by crackpots, and, indeed, the occasional paper from a scientist of reasonable repute who happens to have a bee in his bonnet on some particular issue or another. It became convenient, in such cases, for editors to reject papers on the recommendations of anonymous referees, thereby saving themselves from personal confrontations with angry authors. It was also found that obscurely written papers could sometimes be improved by an author following a referee's comments on presentation. But these relatively minor issues do not seem to me to weigh heavily against the serious objections to the system that can reasonably be raised. Crackpot papers are easy to recognize, only a little trouble being required to remove them. And, since the more clearly a paper is written, the more attention it will receive, every author is under natural pressure to write as clearly as possible.

There appear to me two circumstances that justify the rejection of a paper. One is that it is inconsequential and dull, and the other is that it contains a serious demonstrable error. For a journal like the *Monthly Notices of the Royal Astronomical Society*, which publishes most of the material submitted to it (or, more accurately, which did so in my time at the RAS), I would hesitate to invoke the first of these criteria, since it is always possible for a flower to grow in unlikely ground. But, for a journal like *Nature*—which, for logistic reasons, can publish only a fraction of the material submitted to it—the criterion is valid. Few would question that papers based on demonstrable error should be rejected. The trouble is that most referees who claim demonstrable error do so without demonstrating it. Instead, only opinions are offered, as in the case of the first two referees of Cameron's paper. When I myself recommend rejection of a paper on this second criterion, I feel that it is incumbent on me to demonstrate that what I claim to be false is indeed false, the way one would do in supervising a mathematics student. Nor do I see why, in this role, I need the shelter of anonymity. Rarely when I have followed these precepts have I had complaints from an author. Indeed, I have probably received as many letters of thanks from authors for saving them from the embarrassment of having an error appear in print as I have letters of complaint. No author wants to make a fool of himself, and this is yet another safeguard that would be available in a less restrictive approach to the publishing of scientific work.

I became mentally polarized—which is to say, set in my ways—in the wake of the greatest upheaval there has ever been in science, not realizing that, in the upheaval, the establishments of yesteryear had been toppled into the dust in the process. Consequently, I grew up with the erroneous notion that the scientific establishment welcomes progress, which is the opposite of what is generally true. Progress is equivalent to revolution, with drastic turnovers then occurring in high places. Slender progress means that the sheep cannot be separated from the goats. Nothing happens to threaten existing establishments, as it did with the explosive development of quantum mechanics. When there is near-zero progress, slight steps can be misinterpreted (or misrepresented) as large steps, governments can be urged to throw immense sums of money into the air in the vain hope that something of value will be forthcoming, and, above all, establishments can perpetuate themselves: "That young man of Sir Bradbury Trundlehouse. He seems very sound. I think we should put him on the Arrange Everything Committee." At the present, cosmology is the darling of the scientific establishment, because (with the exception of an inexpensive theoretical step made by Alan Guth circa 1980) cosmology has achieved nothing of real importance in a quarter of a century. Indeed, one might say that cosmology has gone backward, a near perfect situation for the establishment, since great sums must be spent merely to hold the position.

It is against this more perceptive background that the refereeing of scientific papers must be viewed. It is the glory of science that no establishment could withstand the overwhelming impact of Heisenberg's discovery of quantum mechanics in 1925. Once precise numbers can be calculated, the dam wall breaks. But it is rare for the younger generation to make such an immense stride all in one move. Usually, many smaller steps are needed before the decisive step becomes possible, and it is the smaller steps, none of them decisive in itself, that the refereeing system is ideally designed to prevent. Refereeing is a device by which the establishment protects itself. This is why I could make no impression on the system in astronomy, even as president of the Royal Astronomical Society in the early 1970s, and why Hermann Bondi could make none in the late 1950s, despite his being secretary of the RAS. By creating a succession of tumultuous rows as a young member of the RAS Council ten years before Bondi, I may have done better than either of us was able to do in senior positions. I managed to make a fine old row over the rejection of a paper on early ideas about a solar wind by no less a scientist than Sydney Chapman. Chapman was a distinguished scientist of the older generation. The rejection of his paper showed me that such matters are not decided by reputation but by the system, and that when a nation accumulates too many formal procedures, it collapses. But the best story of that epoch was the following.

Ray Lyttleton was particularly irritated by the convention that it was bad form for an observational astronomer to be challenged on grounds of competence, whereas theoreticians were routinely challenged in ways that implied incompetence almost to the point of imbecility. He therefore raised the question of what the comparison really was between observation and theory, if one were to look back in the literature for more than a generation. Which side had really made the most serious blunders? Because he read German, giving him a wider range in the literature, Bondi offered to make the search. What he found was what Lyttleton had suspected, that the worst mistakes of the past had been in observation, not in theory. Thereupon, Bondi wrote a paper to this effect, submitting it for publication to the Royal Astronomical Society.

In those days, the Council of the RAS discussed referees' reports on papers in some detail, and it was the Council that always made the final decision on whether a paper should be published or not. Bondi's paper drew two hostile reports. But the referees, in their prejudice, had made the mistake of claiming that Bondi was in error over his cases—as, for instance, over Walter Adams's erroneous determination of the radius of Sirius B. Because I had already been over every case myself, I knew Bondi was not in error. Contention came easy to me in those days, and I was therefore happy to make a never-ending issue of the thing, which inevitably meant the council meeting in question would drag on until the scheduled time of the main meeting of the society, forcing the president and other officers to miss their tea. The president fidgeted more and more in his chair. His face became puce, and at last he snapped out: "Then will someone recommend the rejection of this paper? Irrespective of its contents." The trouble, of course, was that the RAS establishment did not wish to offend the astronomical establishments in other countries, which, it was thought, Bondi's paper would do. Actually, it is hardly ever possible to prevent a paper written by a competent professional from being published, and Bondi's was no exception. A year or two later, I asked Otto Struve if it had really caused serious offense. Otto grinned in his cross-eyed way and said: "The lucky thing about Bondi's paper was that he never discovered the worst observational mistakes."

Eventually, I came myself to be on the establishment's "Arrange Everything Committee." One day, I was sitting next to my friend Alan Cottrell. I noticed he had a sheet of paper with all the members' names on it. Whenever anyone spoke, he put a mark against that member's name, adding up the number of marks for each member at the end. I took to doing the same myself, with interesting results when young new members (like Sir Bradbury's protégé) appeared. Almost always, they were eager beavers, scoring many marks at each meeting. If they were critical of things, especially of what the chairman wanted to do, you knew they wouldn't be long for the system. If they were obsequious to the chairman, pushing themselves forward to do this or that minor job, you knew they would go far and would end up as one of the pests with which the establishment was liberally infested.

I had learned early, as in my spell on the Royal Astronomical Society Council, that, if at heart you were basically antiestablishment, it was quite useless to attempt to persuade a majority that was strongly pro-establishment. So I always saved what I had to say for the one or two items that were important to me. This actually wasn't bad policy. Lord Alexander Todd, the famous chemist, who was the only person I ever saw with a streak of antiestablishmentarianism in him who was able to beat the system, once confirmed that holding one's tongue was a pretty fair tactic: "Because," as he put it, "if you keep quiet, they all think you must be a pretty deep customer."

Willy Fowler had a sabbatical year due in 1953–1954, and he decided to take it in Cambridge. My memory is that Willy and his wife and two children settled down in Denis Wilkinson's house, Denis being also on sabbatical, but away from the Cavendish Laboratory. Willy bought a soft-topped Hillman 10—but a new one, not like Tommy Gold's specimen, which he had bought for a mere £50. Subsequently, Willy shipped the Hillman back to the United States, and for many years he could be seen in it with the top open on the freeways of Los Angeles, frequently wearing a porkpie hat.

Geoffrey and Margaret Burbidge had recently come to Cambridge, Geoffrey on an ICI (Imperial Chemical Industries) Research Fellowship. They had managed to rent apartments in Botolph Lane, hard by the rooms where the librarian of a certain college lived with a squawking parrot. The librarian, who went around Cambridge in shoes without socks, was much feared in the university library as a purloiner of books. The Burbidges, for their part, were deeply into abundance analyses of stellar spectra, with particular reference to stars of unusual atmospheric compositions. So strange were the results, in some cases, that they succeeded in interesting Fowler, the possibility being that strange abundances might be explained by nuclear processes.

From studies of flares on the Sun, it was known that protons and helium nuclei could be accelerated to energies comparable to those achieved in the Kellogg Radiation Laboratory at Caltech, and the question was what might ensue therefrom. Pretty soon, however, it became clear that neutron processes had to be important, the subject I had earmarked as a follow-up to my paper on the synthesis of the elements from carbon to nickel. I would, therefore, have been delighted to team up with Fowler and the Burbidges already at that stage, in 1953–1954, rather than in 1955–1956, as I actually did, but this was not to be. In 1953–1954, I had a full teaching load in Cambridge, together with the work on type I and type II stars with Martin Schwarzschild, as well as the ill-fated investigation into the synchrotron process in radio sources. There were just not enough hours in the day to take on anything further. Nevertheless, I was able to follow what went on. It was especially interesting when Fowler and the Burbidges got around to calculating the effect of adding neutrons slowly to neon, a supply of, say, ten neutrons to each initial neon nucleus. The result was abundances at odd values of the nucleon number A that were suggestively similar to known abundances. This was for a range of A from 20 at neon to 45 at scandium. Neutron-absorption cross sections were a problem at that time, a problem that inhibited the calculations from going to higher values of A. Published experimental cross sections at high values of A were nearly useless because they were given element by element—which is to say, all isotopes of an element were lumped together. To determine the content of a particular nucleus, both the proton number Z and the neutron number $A - Z$ were needed. The value of Z determined the element in question, and either A or $A - Z$ fixed the isotope. Neutron cross sections with respect to both A and Z were needed, and these did not become available until 1956. Some progress at high values of A could still have been made in the interim, but it is always intimidating to tackle a problem when one knows that essential data are missing.

In the summer of 1955, I had an invitation from Caltech to pay a second visit there in 1956, under the title of Addison Green Whiteway Professor. As I later learned, the story behind the invitation was that a move had been made from Mount Wilson to offer me an appointment. In those days, Mount Wilson worked closely with Caltech. There was a cabal of six or maybe eight who, by the convention of the day, had to agree on offers of appointments. It seemed I had attracted a black ball from someone in the cabal. A compromise was reached—that I would be offered the temporary professorship. Meanwhile, Ira Bowen, who was the director at Mount Wilson, made it plain that the black ball had not come from his side by making me an unpaid staff member at Mount Wilson, with all the usual privileges that went with such an appointment.

I arrived at Caltech to find Geoffrey and Margaret Burbidge already there. Margaret had a temporary research appointment at the Kellogg Laboratory, and Geoffrey had been given a Carnegie Fellowship. We teamed up immediately and, with Willy Fowler, set about the work that was to lead to the review of the entire subject of nucleosynthesis of elements in stars that appeared in 1957 in *Reviews of Modern Physics*, a paper that became generally known as B²FH (for Burbidge, Burbidge, Fowler, and Hoyle). As regards neutron processes, which was then my main interest, it was clear from the start that there had to be two distinct processes, a slow (s) addition of neutrons to some initial seed nucleus, which we took to be iron, and a rapid (r) addition to similar seed nuclei. This could be visualized in the following way. Each individual nucleus could be represented by a point in a diagram, with A (the nucleon number) measured on the horizontal axis and Z (the proton number) measured on the vertical axis. The isotopes of each element would then be represented by several points at the same height (the same value of Z) in the diagram. From the principles of nuclear physics, a curved line could be drawn known as the stability line. Thus, for each value of Z, there was a theoretical value of A giving the nucleus that was most stable.

The stability line was, however, a generality that ignored precise details, which led to many nuclei being stable even though they were not on the stability line itself. But, generally speaking, the further from the stability line, the less stable, with the consequence that nuclei near the line would be permanently stable, whereas, with increasing distance, nuclei would become unstable in shorter and shorter times, until those rather far away would decay in only a fraction of a second. In total, we had about 300 stable points on our diagram. For elements heavier than nickel—the ones in which I was most interested—those to the left of the stability line had usually been produced by the s-process, and those to the right, usually by the r-process. There were a few intermediate cases, and for these we made a half-and-half division between the two processes.

For neutron sources, in the case of the s-process, we had Cameron's reaction, $^{13}C + {}^4He \rightarrow {}^{16}O +$ neutron, operating on a time scale of the general order of 10 million years. Starting from an iron nucleus, and assuming that several neutrons become available for addition in each million years, it was possible to trace how the nucleus would evolve on the Z,A diagram. With fascination, we found it would move along a track that took it through most of the points to the left of the stability line. Moreover, evolution through the isotopes of some elements was considerably slower than through others, and we found that, where evolution was slow, the cosmic abundances of the elements in question were high. This showed that the giant stars in which the s-process occurred had limited lifetimes. Eventually, they either dissipated through mass loss from their extended atmosphere or they blew apart. Wherever in our diagram the evolutionary track had happened to reach at the mass loss or explosion stage, the corresponding nuclei were expelled out into space, and they had been a part of the cosmic heritage ever since.

One day, with the other three away, for some reason, I was sitting in a windowless room in the Kellogg Laboratory, idling away an odd hour until it was time for lunch, when I had the fancy to start multiplying neutron-absorption cross sections by cosmic abundances. This was for *s*-process nuclei in turn. These cross sections had only recently been published by the Atomic Energy Commission of the United States, at the so-called Atoms for Peace conference in Geneva. The cross sections, which had previously been secret, were indeed the data that had been missing in 1953–1954. As I said before, we had seen generally that high cross sections went with low abundances. But now I found that it was nearly exactly so. My products were coming out almost exactly the same. In these days of digital computers, multiplying two columns of awkward numbers would be trivial, but working painstakingly by hand, in those days, through hundreds of numbers, it had taken a while to emerge. As more and more products came out the same, I naturally became increasingly excited, for it proved decisively that we were well set on the right road—at any rate, so far as the *s*-process was concerned.

There were a few discrepancies, however. After a search through the literature, Fowler decided they were all cases in which the experimental results might be wrong. Because it did indeed eventually turn out that our products were right and the experiments were wrong in these cases, one is left with the odd reflection that nuclear security had been pointless—at least as far as neutron-absorption cross sections were concerned. All anybody had to do to obtain even better results than the U.S. Atomic Energy Commission data was to take the reciprocals of known abundances. This was provided, of course, that one had the wit to separate *s*-nuclei from *r*-nuclei using the stability line. It has been experiences like this that have made me generally dubious of the value of security services. Too many times there are the means of arriving at supposedly secret information in ways security agencies do not understand, and such advantages as may arise from time to time are offset by the stultification that security produces within itself. Information given out sparingly on a "need to know" basis can all too often degenerate into self-inflicted disinformation, simply because too few people are able to cast a critical eye over it.

Throughout the work, we were greatly aided by accurate estimates of the cosmic abundances of the elements, which had very recently been published, by Hans Suess and Harold Urey at the University of California, San Diego. Willy Fowler pointed in delight at the three double abundance peaks that Suess and Urey had obtained in the regions of the elements strontium ($A \approx 90$), xenon and barium ($A \approx 135$), and platinum ($A \approx 190$). The three cases arose from the so-called neutron magic numbers at $A - Z = 50$, 82, and 126, the hump at lower A in each peak coming from the *r*-process, the hump at higher A, from the *s*-process. Every day brought new perceptions like this.

We associated the *r*-process with neutron release in supernovae, which set a time scale of the order of a few seconds, permitting another rather precise calculation to be done. If one starts with iron as a seed exposed to a flood of neutrons on such a short time scale, what is the track of the nucleus in the z,A diagram, and what are the relative times spent in each short section of the track? Fowler had decided to review all aspects of the conversion of hydrogen to helium—a time-consuming project, but one necessary for what we were now envisaging as an encyclopaedic article. Geoffrey and Margaret Burbidge were compiling actual evidence of nucleosynthesis from analyses of stellar spectra. So it fell to me to attempt the calculation for the r-process, which I wanted to do anyway. Beta-decay rates were an essential ingredient of the calculation—not known rates, but decay far from the stability line. Both Fowler and I had worked in this field in the late 1930s, and, between us, we agreed on suitable estimates. The results turned out well. When times along the r-track in the Z,A diagram were converted into abundances in a manner similar to the s-track, we had good estimates for all our assignments to the r-track. Moreover, we had estimates for the production of the transuranium elements, which, with some updating, were to serve Fowler and me as an essential datum in devising a nuclear clock for determining the age of the galaxy, a project that had to wait, however, until 1959–1960 for its completion.

The association of the r-process with supernovae appeared to have a dramatic confirmation. Walter Baade had classified supernovae into two types. Type I was characterized by an exponentially declining luminosity, like a radioactive decay of a substance with a half-life close to 55 days. Baade's measurements for a supernova in the galaxy IC 4182 were so strictly like radioactive decay that nobody was in any doubt (except, possibly, Fritz Zwicky) that the energy source supplying the luminosity must indeed be a radioactive substance with a half-life of 55 days. Judge, then, our delight when Geoffrey Burbidge noticed that, in the debris from the U.S. H-bomb test at Bikini Atoll, an isotope of the transuranic element californium (californium-254) had been found to have just the required half-life. Moreover, the decay was through fission, which yields far more energy than alpha emission or beta decay. This appeared to be a twofold triumph. First, we had the energy source for type I supernovae. Second, to produce californium-254, a powerful rapid source of neutrons was needed. Therefore, we argued with great confidence, the identification of the r-process with supernovae was correct.

But this line of reasoning, which looked so certain at the time, turned out to be wrong. We had a hint that it might be so, but we refused to take it. Try as we might, we could not see how the release of radioactive energy in an expanding supernova envelope managed to transfer itself so immediately into the luminous emission. Nor does it, according to the modern view. The cause of the 55-day decline is not radioactive energy as we took it to be but stored radiant energy in the envelope itself. Despite the failure of the argument for the existence of californium-254, however, astronomers have continued to believe in supernovae as the site of the *r*-process (whether rightly or not remains uncertain).

As well as calculating the abundances to be expected from the *r*-process, I recalculated abundances in the iron peak, just the problem from which I had started in 1945. And just as it was then, I found irritating discrepancies in an otherwise satisfactory result. The cause was another error that really was quite a bit worse than the californium-254 error. At temperatures of about 2 billion kelvin, oxygen-burning produced such common elements as magnesium, silicon, and sulfur. With still rising temperature, even the most stable of these elements, silicon, became susceptible to break-up by gamma-ray absorption. Because silicon was slightly more difficult to break up than magnesium and all other lighter elements, I argued that, once silicon succumbed to gamma-ray absorption, there would be a general approach to statistical equilibrium. If I had troubled to calculate the balance of ^{28}Si + gamma ray $\leftrightarrows$ ^{24}Mg + ^{4}He, I would have seen that, because of the rapidity of the return reaction ($\leftarrow$), the silicon could survive many gamma-ray disintegrations. Mostly when silicon-28 was disrupted, the addition of an alpha particle turned the magnesium-24 back into silicon-28. The extra persistence that this confers on the silicon permits it to exist to higher temperatures than I thought possible—indeed, to temperatures such that chains like ^{28}Si + ^{4}He $\rightarrow$ ^{32}S + gamma ray, ^{32}S + ^{4}He $\rightarrow$ ^{36}Ar + gamma ray, and ^{36}Ar + ^{4}He $\rightarrow$ ^{40}Ca + gamma ray become important. In effect, the iron peak is built as before, but by reactions that are out of statistical balance. The peak is still centered at $A = 56$, but the relevant nucleus is that of nickel-56, not iron-56. This might seem a mere detail, but it is technically important if one is to get rid of the irritating discrepancies I always found for the statistical calculation.

I must shoulder the blame here, and doubly so, for the German physicist Hans Jensen from Heidelberg, who was then staying for a term at Caltech, kept telling me that the crucial nucleus had to be that of nickel-56, because, with $Z = 28$, $A - Z = 28$—it was "magic," both in neutrons and in protons. This was Jensen's instinct, and he was absolutely right.

It may seem strange, but I have few memories of daily minutiae over the six months from January to June of 1956. There was a hot afternoon on the beach at San Juan Capistrano, a laboratory outing at a park in the San Gabriel Mountains somewhere behind Mount Wilson. Yet I remember in startling detail a week in June spent in the high desert of northern

Arizona and southern Utah with Bill Miller, who ran the photo lab at Mount Wilson. I remember an apparently ideal site at a place called Elk Ridge. We slept in the open on collapsible camp beds, to be wakened at dawn by streams of trucks throwing up huge clouds of dust. This was the time of the uranium bubble in the United States, when fortunes were made by buying and selling deposits of low-grade ore that subsequently collapsed into near worthlessness. There was the spectacular penetration by jeep into the almost inaccessible (in those days) land of standing rocks. There was the ferry across the Colorado River at Hite, Utah, where the temperature fell to only 100°F, but where the high humidity made conditions more oppressive than 130°F in the generally dry heat of the desert. It was 140°F at Dug Out Ranch, where there were irrigation-supplied fields of alfalfa, and where the most important thing in the world was the sack of fresh grapefruit that Bill had brought along. There was the camp by the Totem Pole in Monument Valley, Arizona. We did our exploring mostly in the hours immediately after dawn. By noon, we had sought out a cliff that would give shelter from the Sun in an hour or two. The cliff would also be important to our comfort later on, because it would continue to radiate heat during the cold of the nights. There was a long lone traverse of Navajo Mountain (in southern Utah near the Arizona border), where, because of heavy tree cover, I had all sorts of difficulties in finding my way in the descent. There still remains a kaleidoscope of memory of that desert trip, whereas the day-by-day details of life on the campus at Caltech have by now flown far away.

I returned to Ivy Lodge in Great Abington at the end of June 1956, leaving my colleagues with a huge welter of notes and calculations to be disciplined into a publishable manuscript. Since I had consumed all of the sabbatical leave due to me, there was no hope of my returning to Pasadena to shoulder my share in the writing of B^2FH, unless I could make a special arrangement with the university. I had one lecture course scheduled for the autumn term, which there was still time to swap with someone for an additional course in the Lent term of 1957, provided the university would grant me a leave of absence. I visited the secretary of the faculties, who agreed to put the matter to the General Board on the basis of my being granted a leave without pay.

Meanwhile, St. John's College had at last made some three acres of land available for building in Cambridge. It was an infill between existing houses in Clarkson Road, a cul-de-sac that the college had decided to name Clarkson Close. There were to be four building plots, the understanding being that fellows of the college had first preference over others from outside St. John's, with preference among fellows themselves going according to seniority. There was to be a £500 down payment as key money, and a ground rent on a 100-year leasehold. Most fellows felt the terms too stiff, and, in the event, only two of us applied. As I happened to be the more senior, I got the plot we wanted. Viewed from Clarkson Road, it was the far plot on the right, with the Trinity playing fields straight ahead and a bird sanctuary on the right.

So there were discussions and contracts to be signed on that score, after which the entire family was off with a car and caravan. Normally, we went in the summer to Cornwall, but, after hearing a panegyric from the historian Michael Oakeshott (who had bought my old papier-mâché DKW car) on the beach at Marloes in Pembrokeshire, we went there instead, obtaining an excellent site for our caravan on a farm immediately overlooking the path down to the beach. Once again, I remember far more about the five weeks we spent in western Wales—the sand, the sea, the cliffs, and the birds—than I do of the six months at Caltech.

In late September 1956, I returned to Caltech, now accompanied by wife and daughter. We stayed through Christmas until after the Rose Parade in Pasadena, and then we returned to Cambridge to a house that would shortly begin construction and to a hard term of heavy teaching, leaving behind my written contributions to B^{2}FH.

In many ways, the most spectacular international conference I have attended took place in mid-May 1957 at the Vatican. My memory is of a small group accommodated together in a hotel about three miles from the Vatican, a group that had collectively accumulated so great a wine bill that, in the end, we welcomed with relief the news that Pope Pius XII had decided to cover it. Aside from an even smaller group of the Vatican's own astronomers and guests, there were, I think, seventeen of us: Allan Sandage and Walter Baade from Mount Wilson, Willy Fowler from Caltech, George Herbig from the Lick Observatory, Bill Morgan from Yerkes, Martin Schwarzschild and Lyman Spitzer from Princeton, and Ed Salpeter from Cornell to complete the American contingent—except for the Greek-born J. Jason Nassau, then from Cleveland, Ohio, who said he was there to make certain we took matters seriously. The European contingent comprised Bertil Lindblad from Sweden, Bengt Strömgren from Denmark, Otto Heckmann from West Germany, Jan Oort and Adriaan Blaauw from Holland, D. Chalonge from France, Georges Lemaître from Belgium, David Thackeray from South Africa, and me from Britain. The conference was on population I and population II stars, which meant that it was concerned with stellar evolution, displaying a leaning towards nuclear reactions in stars, as well as towards the distribution of stars in our galaxy.

By now, my wife and I had graduated from our old prewar DKW and had acquired a tough, largish device called a Humber Hawk. Without it, we would not have been able to tow a caravan to Marloes the previous summer. We decided to drive to Rome, thinking (correctly, as it turned out) that the Humber Hawk would be a match for the Fiat 500 cc, "the Mickey Mouse," except possibly in the press of Roman traffic, which has eased not a jot since classical times. But there was no traffic to speak of in the center of Chartres. After an hour in the cathedral there, with the car parked immediately outside, we had tea and cakes in a nearby café. The café is, I think, still there, but you can't park anywhere near it nowadays. Our way through the Massif Central took us to a small place called Les Halles, where we found a marvelous little hotel. The Italian Riviera went, in those days, through a seemingly endless succession of small coastal towns, which was made acceptable by the great masses of flowers that were then in bloom. We took the coastal route through Pisa, intending to visit Firenze on the return. Of course, we padded up Galileo's famous leaning tower. But there was a phenomenon in the nearby basilica that was more interesting. Sound damping in the big dome was so minimal that a long note sung by the human voice would persist while a second note was sung, and so on, permitting someone with a true voice—which the guide had, this being a requirement for his job—to build up an entire chord. With the guide, it was the common chord, but all sorts of more ambitious harmonic structures could have been attempted. In sum, it sounded as if a whole choir had given tongue, making me wonder what happened when a whole choir really did give tongue in such a place.

Since the invitation had extended to a lady companion—wives may have been specified as the only category of lady, but of this I can't be sure, marriage certificates not having been required—a special ladies' program had been arranged, which took them to the very best that Rome had to offer. Thereafter, in our spare moments, my wife insisted on revisiting the places and museums she felt I would enjoy most. The Palatine Hill in the Forum was where we most often ended our personal tours. We were also taken on highly special tours within the Vatican itself. I remember looking down in some awe at the signature of Frederick Barbarossa, and I remember Walter Baade muttering in my ear, "The art and antiques market is going to collapse, if they ever decide to release even a tenth of all this."

Two summaries of the conference were given, one by Jan Oort, from an astronomical point of view, and I was entrusted to give the other, from a physical point of view. This led Father O'Connell, the Director of the Vatican Observatory, to ask Oort, Baade, and me to prepare a list of suggestions for future work, suggestions that were to form the last section of the book on the conference. Rather than stay in Rome for this, the three of us decided to meet at a hotel in Amalfi during the week following the conference. Hearing that my wife and I were driving south of Naples, Georges Lemaître asked to come along.

I think the autostrada from Rome to Naples was the only special road in Italy in those days. From Naples, we made our way past Vesuvius to Sorrento, where Willy Fowler and his wife had gone. Willy was in the highest spirits because, on the rapido from Rome to Naples, he had somehow got himself into the driver's cab and had even more improbably managed to take over the controls for a while. What intrigued Willy about that hotel in Sorrento was a boy of ten who functioned as a computer. His job was to know where all the guests were at all times. Whoever you wanted to see, you just asked the boy and he would tell you exactly where to go and look. From Sorrento, we went across the peninsula to Amalfi, where, in a couple of days, Baade, Oort, and I finished up the conference recommendations, greatly assisted by Blaauw.

On the journey north, Lemaître had a seemingly endless list of special places he wanted us to visit. At first, we thought Georges's itinerary, which he seemed determined to follow, was perhaps not quite where we wanted to go. But, in the event, he took us to so many places we would never have found for ourselves that this too proved to be a joy, like the Sunday morning band in uniform playing Verdi's march from *Aida* in the square of a small hill town in the Appenines. Only in one respect did Georges's presence raise an issue, and this was over lunch. I always wanted a light lunch, so that I could continue driving in the afternoon, whereas Georges wanted a heavy lunch with a bottle of wine, so that he could sleep in the afternoon. We compromised by allowing Georges to sleep in the back of the car, which, unfortunately, led to his awakening almost always with a shocking headache.

Lemaître was a round, solid man, full of jokes and laughter. There was only one moment when Georges and I came at all close to a cross word, and it wasn't over either cosmology or religion. True, as lunchtimes extended beyond the two-hour mark, I used to become fidgety; but, because I knew Georges would shortly be suffering from a headache, I could never bring myself to become cross over lunchtime. No, it was at dinner, and it was because of a fish, and it happened on a Friday at an old-fashioned hostelry in the town of Landeck in the Austrian Alps. At dinner, Georges ordered fish and I ordered steak. When the fish came, it was enormous, while my steak was naturally much smaller. So, seeking to keep the atmosphere genial, I said, "Now at last I see, Georges, why you are Catholic." Lemaître responded with uncharacteristic sharpness. The sharpness, I immediately realized, had nothing to do with my religious skepticism. It had to do with his wanting my steak. Being aware, from my remark, that I thought the fish looked pretty good, it had to do with

Georges's inability to offer a swap, the day of the week being Friday, as I said. So he had to plow his way gloomily through the enormous fish, an unhappy situation, slightly ameliorated by an equally large bottle of wine.

We left Georges at Brüningen in Switzerland, at a hotel where he had stayed before and of which he was fond. Georges's presence in the car had prevented us from getting around to a pressing subject that now had to be faced. By attending the Vatican conference, I had pretty well broken the bounds of my job in Cambridge. It will be recalled that, to do my share in the writing of B²FH, I had been obliged to ask the university for a term's leave of absence without pay, and, to attend the Vatican conference, I had again been obliged to repeat what was already an imposition on the university system—and, of course, on the patience of St. John's College. But as the only representative from the United Kingdom, at a conference that included a good half of the world's top astronomical brass, it had been impossible for me to refuse to attend. With the evident possibility that the situation was only too likely to become still more contradictory, I felt that some action had to be taken—the question, of course, being *what* action?

The simplest solution would have been to follow Tommy Gold, who had left the Greenwich Observatory in the previous year to take up a professorship at Harvard. I don't mean specifically at Harvard, although Tommy was forever suggesting that I should consider the possibility, but at some American university or observatory. There were several obvious leads that could be followed in those directions, but, for one thing, my wife did not wish to emigrate. Unlike the wives of many academics, who quickly become enamored of the higher American standard of living, she had not. Because we hadn't changed our lifestyle at all, following my slice of luck with the BBC in 1950, we were not under any financial imperative to improve our lot. The question, therefore, resolved itself into what was the least drastic step to resolve the contradiction that had been unavoidably set up.

My own position was indecisive, and, if I were taken back magically to that moment in 1957, I would still be indecisive, because, where research is concerned, there is no ideal way to associate oneself with any university or independent research institution. For more than a decade, my job in Cambridge had come close to the ideal, so that it would take something unusual to be an improvement. Even to stay the same would not be all that easy, because, by 1957, there were ominous signs that business in the university would infringe more and more on the sanctity of the vacation intervals between teaching terms, a trend that was indeed to accelerate in the 1960s.

The other sixteen who had attended the Vatican conference were able to do so as a matter of course. In effect, all were working in astronomy full-time. A full-time post in astronomy, like the others enjoyed, therefore seemed the easy solution, and, if a congenial possibility had been offered, I would doubtless have taken it. But I would have taken it with certain trepidations, for I have always had a near-horror of being expected to produce worthwhile research. Some years later, I had what objectively seemed a fantastic offer. In 1965 money, it was an offer of $100,000 per annum, to cover salary and research expenses, to be decided entirely by me. Tommy Gold always maintained that one dollar without strings, such as this, was worth eight dollars with strings from a government agency. On that reckoning, I was offered, on a personal basis, the equivalent of a present-day agency grant of several millions of dollars per year—and that ain't hay, as Olin Eggen (who had become chief assistant at the Royal Greenwich Observatory) was fond of saying.

It would have meant a fine office, but I much prefer to work at home. It would have meant a superefficient secretary, but you have to work quite hard to keep an efficient secretary busy, lest boredom becomes unhappiness. It would have meant a fine house, but would this give us any more pleasure than what we were deriving from planning the much smaller house in Cambridge? It would have meant driving around in a newish Cadillac, but would this give us more pleasure than the old DKW had done, or even the £5 Singer? And standing over me at all hours of the day (and of the night, likely enough) would be the expectation of those around me that I produce world-shaking research. So I had the sense to refuse the offer. Nothing that is much good can, I believe, be produced in the face of great expectations. I doubt that even Mozart could have overcome this rule. It was indeed just because he was regarded as a funny little oddball that Mozart was able to rise so consistently to the greatest heights.

My observations of organizations devoted exclusively to research are not so bright either. The pressure of expectations affects such organizations in diverse ways. As my friend Don Clayton once observed, most competent scientists have the luck to hit a profitable line at some stage of their careers. Having attracted attention when the roulette wheel of scientific discovery stops providentially at one's own number, and having accepted the world's commendations, the restful thing for a person engaged in full-time research to do is then to stick with that line. It was the same with some in the university. Having arrived at a satisfactory course of lectures, they would remain content to give just the same course year after year, for ten years, for twenty, for however long it takes to reach retiring age. When a research organization is of this kind, it is characterized by each of its members being immensely competent in one particular specialty, but a specialty that is mostly past its day, a specialty that, in many cases, becomes almost outré in its old-fashionedness. Such organizations are easily recognized, for they have a clear but undefinable smell of intellectual mustiness about them.

The other brand of organization is totally different. It is characterized by the loud hum of an impressive intellectual dynamo. Everyone there is working on the very latest thing in every field covered by the organization's activities, with a brilliant air of immense competence apparent on every hand. This fervent state is achieved by the determination of all organization members to jump, at the earliest opportunity within their purview, onto every bandwagon that rolls. The entire organization is at the frontiers of knowledge, wherever the fighting is heavy. The problem, in such places, is that everybody is so avid to jump onto someone else's bandwagon that nobody in the group ever makes a primary discovery. The entire organization is secondary and largely irrelevant, although such places habitually impress government funding agencies and, consequently, flourish like green bay trees.

Our problem therefore unsolved, my wife and I returned home to Ivy Lodge in Great Abington and to the still unfinished house in Clarkson Road. The completion of the house was likely to be delayed by two months until late September or early October. We had given notice on Ivy Lodge with a quit date at the end of June, so, for a while, we had to live in our caravan, putting our furniture into storage. We found that we had a marvelously cheap storage space in our newly built garage, but, since quite a few of our valued possessions disappeared somewhere in the transition, it didn't turn out to be so cheap in the long run. Because we had to live in the caravan anyway, as soon as school holidays started for our children in July, we hitched up to the Humber Hawk, driving north up the ancient A1 road to Scotch Corner, to Penrith, to Dumfries, over the Stranraer-to-Larne ferry, through Northern Ireland, through Eire to Kerry and Dingle, and over a low pass to Dunquin, where, for six weeks of glorious weather, we parked down a narrow lane winding between stone walls on a site hard by Paddy Browne's cottage, sited some 400 feet above the sea. When the real Atlantic storms raged, Paddy used to complain of the wave tops coming down through a leak in his roof, in proof of which he claimed that the water tasted salty. Maybe.

We swam on beaches where seals came close in to watch us, and, after the evening meal, with the children in bed, we visited Paddy's study, where he regaled us with tales of Ireland long gone by as well as with Power's whiskey. Paddy was in excellent form that summer, for he had recently finished translating Homer's *Iliad* from the Greek into Gaelic. At other times, he would read from Erwin Schrödinger's poetry, always with unsuppressed gales of laughter. This was just the sort of thing that is generally missing from research institutions, especially from those who work in a perpetual frenzy on other people's bandwagons. We hired a fishing vessel and visited distant islands, thinking, all the while, that somewhere out there to the west lay *Tír na nÓg*, the legendary land of youth.

We moved, in September, into the new house, and we settled down, as the new academic year began, to enjoy a quiet existence, except now we could hear the roar of the Saturday crowd from the university rugby ground in Grange Road. In December, however, I was away to Caltech again, returning in January to be told by my wife that, in my absence, a letter had come from Lord Adrian, then the vice-chancellor of the university. The letter was to ask if I would be interested in the Plumian Chair. This being the academic year in which Harold Jeffreys retired, the university was moving to appoint a successor to take up the chair in October 1958. My wife had replied for me, saying that I was away and unavailable but that she thought I would be interested.

Although the move from Great Abington to Clarkson Road had been an improvement in convenience, we both felt the loss of the sense of security one feels as a result of living for a decade or more in a village. Nostalgically, we were given to returning to the section of the ancient Roman road behind Great Abington, the section of it from which I had seen the naked-eye sunspot ten years earlier. On a Saturday morning immediately after the beginning of the Lent term, we went out there with sandwiches for lunch. It was a special day, with snow, blue sky, and bright sunshine. We drove back home for tea. Shortly afterward, there was a ring at the door. It was a university messenger, the kind of chap who would run through the streets with *non placet* leaflets. He handed me a small letter. It was from Adrian, with the statement that I had been elected to the Plumian Chair. Would I accept? Vividly conscious that I was crossing a watershed, where everything would now flow to an unfamiliar ocean, I replied that I would. As the messenger left, walked down the path, and closed the garden gate, I had the inevitable sadness one feels when another chapter of one's life has passed away into memory.

PART THREE

Home Is Where the Wind Blows
1959–

Chapter 21

A Vintage Year

ELECTION TO THE Plumian Chair in Cambridge took me, in one bound, to the top of the British astronomical establishment. It was with that event that the most curious period of my life began. All sorts of humorous stories remain to be told about what happened in the ensuing fourteen years. But first there was an interim period of nearly a year that would elapse before I officially took up the appointment, on 1 October 1958.

The Solvay Conference for 1958 was held in the late spring, at the opening of the Brussels World Fair. The pavilions of the nations told you a lot about what was going to happen in the second half of the twentieth century. I left it to the American attendees at the conference to complain about the U.S. pavilion, concentrating my attention on the British, which was decidedly peculiar. To start with, the British designers seemed to have their time axis the wrong way round, pointing backwards in time instead of forwards, the British pavilion being shaped like a medieval tent. On the other hand, a lesson had been learned from the London Exhibition of 1951, held at the century jubilee of the Great Exhibition of 1851. It had been found that the stands that most interested the public were those in which something moved. So everything in the British pavilion moved. It didn't matter how—up-down, left-right, backwards-forwards, in-out, underhand-overhand—so long as it moved. Because of this, the designers got away with their skins, the pavilion being generally judged for merit to be among the first half dozen.

The Norwegians had little money for their pavilion, but to help out, they had a good idea. On a patio beside their exhibits, they provided tea in huge pots. Since everybody sooner or later was going to become parched and weary of foot-slogging, the Norwegians ensured that most of those who visited the fair became aware of their presence. Likewise, the Hungarians sold an immense quantity of goulash.

The Soviet pavilion was something to conjure with. It was a colossal rectangular box, easily dwarfing the American pavilion, which was why the American attendees at the Solvay Conference criticized it. Actually, I thought the American design imaginative, light, and airy, in favorable contrast with the Soviet box. It was the contents of the American pavilion that left something to be desired. The Soviet contents consisted of space gear, tractors, and coal-cutting machines, with vast wall paintings of happy workers waving great banners as they marched in a long line, a so-called crocodile, through glowing fields of corn. My sociological knowledge was primitive in those days—otherwise, I would have seen from those wall paintings that, sooner or later, the Soviet Union was destined to collapse, a tragic husk of a nation. Much the best pavilion in all respects was the Czech, from which, it was clear, the Czech people would really be on the march, once they managed to rid themselves of the political millstone presently hung around their necks. Neither the Germans nor the Japanese were represented in those days.

Wolfgang Pauli was the president of the Solvay Conference. One day he asked my wife and me to have lunch with him and his wife. "Ah ha!" he cackled, "I have just read your novel *The Black Cloud*. I thought it much better than your astronomical work." Pauli told me he had studied *The Black Cloud* in some detail, together with the psychologist Jung, and, indeed, that Jung had written a critical essay on it. I didn't have the temerity to explain that I thought I was only writing a story. But I had an intelligent life form in the story that didn't think in words, a form that had to learn words before it could communicate with man. Pauli knew all about Schrödinger's cat, about arguments over the origins of mathematics, while Jung knew about human emotions. So it was evidently the problem of what lies behind words that had been occupying them.

Frankly, I was less interested in gathering plaudits over *The Black Cloud* than I was in finding out if a story from the 1920s told about Pauli was true. Those were the days in which the young German was supposed to know his place among elderly bearded senior German professors, when even lecturers knew their places precisely. At important colloquia, the professors would sit on the front row and the younger people behind, first lecturers, then the assistants, and finally the students. Except for Pauli. Pauli sat on the front row, in the middle, dressed, quite likely, in Tyrolean leather shorts. So much is known and well documented. The occasion in question was a seminar given by Einstein, at the end of which there was a hushed silence among the bearded professors, unsure of which of them should begin the discussion, each anxious to get the order of precedence right. It was then that Pauli half turned to those behind him and began: "What Professor Einstein has just said is not really as stupid as it may have sounded."

This was what, in 1958, I wanted to ask Pauli about. He began to recall the occasion, and then he collapsed into helpless laughter, rolling in his chair like the ball in Galileo's famous bowl experiment. So I never quite had it from Pauli personally that the story was true, but those who knew him well assure me that it was.

Nick Kemmer told another good story about Pauli, also from the 1920s. Paul Ehrenfest, who was a close friend of Einstein, had sought to put down the young Pauli in print, which had led to acrimonious interchanges at a distance. Eventually, they came together at some conference, when, to the surprise of other attendees, they seemed to get on well together. Towards the end of the conference, in front of attendees, Ehrenfest, to gild the lily, said to Pauli: "Well, Pauli, I used to think you something of a bounder. But now I see I was quite wrong. Although I used to think highly of your work, of course." To this, Pauli cackled in a flash at the limit of his vocal range: "Ah ha! With me, it was just the other way around."

After the Solvay Conference of 1958, I did not see Pauli again. At the next Solvay Conference I attended, the president would be Robert Oppenheimer, and the four wines Pauli insisted on at lunch would be cut down to a more abstemious two. My abiding memories of Pauli are of his helpless laughter as the youthful remark about Einstein came back to him, and his rolling back and forth, a nearly perfect sphere, on the squeaking board in Room A of the Arts School in Cambridge.

The span of our lives is surprisingly great, provided you take account of contacts with old people you knew when you were young and with young people you know when you are old. Reckoning this in my case, the span is about 200 years, say 7 generations, half the time back to Shakespeare, a little less than a quarter back to William the Conqueror, a tenth back to Roman Britain, and a fourteenth back to Homer and the heroes of ancient Greece. But there were 10,000 generations in prehistory—immense indeed, compared with our personal experience. It should be no surprise, therefore, that the way we feel about what is important in life should have been molded far more by the long-rolling generations of prehistory than they have been by the social churning of the past 2000 years. Any politician who understands this will go far, and any social system that incorporates it will be successful. The essential point—the overriding point— is that the number of people with whom we need to interact in our daily lives should not exceed about a hundred, and preferably, on any enterprise of difficulty, not more than twenty-five. This is because twenty-five was the typical size of the hunting parties of prehistory. It is the scale of the medieval village, the scale of the modern cabinet in government, of the

Politburo, the scale of an army command, the scale of the players in a football game, the scale of the command team in Houston during the NASA landings on the Moon, the scale of committees, the scale of a company of actors, the scale of an orchestra, the scale of a family and relatives, and the scale of a dynasty. More or less everything that lies within it will be successful, and more or less everything that lies outside it will not.

Civilization is a device to maintain this principle in appearance while not maintaining it in fact. This is done by the process of subdivision. A century in the Roman army consisted of a body of 100 men, the officer in charge being the centurion. Take 50 centurions together, and you have a legion with its own commander. Take ten or more legions, and you have the Roman army, in which no man had to be in personal contact with many more than 100 of his fellows. It was this idea of using subdivisions to maintain the necessary scale in the daily lives of every man that was at the foundation of Roman power, a power that lasted successfully for more than half a millennium. So strong does the century idea still remain today that the batsman in cricket who makes 100 runs stands there acknowledging the applause and the congratulations of his fellows, whereas he who is out on 99 (essentially just as good, from the point of view of the game itself) walks off in comparative silence, feeling he has failed at the last gasp.

The same principle of subdivision applied within Cambridge University. The university staff of perhaps 800 was divided into faculties in such a way that each faculty had a scale that fits our rule, while the General Board, which controlled the university, was like a meeting of centurions, and it too fitted the rule of scale. When subdivision is used in a society, the rule of scale is always applied at the top of the pyramid, whether in a dynasty or in modern governments. What really matters, however, is the extent to which the right scale applies all the way down the social pyramid. If it goes all the way down, as it did in the Roman army and as it does in the modern British army, the outcome will be successful. But, if the rule of scale fades out some way before the bottom of the pyramid is reached, the society will not be permitted to last for very long. If an army fails to satisfy the rule at the bottom, it will be a rabble and will soon be defeated. If the rule fails in an economy, with the bottom of the pyramid described, in Marxian terms, as "the masses," it will also collapse ignominiously. Karl Marx failed in his economic analysis more drastically than other economic analysts of the nineteenth century because of his insistent concentration on the top of the pyramid, which was to be the full 180 degrees away from the correct position. Lenin was another top-of-the-pyramid man, with the conviction that, if the top was right, the bottom could be kicked around in any manner he pleased. This was the beginning of the route to disaster for the Soviet Union. This was why the blood congealed at the Brussels Fair of 1958, when one saw the endless crocodile of workers parading under banners through the cornfields of Mother Russia. Instinctively, one knew it was all wrong, but, at the time, I didn't realize exactly where. It was just the same story when the International Astronomical Union held its meeting for 1958 in Moscow.

We were all housed in a truly monstrous structure of a hotel called the Ukraina, from the front of which the tiles were falling off in quantity. As George Orwell perceptively remarked about buildings in the USSR, the place smelled, into every nook and cranny, of boiled cabbage. Tommy Gold spent half his time trying to discover the multitudinous electronic bugs with which the place was infested, as our tents had been infested with midges in a warm August in Skye in 1936. In this, Tommy was following a not too difficult line of investigation. More revealing were the boiled eggs in the hotel restaurant. I hit on them one morning at breakfast when I found that, of the two I had been served, one was boiled as hard as a stone and the other was so soft that, when I cracked it open and held the two halves up in an inverted position, the contents ran out freely onto my plate. How was this amazing feat performed by the kitchen staff? Investigation showed that they had great vats of boiling water into which eggs were tumbled in enormous number. Whenever two were needed, they simply fished them out at random, so that, if you were exceptionally favored, as I was, you got one that had been tipped into the vat thirty seconds ago and another that had been in there for an hour. Nothing more than that was needed to convince me that the Soviet Union was destined to experience a bad future. Once the bottom of the pyramid is ignored to that contemptuous extent, the case is terminal.

Western capitalism is not so bright either, but it is not terminal, as its survival has proved. Democratic politics emphasizes the bottom of the pyramid—at least in theory. At election times, politicians at the top bow rather unconvincingly to the bottom so as to be reelected back to the top. But at other times, the crucial problem of scale at the bottom hardly seems to be addressed. Instead, there is an illusory concentration on material standards. As governments inevitably discover, no sooner is an economic standard achieved than more is demanded—as it always will be, so long as the essential problem remains unsolved. Japanese industry has been more successful at dealing with this point than American industry. By taking notice of personal scale at the bottom, the Japanese have outstripped the rest of the world in the reliability of their products. Introduction of Japanese methods into Britain over the last decade has miraculously transformed manufactured goods from being among the least reliable in the industrial world to being among the most reliable.

By the time of the Moscow meeting of the International Astronomical Union in August 1958, I had discovered a mistake in the physical summary I had given at the Vatican Conference in May 1957. I had emphasized the need for a concerted effort in developing computer models for evolving stars. In B²FH, the nuclear physics had been handled more accurately than had the structures of the stars in which nucleosynthesis took place. There was an evident need to improve the quality of computer models to match the precision of the nuclear physics, and, indeed, from the time of my joint paper with Martin Schwarzschild in 1954–1955, I had been endeavoring to do just that. As the problem developed, especially in 1957–1958, I found the situation to be much more complicated than I had expected. My estimates in May 1957 of the effort that would be needed could now be seen to be far too modest. It wasn't so much a year or two of work that would be required as it was the work of a generation of theoreticians. When establishment problems began to occupy a good deal of my time, beginning in October 1958, this matter of computer models was the first of my research efforts that unfortunately had to be shelved—but not before I managed to distil a significant result out of what I had got thus far.

In 1958–1959, I found the ages of the oldest stars in the disk of our galaxy to be about 12 billion years old, a really big increase on Edwin Hubble's old estimate of 2 billion years for the age of the entire Universe. Then, in 1959–1960, Willy Fowler and I found a closely similar result from what we called nucleochronology. This involved calculations of the primeval abundance ratios of thorium, uranium, and other transuranic elements, with geochemical estimates of the general thorium-to-uranium ratio in terrestrial rocks being also required. We were able to show that, subject to certain assumptions about the rates at which stars had formed and evolved, the age of our galaxy lay between 10 billion and 15 billion years, with 12 billion again being the most likely value. Subsequent work by Fowler has not shifted this estimate significantly, while more extensive later work on the evolution of stars has also given much the same 12 billion years for the age of the galactic disk, but it has also given rise to estimates of 13 billion to 18 billion years for the ages of the stars in the halo that surrounds the disk. The tendency in recent years, however, has been to revise the latter estimate down to the lower limit of 13 billion years, a value of great importance in modern controversies over cosmological theories.

Chapter 22

Droll Stories

T HE ISAAC NEWTON Telescope started with a proposal immediately after the war, in which Harry Plaskett, Professor of Astronomy at Oxford, played a leading role in his capacity then as the President of the Royal Astronomical Society. Parliament responded graciously and almost immediately by voting the necessary money. The correct and obvious move to follow would have been to instigate a search for someone in the astronomical community endowed with at least a little common sense, and to charge this person, when at last discovered, with making an approach to the staff of the Mount Wilson Observatory, which then knew more about making 100-inch telescopes (the aperture proposed for the Isaac Newton Telescope, or INT) than any other institution in the world. A design improving upon Mount Wilson's own 100-inch telescope could have been arrived at very quickly. Glass of good quality could have been ordered either from Corning or from Pilkington; American help could have been obtained in the optical work; the mounting could have been ordered either from Vickers or from Grubbs; and, in the meantime, a suitable site out of Britain (where the sky is never really suitable for optical astronomy) could have been decided, so that, by the early 1950s, the INT could have been operating well and efficiently.

Instead, committees were set up, both by the Royal Astronomical Society and by the Royal Society, and a general request was put out for proposed designs. Every amateur telescope designer in Britain was agog with some suggestion or another, while the Royal Society hunted avidly among its engineering fellows also for suggestions. The outcome was a gargantuan pile of proposals, of which the most extreme I can recall was one that made the telescope tube out of floppy material so that the whole affair could be packed away in a box, the idea being to stiffen the tube when in operation by pumping fluids through its members.

Meanwhile, Harold Spencer-Jones, the Astronomer Royal of the time, was busy in the United States—but not at Mount Wilson. R. R. McMath was then the Director of the University of Michigan Observatory. He had acquired a 100-inch glass blank that had been cast in connection with the building of the 200-inch telescope on Palomar Mountain, and it was this that Spencer-Jones coveted. No one can really say how he acquired it. Spencer-Jones said it was offered as a gift. A well-known American astronomer, who was there at the time, said that McMath was put in a position from which he could not refuse to make the gift. Maintaining his point, this well-known astronomer thereafter always referred to the unfortunate Sir Harold as Scrounger-Jones. To avoid controversy, let us say that it was a swap, that McMath gave the glass disk and Spencer-Jones gave, in exchange, his famous lecture on time. And let us say that neither got quite what he expected.

Spencer-Jones's lecture on time was a fearsome happening. Time is concerned with what you care to make it. Mathematical, physical, or philosophical subtlety, if you like, or just plain clocks. A metronome can be said to be a musician's type of clock. So, at a practical level, you could say that time is concerned with metronomes, and this was the way Spencer-Jones interpreted it. He would show a slide of a metronome disassembled into its components, then a slide with the same metronome assembled and, with its upright tongue wagging from side to side, making the loud ticks that keep musicians in tempo. Then he would show a third slide, with the instrument stored away inside its case, from which you could see nothing at all. It might take three minutes for all that; multiplied by, say, fifty different kinds of metronome, it comes to just two hours and a half, which is exactly the time required by Spencer-Jones to deliver his lecture on time to an unsuspecting audience. I never see marathon runners starting off but what I think that even the slow ones among them are going to cover the 26 miles plus in less time than it took Spencer-Jones to cover that lecture on time, which was all that McMath got in exchange for his 100-inch glass disk.

Spencer-Jones wore thick, black, tortoiseshell-rimmed spectacles, such as those the comedian Harold Lloyd had adopted for the humor of it, but that had unexpectedly become popular in London society. Spencer-Jones was far-sighted, so, at the beginning of showing each metronome slide, he would put on the big tortoiseshell glasses to read from a sheet of paper to find out what it was, whether it ticked in crochets or minims, and then he would take the big black spectacles off as he looked up at the slides. This made for a good deal of putting on and taking off. Indeed, whenever he wanted to emphasize some particular point, he would fold the spectacles carefully in his hands and would smile confidently at his audience.

You could say that, if a lump of old glass has scrap value, the way old metal has scrap value, Spencer-Jones got the better of the swap; but, to my knowledge, neither pawnbrokers nor scrap yards have much use for lumps of glass by themselves. So it may actually have been McMath who did best, for the disk was so full of bubbles and striations that, to figure it at all, a lot of glass had to be taken out of the middle. This meant that the eventual paraboloid into which the surface of the glass was ground had to be deep in the middle, leading to an unusually short focal length. It meant that the plate scale was poor, the tube length was short, and the dome housing the INT was small, which explains why the unfortunate Sir Isaac's memorial telescope is a little runt of an instrument.

It was an important characteristic of Spencer-Jones that you felt very much at ease in his presence. If the world has dark aspects, you didn't see them when you were talking to him. In contrast, his successor as Astronomer Royal, Richard Woolley, was someone with whom I never felt at ease. So long as he was at center stage, Woolley could be charming—charismatic, even—but, at all times, there was an insistent psychological demand that he be at the center. Spencer-Jones was a better source of droll stories than Woolley. Not long after the end of the war, he persuaded the Admiralty to move the Royal Greenwich Observatory from its traditional site to Herstmonceux Castle in Sussex, the RGO being administered in those days by the Hydrographer's Department of the Admiralty. The reason given for the move was bad observing conditions in eastern London, which was true, the pertinent question being, of course, How much better were they at Herstmonceux, a few miles inland across marshes from the sea?

The first requirement for an astronomical observatory is a general absence of cloud. On this score, there was obviously little to choose between Herstmonceux and Greenwich. The next point is an absence of air movements, particularly at heights from 10,000 to 30,000 feet. It is this that mostly decides what astronomers call the "seeing," measured by the radius of the apparent patch into which the atmosphere spreads the light of a star of some standard brightness, a phenomenon popularly known as twinkling. It came as a surprise, in the 1960s, to discover that the whole of western Europe is just about equally bad in this crucial respect. Italy and Greece were found to be essentially as bad as Britain. Nothing that was good was found even in the huge territory of the Soviet Union. Only in Spanish territory, especially in the Canary Islands, were sites found that were comparable to those in California and the Hawaiian Islands, and for just the same climatic reasons.

In addition to these objective factors, astronomers can all too easily inflict troubles on themselves. Telescopes placed where the sky is bright from city lights is an example, and, in this respect, Herstmonceux was certainly better than Greenwich. But the unsuitability of the higher atmosphere remained just as serious after the move of the RGO as it was before.

Another problem was the generation of air movements close to the telescope itself. This happens because of temperature variations between day and night. A telescope and its dome heat up during the day and cool off in the night, the cooling generating local air movements. There was a suggestion made by French astronomers that the problem might be cured by installing air blowers in the dome, but nobody had the courage to use strong enough blowers until Vince Reddish did so years later in the dome of the British 48-inch Schmidt telescope at Siding Spring Observatory in New South Wales, Australia. Improvement was remarkable, causing a rapid reassessment of the problem, with strong blowers becoming ubiquitous in large telescope domes thereafter. In earlier days, however, only indirect solutions or partial solutions were available. On a fraction, perhaps a fifth, of days and nights, it happened at Mount Wilson that there would be only small temperature variations, and these would be the occasions on which the best results were obtained. The simplest palliative was to keep the dome interior as cool as possible during the daytime. This is why telescope domes are painted a dazzling white, to reflect as much sunlight as possible, and the paint should have a base that radiates heat as strongly as possible.

When Spencer-Jones moved the Greenwich telescopes to Herstmonceux, he did things the wrong way round, counteracting the advantages of the move from Greenwich. He had the domes made of copper, which ensured a maximum absorption of sunlight, making the daytime temperatures as high as possible, the way it is with car door handles on sunny days. In fact, car door handles can get so hot that they practically boil your blood. I don't know whether the inhabitants of Yuma, Arizona, still use oven gloves when they open their cars around midday, but that is just what they used to do in the 1950s, when I drove through the town on a number of occasions.

Herstmonceux Castle wasn't an old castle. It was an imitation castle built circa 1908. Yet it had a supposed ghost room, better than most genuinely old castles, where they used to put up special visitors. When I became a special visitor, that was what they did to me, and I had to wrestle with the ghosts through many a long, hair-raising night. On one occasion, my wife was there with me, and I thought that, with the two of us together, we might manage very well. It chanced that, because of a discussion with Richard Woolley, I was half an hour behind my wife in arriving for bed. As soon as I appeared, my wife exclaimed, in a disturbed voice, "This place is haunted." And so it seemed to be, by the unpleasant mind of the person who built the imitation castle. Every aspect of the shape of the octagonal ghost room was wrong. Every sound echo was wrong. When falling asleep over the 10,000 generations of prehistory, humans must have developed the ability to recognize warning signals subliminally. In that room, you got warning signals all the time. The bed was on a raised dais with narrow steps all round it to the floor level below, so that, if you were to waken in terror in the dead of night and seek to make an impulsive dash for it, you would trip and end your days on the floor with a broken neck. The walls had long strips of musty tapestries, which gave the room an authentic smell, but, if you tried to get the smell out a bit by opening the medieval-style windows, bats would fly in. The creatures would fly around ceaselessly over your head, giving the place an apparently genuine vampire atmosphere.

I dearly wish I had a videotape of a Royal Astronomical Society meeting, held circa 1948, to discuss the siting of the Isaac Newton Telescope. The inevitable proposal occupying most-favored status was to site the INT at Herstmonceux. It was also inevitable that a few wisps of sanity would float around, delicately suggesting that perhaps some other site might be more appropriate than the Pevensey Marshes. My videotape would concentrate on a senior establishment figure (not Spencer-Jones) who rose up like the morning Sun to dispel the few wisps of sanity. The picture he painted for us was this: A nation has one large telescope and several smaller ones, and it has several sites, one poor with much cloud cover and the others somewhat better. So how does it dispose its telescopes? Obviously, this glowing morning Sun explained in a spluttering manner to those of us attending the meeting, the best telescope goes to the site with the most cloud, because the fastest-operating instrument is needed to take best advantage of the rare moments when the clouds happen to clear. The better sites can manage with the smaller telescopes, obviously, he spluttered, and then burst into a great guffaw. Not at the absurdity of his argument, but at his imagined skill in trumping the trick of the few RAS members stricken with common sense who happened to be around in those days. When I asked Walter Baade what he thought of Herstmonceux as a site, he grinned in his pixieish way and said: "It is like building a submarine in the Arizona desert."

By the mid-1950s the British Treasury wanted to know how the telescope it had agreed to fund a decade earlier was getting on. Finding that nothing had happened, there was an inquiry, as a result of which the project was abandoned. Then, in 1956, came Richard Woolley, finding himself in a situation that brought out in him both what was best and what was worst. What was worst was a near-zero store of common sense. What was best was his courage and tenacity. Displaying the best, he persuaded the Admiralty to revive the INT project with the Treasury. Then, in 1958, I came. I was no sooner installed at Cambridge than antiestablishment figures taunted me with the INT. After being a colleague of theirs for so many years, was I going to be a party to the INT, joining the establishment for a few pieces of silver, the way it was said of Wordsworth?

Woolley was willing to listen when I drove to Herstmonceux to discuss the fate of the INT. He half admitted that he had been sucked into a mistake, and he was charming about it. He had made other mistakes, he recounted, that he also regretted, the best known being on the occasion of his arrival from Australia. After the fatiguing journey, as it was even by air, in those days, he had been waylaid by hungry reporters. Asked to state his views on space travel, then much in the news among the feather-headed, he robustly observed that space travel was bunk. Then they got him on television and embarrassed him by asking how it felt to be just an ordinary person living in a castle. He muffed that one too, so that, by the time I visited him, he was much in need of an ally. The correct answer to have given on TV was that he very much enjoyed living in a castle, bats and all, so much so indeed that he was sorry for all the unfortunate, ordinary people, like the TV interviewer himself, who were not able to live in castles.

Was it better, Woolley asked me, to have some sort of a telescope or none? Stop it now and the double indecision would make it certain that astronomy would be barred from all major support by the government in the future. So the situation was this. Eggs have been dropped, and their contents are all over the floor. Should we scoop up the mess and make out of it the least unpleasant omelette we can, pretending thereafter to enjoy the eating, with the prospect of receiving more eggs? Or should we sweep the lot down the drain with the prospect of receiving no more eggs? Well, since it was Woolley who would be eating the omelette, not me, I felt it would have been churlish to disagree with him, even though my antiestablishment friends subsequently thought I had indeed sold out for the pieces of silver.

In physics, the word "power" has a precise meaning. The word circulates in general use, in politics, in commerce, as if it also had a clear meaning, which it hasn't. According to some, money is power, but I would rather say that money gives "potential," the potential to buy an article or a house. By power, I mean the ability to change the course of events in the world. True, if a wealthy man offers me x pounds to spend a week in Paris, nothing else being required, and if x is large enough, he changes the world to the extent that I will spend the week in Paris. The last person to hold power in any real sense in Britain was Henry VIII, and even Henry's power was insufficient to alter, in any easy way, the terms of his marriage to his first wife—a pretty small affair, one might think. Continuing to define power as the ability to change the course of events in nontrivial ways, I have to doubt that anybody in the British political system really has much power. Where people go wrong on this matter is to think that a big cog wheel in the system is somehow powerful. Big cog wheels are just big cog wheels that fulfill a big function, but they rarely alter anything outside an inevitable course set by the relation of the machine as a whole to the rest of the world.

The University of Cambridge can be viewed as a state, an examination of its structure providing a fascinating glimpse of the functioning of a complex organization. From inside the university, every individual had the impression that such-and-such a person or committee was powerful, but an examination of the whole structure showed how very carefully things had become adjusted so that this was quite untrue. It was simply a case of big cogs and little cogs, all grinding away to an end that was outside the control of any individual or, indeed, of any limited group of individuals.

Like a state, the University of Cambridge had a book of laws, the *Statutes and Ordinances*, a statute being a law that cannot be changed, except by outside permission from Parliament, and an ordinance being a law that can be changed from within. The laws defined the various faculties—English, Law, Physics and Chemistry, History, Mathematics, and so on—and how the boards of faculties were to be elected. They permitted the boards certain powers in respect to the content of lecture courses, but not in the form that examinations should take. The powers of a General Board were defined as well as the various ways in which its members were to be elected. The essential power of the General Board lay in proposing new laws or changing old ones—proposing, but not enacting. For a proposal to become a law, it was necessary to go through a procedure in the Senate House in which the Head Proctor went through an arcane business with his square, a procedure I believe of Italianate origin. Any member of the university could then challenge any proposed legal change by a simple declaration of *non placet*, "I don't agree." The freedom to declare *non placet* was substantial and much feared by the higher committees of the university, but it was not a power special to any individual. We all had it.

The *Statutes and Ordinances* specify how the University of Cambridge shall be governed with a minute accuracy designed to avoid all ambiguity. The *Statutes and Ordinances of the University of Cambridge* could exist perfectly well, however, even if there were no actual University of Cambridge. The *Statutes and Ordinances* are simply a logical blueprint for operating a university. To obtain an actual university, something else is needed. The *Statutes and Ordinances* might mention the Director of the University Farm, but it does not say who the Director is to be. Likewise, the identities of the nearly 1000 officers of the university are not specified. So how did the huge and ponderous system cope with the hall porter who insisted that he was the Head of the Department? Ray Lyttleton's grandfather had been a famous lawyer in Ireland, and his talent for the ludicrous had carried through to his grandson, for this was exactly the kind of problem Ray used to delight in. There was no way by the rules and regulations of which the university was so fond. At the end of the day, a welcome entry had to be made by common sense. By precedent also, since it is hard to think of any purely written system that does not appeal to precedent at some point. According to the laws, electors on professorial boards were appointed for periods ranging up to five years. Electors came in various ways, some from the faculty to which a professorship related, one or two from allied faculties, one or two from the General Board, and one or two from the high level of the Council of the Senate. It was also a law that at least two electors should be from outside the university. There was no law, however, that prevented electors from being changed at the end of their periods of service. It was here that precedent came in, for precedent had it that electors—especially external electors—were not changed except for some understood reason, usually retirement or the wish of an elector to be relieved of the duty.

There was room for some circumvention of the system in the way papers were circulated. Obviously, it was impossible for all papers to be circulated to everybody. Restriction was essentially on a need-to-know basis. You received those of the committees to which you belonged but not those of other committees, which was fine so long as other committees were not deeply into your business, as happened for me in 1963. By a curiosity, however, the relevant papers on that occasion came into my hands later, and pretty drastic they turned out to be. The affair went like this: Lyttleton became involved, together with an influential group of other scientists, in changing the rule for entrants to the university to have obtained a qualification in the Latin language. Lyttleton used to ask, What use was it to a mathematics student to have read a lot of stuff about golden bulls? Actually, I didn't agree with him. I would have been glad to have had a golden bull at the bottom of my garden, or even inside the house. Anyway, an influential scientist instructed a secretary to send papers from a certain filing cabinet to Lyttleton, papers concerned ostensibly with the issue of the Latin requirement. By a droll mistake, the secretary sent him highly sensitive material, including the shuttlecock that the university had made out of my business.

The only person I ever saw who managed to change the *Statutes and Ordinances* singlehanded in a significant way was George Batchelor. The effect was to split the Faculty of Mathematics into two departments, each with a head, one supposedly rotating and the other nonrotating. In place of a previously united free organization, mathematics became like a tent at an old-fashioned fair, with two heads outside grimacing at the passing crowd in different ways.

The potential for a split between the pure and applied sides of mathematics had always been there, a potential that arose from a genuine uncertainty in how theoretical physics—properly speaking, the major part of the applied side—should be organized within a university. The laws of physics are highly mathematical, which suggests an association with pure mathematics. On the other hand, advances in the formulation of the laws come primarily from experimental physics, so theoretical physics can be organized alongside experimental physics with profit to both, as it is in most universities throughout the world, or it can be organized alongside pure mathematics. Traditionally, Cambridge had always done things in the latter way. Either made sense. What did not make sense was to put theoretical physics into limbo, which was what Batchelor managed to achieve in his Department of Applied Mathematics and Theoretical Physics. The term "applied mathematics," as distinct from theoretical physics, largely meant fluid mechanics (or continuum mechanics, as some liked to call it). The implication of the title, that continuum mechanics had an importance to science comparable with that of theoretical physics, was an elaborate pretense that did not augur well for the new arrangement.

If it be asked how these mistakes came to pass, the answer is—as it always is, when a free society declines—through its members giving away their freedom. Batchelor collected a formidable list of signatories for his proposal from lecturers and junior lecturers on the applied side. The problem, as it was presented, was to obtain material facilities from the university—paper, envelopes, stamps, more secretaries, desks for research students, a building. Such facilities were available to departments who pushed for them, but (traditionally, again) they had not been provided to free democracies, largely because the democracies preferred to ask the university for other things, which might be more faculty positions or a larger library.

Once Batchelor had his impressive list of signatories, the movement towards a divided faculty was essentially impossible to stop. So the next question to arise was, Who should be the head of the new department? Paul Dirac and I were the only two professors then available, and we were both occupied in what we saw as the proper activities of professors— doing research, not organizing an unnecessary paper chase. Besides which, Batchelor drew the attention of the Faculty Board of Mathematics to the way things were done in the United States, where the job of department chairman would be rotated. Each of a number of people in a department would do a stint of a few years, passing the chairmanship then to someone else with a sigh of relief, their duty done. All this was true, and it was difficult for the cynic to argue against it. In the United States, department chairmen are strictly pro tempore, and the system there works very well. So, it was decided, the new department would have a rotating head, nobody realizing that, in rotating problems in mechanics, there is usually a particular solution in which the rotation remains zero. When the proposed addition to the university ordinances defining the new department was being drawn up by the Faculty Board of Mathematics for submission to the General Board, it was suggested that the need to rotate the head of the department at the end of each five-year period be specified in the new ordinance. But it was pointed out that, by doing so, a change at each five-year period would be forced, whereas, in exceptional circumstances, the general view might be that the incumbent head should continue for a further spell. So why not simply require that, at each spell, a department head should be elected, irrespective of who had happened to be the head during the previous spell? It was difficult to argue against this logic, put forward somewhat naturally by George Batchelor. This was all in 1958–1959 and not very droll in itself. The droll part comes when we move along to the 1980s and find George Batchelor still there, still head of the department, the head that didn't rotate, "rusted in," as Lyttleton used to say.

I have written about the division of the Faculty of Mathematics at some length because it had long-term consequences that, I must admit, I had not perceived at the time and for which I must really take some blame. So far as examinations were concerned, the pure and applied sides were split even before I joined the faculty in 1945. In part I of the mathematical tripos, examiners routinely checked all questions, but mostly not in part II. Of the six examiners in part II, the three on the pure side would check each other and the three of us on the applied side would do the same, whereas earlier, in the 1920s, all examiners checked everything, a safe-guarding practice that ensured that part II did not become too difficult. By 1945, this restraint had gone, and a first step towards increasing difficulty had been made, because, of course, the less an examiner has to cover, the harder he can make it. My last experience as an examiner, in the early 1960s, was as chairman, which gave added responsibility for judging the overall conduct of the examination. I was disturbed to find that, in less than five years since the Department of Applied Mathematics and Theoretical Physics had emerged, applied mathematics (continuum mechanics) had become largely disjoined from theoretical physics, in the respect that an examiner in the one seemed disinclined to become concerned with the other, a further peg on the ratchet of inevitably increasing difficulty. If one looks at today's examination papers, not only for part II but even for students in their second year, the situation is so highly specialized that a rubric in bold type is attached to every question so as to inform the student to which specialty it belongs. What I feel is needed today by Cambridge mathematics is an examination of its own examiners. Take them off their own specialty and see how well they fare on other specialties. Test them on unseen questions, and discover whether they are really the Gauss-like geniuses they would need to be in order to perform well on what unfortunate students are routinely asked to do.

The rate of entry of mathematics students into Cambridge is, I imagine, somewhere over 200 each year, the competition for entry being so intense that it would not be far-fetched to suppose that Cambridge has access to half of the young mathematical talent of the entire nation. My judgment of the examination papers is that perhaps ten students each year profit from their time at Cambridge, perhaps another twenty to thirty survive, while the remaining 170 plus are put under strain by an unacceptable level of difficulty in the examinations. If the majority tells me I am wrong, I will withdraw my criticism gladly. But I do not think I am wrong.

It was in the early 1960s that I had a call from a secretary at the Greek Embassy in London. King Constantine and Queen Fredericke would like to visit Cambridge to talk to Martin Ryle and me. Would I arrange it? And would I also arrange the lunch? Arranging lunch for visiting royalty is not as hard in Cambridge as it might sound, for the colleges do that kind of thing very well. I had only to call the Master of St. John's, my own college, and all would be taken care of. But, since Ryle was equally involved in what the royal visitors wanted to do, I thought it would be a courtesy to call the Master of Ryle's college, Trinity, instead. Lord Adrian agreed immediately to take responsibility for the lunch, and excellently well he did it. After I had called Adrian, I rang Ryle, who suggested that, after the lunch, the party should proceed to Lord's Bridge, where the discussion should take place. Lord's Bridge was the site, four or five miles out of Cambridge, where Ryle had the university's radiotelescopes. Since I had no place to offer that would be any more suitable for a discussion, I agreed.

Adrian had invited Nevill Mott, the Cavendish professor, and our wives. Matters proceeded according to arrangements, with me and my wife arriving home in Clarkson Road sometime between 4:00 and 5:00 P.M. Immediately, my wife burst into tears: "It was the hostility of it," she exclaimed, meaning that of the university members of the party. "If anybody wants to see you from now on, and I don't care who, they can come here to the house. That's how it's going to be from now on." And that is just how it was from then on.

Nothing droll thus far. The droll bit came a few days later, when I happened to be in the Cavendish Laboratory, then on the New Museum Site behind Barclays Bank. Nevill Mott, head of the laboratory, asked me into his office and said, in a gloomy voice, "I'm afraid Martin Ryle thinks you acted in a very high-handed way over the Greek visit." In view of my blank face, he continued: "He thinks you should have contacted him before Adrian." My face must have remained blank, for Mott continued yet again in a voice of some desperation: "Why is it, Fred, that you can't get along with Martin Ryle?" I am quite bad at answering that sort of question. So many possible answers jump into my mind that I can't think which of them to choose. So I simply replied, "It must be because I have no sense of humor."

But I mustn't give the impression that life at that time was only a sequence of downbeat situations. There were successes as well, and the discovery of quasars was one of them. In 1959, Pat Blackett, who was Chairman of the Research Committee of the Department of Scientific and Industrial Research, later to be the Science Research Council, asked me to become chairman of the astronomy subcommittee of the main Research Committee, on which I would also sit: "Because," as Blackett explained, "I want someone who isn't after the money for himself."

It is interesting to reflect that a really big discovery came out of an expenditure of a relatively small sum, £50,000. The proposal came through Bernard Lovell, from Henry Palmer at Jodrell Bank. Bernard, together with Martin Ryle, had already exhausted the funds available to my committee, leaving the astronomy cupboard bare for the unfortunate Palmer. So it was to the contingency of the main Research Committee that I had to address myself. Supported by Harrie Massey and Alan Cottrell, I got the £50,000, with Blackett nodding sternly from the chair and saying: "Remember, Hoyle, that's all you're going to get." So let me tell the story of Henry Palmer and of his colleagues L. R. Allen and R. Hanbury-Brown, and of just what it was that came out of that £50,000.

The proposal was to build what, for the time, circa 1960, was an exceedingly wide interferometer. One arm of the interferometer was to be the 250-foot Jodrell Bank radiotelescope, the other, a movable telescope going to about 100 miles from Jodrell. It was the latter that needed the £50,000. The interferometer pattern would have its maxima and minima at far smaller angular separations than anything in radioastronomy to that date. The rotation of the Earth caused the interference pattern to move across the positions of radio sources in the sky. Should a source be a literal point, the combined signals from the two arms would give an output, after suitable balancing, that fell to zero as the minima of the pattern crossed the source. But, for a source of appreciable angular size, there would be a smearing effect between adjacent maxima and minima—indeed, there would be little or no oscillation of the combined signal for sources with angular sizes sufficient to cover several adjacent maxima and minima of the pattern. The basic idea, therefore, was to use this dependence of the signal on angular size, together with a variation in the spacing of the two arms of the interferometer, to determine the angular sizes of the patches on the sky that constituted the radio sources. It was an idea that was destined to develop into what has been perhaps the grandest concerted program in radioastronomy, a program that would eventually examine the finest details within individual sources by means of interferometers with a base line not of a mere 50 miles or so but comparable to the radius of the Earth, with thoughts for the future that would establish still larger base lines in space beyond the Earth.

First results were available by 1961, when my student Jayant Narlikar and I visited Jodrell Bank as guests of Robert Hanbury-Brown. Many of the radio sources from a catalogue compiled by Ryle's group, known as 3C, had been investigated and had proved amenable to the technique. Their angular sizes had been determined. But there was a residue, a special class of "chaps," as Hanbury-Brown put it, for which the technique had not succeeded. Their angular sizes had been too small to be measured. This residue of the "chaps" turned out to be the quasars.

The small angular sizes of quasars could thus be distinguished already, by 1961, from the larger sizes of radio galaxies, showing that the entire class of radio sources that were isotropic with respect to the Earth could be divided into two groups in a ratio of about 1 to 2, respectively. The discovery of a small-size group, constituting about 30 percent of the radio sources, suggested to some people that both sides in the Massey Conference held a decade earlier at University College, London, had been right after all. Gold and I (the "theoreticians") had been right about the radio galaxies, but Ryle and the establishment behind him had been right about the group with small angular sizes. The consensus took them to be local stars, the old radio stars, an idea that was to prove as fruitless on this second occasion as it had on the first, an idea that was to prevent the consensus from making progress until early in 1963.

The problem with the scientific establishment goes back to the small hunting parties of prehistory. It must then have been the case that, for a hunt to be successful, the entire party was needed. With the direction of prey uncertain, as the direction of the correct theory in science is initially uncertain, the party had to make a decision about which way to go, and then they all had to stick to the decision, even if it was merely made at random. The dissident who argued that the correct direction was precisely opposite from the chosen direction had to be thrown out of the group, just as the scientist today who takes a view different from the consensus finds his papers rejected by journals and his applications for research grants summarily dismissed by state agencies. Life must have been hard in prehistory, for the more a hunting party found no prey in its chosen direction, the more it had to continue in that direction, for to stop and argue would be to create uncertainty and to risk differences of opinion breaking out, with the group then splitting disastrously apart. This is why the first priority among scientists is not to be correct but for everybody to think in the same way. It is this perhaps instinctive primitive motivation that creates the establishment. Its one formidable defense lies in its ability to forget its mistakes and, indeed, to remove all memory that they ever existed. Whence we see it is the rationale of 100,000 years ago that we find on all hands today. And why not? the establishment man might ask. Because, says the heretic, if prey is not found to lie in a particular direction, it is because prey does not, in fact, lie in that direction. If the scientific establishment does not solve a problem by moving in the direction consensus has chosen, with all the advantages of state aid and the like, it is because the solution does not lie in that direction.

Willy Fowler had sabbatical leave from Caltech in the year 1961–1962, which he spent in Cambridge. Our main production was a large paper on nucleosynthesis in supernovae. This paper had a vogue in the 1960s. But of more interest, in the long run, was a smaller investigation we did on the genesis of the immensely powerful outbursts of radio galaxies, which energies Geoffrey Burbidge had calculated to be on the order of 10^{57} ergs in some cases. This was a hundred thousand times more powerful than a supernova, which had led Burbidge to the idea of triggering a chain of supernovae in some way, like a row of falling dominoes. Fowler and I considered this idea, and, in the spring of 1962, we had the thought of putting all of Burbidge's supernovae together into what we called a single massive object, the mass being upwards of a million times greater than that of the Sun. It was our natural style to attempt to power our massive objects by nuclear energy, but, by the late autumn of 1962, it had become clear that nuclear energy simply would not do. Surprising as it seemed to us, the only sufficiently powerful source of energy was gravitation, which required our massive objects to be very compact—"collapsed massive objects," we called them, but they are known nowadays as "black holes." By December 1962, we had also begun to consider hunting around to discover if our theoretical deduction of collapsed massive objects might actually correspond to existing objects. As it was soon to turn out, they did. They were the quasars.

By early 1963, much water had flowed under the bridges since my 1961 visit to Jodrell Bank. The issue of whether the radio sources of very small angular size that were discovered by Palmer, Allen, and Hanbury-Brown really were the elusive radio stars could not be settled until highly accurate positions on the sky had been obtained for at least a few of them. It was realized that, in the course of its variation from month to month over a period of about 18⅔ years, the Moon's path in the sky crossed over several of the "chaps." Cyril Hazard also realized that, by determining the exact moment at which the Moon covered a source, and from the *known position* of the Moon at that moment, it would be possible to obtain the position of the chap in question to a far greater measure of accuracy than had ever been achieved before. Hazard's first success at Jodrell Bank was for the source 3C 212, but this did not prove adequate to crack the nut. Hazard's second choice, 3C 273, was to prove very different.

With the Moon's passage over 3C 273 in the offing, Hazard left Jodrell Bank to take a job in the Physics Department of the University of Sydney, Australia, where there was no suitable radiotelescope for observing the coming occultation of 3C 273. The only suitable telescope was in the possession of the Commonwealth Scientific and Industrial Research Organization, and between CSIRO and the Physics Department of the University of Sydney, relations were strained to the point at which, on a visit to Australia in the autumn of 1962, I could maintain peace only by visiting the two organizations in separate weeks, not even on alternate days. This was the droll problem on which the discovery of quasars really turned. Only someone with Hazard's amiable, somewhat vague manner could have solved it. A more eager, more acute sort of person would never have been able to do so. The solution arranged with John Bolton, then Director of CSIRO's 210-foot radiotelescope at Parkes, New South Wales, was that Hazard would work on 3C 273 together with M. B. Mackay and A. J. Shimmins of CSIRO.

The Moon duly occulted 3C 273 without the need to make any such personal arrangement. The occultation was observed from Parkes, and an exceedingly accurate position (by the standards of the day) was obtained.

Bolton communicated the resulting position of 3C 273 by letter to Tom Matthews, head of radioastronomy at the California Institute of Technology. Matthews, however, was not permitted, by house rules, to observe 3C 273, either at Mount Wilson or at Palomar Mountain, so he was obliged, if anything was to be done at Caltech, to pass the position to Maarten Schmidt, who was the accredited observer in such matters. If Matthews had instead telephoned Margaret Burbidge in La Jolla, the position would again have been different, in that Margaret would then have had the first chance at the discovery.

For the reader to appreciate more fully how it was at Caltech and Mount Wilson, it should be said that one of the "chaps" in the Jodrell Bank list, 3C 48, fell in a region of the sky where there was a sufficient dearth of ordinary stars for one particular starlike object to be suspected as the radio source. An optical spectrum of the starlike object had been obtained by Allan Sandage a year or more before the position of 3C 273 was received at Caltech. The spectrum had a large number of emission lines in it, which those who were privileged to see the spectrum could not identify, essentially because they attempted the identification within the wrong framework of the radio-star hypothesis. The spectrum of 3C 273 obtained by Maarten Schmidt in early 1963 admitted no such mistake. It had three dominant emission lines with the wavelength spacings of the lines Hβ, Hγ, and Hδ of the Balmer series of hydrogen. (The spectrum of 3C 48 turned out, in contrast, to have a complex mixture of lines of hydrogen, helium, oxygen, neon, and magnesium.) The surprise was that the measured wavelengths of Hβ, Hγ, and Hδ were increased by 16 percent from their laboratory values, giving 3C 273 a red shift of 0.16. Interpreting the red shift cosmologically, so far from being a local radio star (as the "chaps" had been believed to be only a month or two earlier), 3C 273 was at a distance of about 2000 million light years, and it was therefore enormously luminous.

The work of Hazard, Mackay, and Shimmins, and that of Schmidt, appeared together in *Nature* about two months after the appearance in the same journal of the prediction of the existence of massive objects by Fowler and me. The strength of the prediction was that it told one something of the basic nature of quasars—they are massive and they are highly collapsed. The weakness was that we had no way to decide if such objects exist. The strength of the observational work was that it showed that quasars exist, and it told one something about their outer form.

From the circumstance that the theoretical prediction appeared only two months before the observations, and in the same journal, even a careful science historian might think some kind of immediate connection existed between the two. However, so far as I am aware, there was none. It was just that both were derived from the same Jodrell Bank list of "chaps," and it happened that both took a similar length of time to arrive at similar conclusions. Thus, the original source of the discovery of quasars was Henry Palmer's proposal for a new and more powerful interferometer, and the decision that made the discovery an eventual reality was the voting of the necessary money by the Research Committee of the British Department of Scientific and Industrial Research.

I had first encountered Alexander Todd on the Faculty Board of Physics and Chemistry in the late 1940s. Todd, in his capacity as Head of Chemistry, was often involved in attempts to modify university rules affecting his department, which he considered to have been drawn up in antediluvian times. Often enough, he would be opposed by the physics interests on the board, and I usually found myself agreeing with him. So it came about that I voted with Todd on such occasions, without any thought of his becoming favorably disposed towards me in the future. In the period 1953–1964, Todd, first as Sir Alexander and then as Lord Todd, was Chairman of the Government's Advisory Council on Scientific Policy. In that capacity, I had a communication from him in 1959, asking if I had proposals for improving the national situation in theoretical astronomy. At that time, I had already formed the perception that the optimum size of an interacting group was about 25 people, the size of a hunting party. No single university could support such a number of theoretical astronomers. My proposal then was that government assistance be made available to establish such a group at a university, perhaps Cambridge, to be agreed upon by the members of the group. The proposal was clear-cut, and it was modest, compared with what others were suggesting, so it received immediate support from Todd's council. The proposal even appeared in print in a published government paper, a step that usually indicates that something is really going to happen.

Therefore I expected that, from 1960 onwards, I would be involved in such an active development. But, instead of active development, I was asked to refer the matter for discussion by the National Committee for Astronomy, which operated under the aegis of the Royal Society. Todd later apologized for this stopper, telling me that William V. D. Hodge had been there on his council as an assessor for the Royal Society. Hodge had come in with the suggestion that an opinion from the Royal Society be sought. Todd told me he had been determined to push the matter through from the chair, but that Hodge's intervention had made it impossible. It was this kind of situation that Alan Cottrell had in mind when he told me that Whitehall was a place where nobody breathed the air freely. Everybody sat around in face masks, nobody controlling his own oxygen, the only freedom anybody had being to turn off somebody else's supply.

Richard Woolley was the Chairman of the National Committee for Astronomy. He refused to call a special meeting to discuss the issue, sensing, I suppose, that, if things went smoothly ahead, I might become too "powerful." So I had to wait six months before the referee even blew the whistle for the kickoff. The National Committee appointed a subcommittee to report, which subcommittee had to meet three times before it felt able to report. The three meetings, involving theoretical astronomers from all over Britain, took up the best part of a year. The subcommittee upgraded my ideas by a large margin, to an Institute of Astronomy with an estimated cost of £3 million—which wasn't exactly hay, being equivalent, in present-day terms, to perhaps £25 million. When a committee appoints a subcommittee to make recommendations, it, in effect, loses control of the situation. Subcommittee members will be far better primed with the details of what it has considered and so will be able to outargue other committee members; and, by asking a subcommittee to do the work on a tricky matter, the main committee acquires a kind of moral commitment to pass through whatever is recommended. Never once, that I can recall, did I seek to appoint a subcommittee on any committee of which I happened to be chairman, since, in my opinion, it is simply a stupid thing to do, unless you don't give a rap about the situation in hand. But, unfortunately for him, Hodge did give a rap, a big rap. Now he had to sit and watch a colleague—a not-so-dear colleague, I suppose—from the Faculty of Mathematics at Cambridge being given the imprimatur of the Royal Society for a project of truly astronomical proportions; and, vaulting upwards to Todd's council, he had to watch the same thing there.

Thus it came about, in 1963, that I was sitting with the prediction of the existence of quasars in one pocket and £3 million in the other pocket, at the oldtime value of the pound sterling (£1 = $2.8). Now all I had to do was to convert a promissory note into hard cash. Since the institute was to be in Cambridge and attached to the university, I had to obtain General Board approval for the matter to be written into the *Ordinances*, and these would need to be "graced" in the Senate House. The danger of the latter procedure attracting a serious *non placet* seemed essentially nil. This was at a time when the government was starting a host of new universities, handing out packets of money in every direction, except to Oxford and Cambridge, a situation that caused annoyance at Cambridge, meaning that my colleagues generally in the university were only too likely to applaud anyone who had managed to get a little of what was going so quickly elsewhere. What it came down to, therefore, as the last hurdle, was the General Board. So it came about that I took all the papers to the Secretary of the General Board, H. M. Taylor, and left the affair in his lap.

When two months had passed without a response from the General Board, which met weekly, my internal antennae began to quiver. When I wasn't called to appear before the General Board to explain the matter, the antennae twitched more violently. Eventually, I had a call from H. M. Taylor, asking me to his office, where he told me in sepulchral terms (which implied he recognized that it was an issue over which I might reasonably contemplate resignation) that the General Board would have no part of it, not in any way, not nohow.

I described before how, in an issue over the abolishing of Latin as a qualification for university entrance in science and mathematics, the golden-bull affair, Lyttleton had been sent a batch of wrong documents. H. M. Taylor had been very careful to tell me nothing of what went on when my business came before the General Board. But now, as if the secretary's mistake had been supernaturally inspired, there were the General Board's minutes describing it all. Although neither Todd nor I had been called to appear, W. V. D. Hodge appeared like a rubicund uncle. If you are ever to understand the establishment, you have to be clear about the impeccable logic of what happened. As an interested party, I could be called or not, as the board saw fit. Courtesy would suggest that I should have been called, but there was no requirement for courtesy. Todd could *not* be called in his governmental capacity—that would be improper. Todd could have been called in his capacity as a Professor of Chemistry in the University of Cambridge. But what has chemistry to do with astronomy? Nothing, in those days, although quite a lot today. But Hodge was Lowndean Professor of Astronomy and Geometry. No matter that he was all geometry, never having contributed anything of a technical nature to astronomy, that was what the title of his chair said. So Hodge could be called, and he was.

I had occasion before to refer to Hodge's conduct on committees, which previously had seemed more amusing than objectionable. But now it was clearly objectionable. I noticed doubtful behavior already when, as a very junior person, I first became a member of the Faculty Board of Mathematics. At that time, Hodge was also a member of the General Board, a more formidable body than the Faculty Board, in the sense that the General Board could legislate, whereas we could only suggest. What Hodge did perpetually was to make knowing asides about General Board business in order to sway the decisions of the Faculty Board of Mathematics. It was a tactic he had long since perfected when, fifteen years later, he came to discuss my business before the General Board. He was now Physical Secretary of the Royal Society, and he sat on the Government's

Advisory Council for Scientific Policy, powerful cards indeed to wave in front of the members of the General Board, who might very well be from history or the classics, and who would therefore know little of the extraordinary shenanigans that went on in scientific circles. What Hodge told them, with the weight of apparent authority behind him, was that, in his view, my project would ultimately lead to the University of Cambridge being required to fund the proposed institute out of its own resources. With that advice, the General Board had little option but to turn the offer down. If I had been from history or the classics, I would have done the same. Fairness compels me to say that, in the very long run, Hodge might well have turned out to be right. But, in the early 1960s, there was no way to guess at the financial stringencies of the 1980s. With technical colleges being upgraded to universities as well as many entirely new universities springing up like mushrooms in the 1960s, there did not seem to be the smallest cloud on the educational horizon.

If my aim in establishing an Institute of Astronomy had been to build a so-called power base for myself, then doubtless I would have resigned there and then and gone somewhere else to build my power base. But I have never had any interest in "power." People keep on almost endlessly talking about it but, frankly, I don't believe it exists. My clash with Hodge and the General Board was over what a later American president was to call "this vision thing." In 1911, the university had judged astronomy to have become a moribund science, when it transferred the occupant of the Lowndean Chair of Astronomy and Geometry from its primary title of astronomy to its secondary one of geometry. Hodge's behavior over many years showed that he still lived with this conception, although Eddington had clearly exploded it already in the 1920s. For my part, I had the vision—correct, as it was to prove—that astronomy was to emerge as one of the great sciences of the twentieth century. And the mistake was that the flaccid mode of governance of the university preferred to look backwards rather than forwards.

It was not long after this affair that the matter of George Batchelor's zero rotation appeared, which I actually disliked more than the General Board debate. I went to see the vice-chancellor about Batchelor's non-rotation, taking along the minutes of meetings from 1959, in which Batchelor had emphasized the joys of rotation. The vice-chancellor looked serious and told me that, while he understood and sympathized with my point of view, nothing that contradicted the *Ordinances* had been done. Despite the gravity of an occasion when one visits a vice-chancellor, I disliked the attitude that clear verbal agreements could be ignored. It must have had to do with my boyhood, when business in the cloth trade of West Yorkshire proceeded on a handshake, without legal documents being required. If businessmen with little education could stand to their word, I thought, those who operated an ancient university should do likewise.

We now have two issues, sufficiently close in time to be taken together: the General Board debate and the Batchelor nonrotation. The world would think the first an issue over which resignation was understandable, which I didn't, and the world would not think the second a resigning issue at all, which I did—a slightly droll situation. It occurred to me, towards the end of the term following Easter of 1964, that, by taking both issues together, all parties might be satisfied. I therefore determined to resign at the end of September 1964. The sensible procedure seemed to be to use the months from June to September for thinking over what other post I would like to try for. At the end of September, I kept to this plan by sending off my resignation to the vice-chancellor. My wife and I set off immediately for a week's holiday in the Lake District. This was a mistake. We should have gone away for a month.

Chapter 23

The Munros of Scotland

TROUBLE STARTED straightaway after my wife and I returned from the Lake District in early October 1964 to find a letter from the vice-chancellor acknowledging my resignation. He hoped I wouldn't mind his leaving it in abeyance for a short while, since he would not be able to report to the General Board of the university until the first meeting of that body—I believe it was on 12 October. More ominous still, I was to learn that the phone had been ringing all week, and would I please call a certain number. Calling it, I found I was speaking to Lord Todd's secretary at the Department of Chemistry in Lensfield Road. Could she fix a time for Lord Todd to come round to see me? Although I was leaving soon for a two-month stay in the United States, I could hardly refuse, since, in the period 1960–1963, Todd had not wavered in his support for the ill-fated institute project, which was struck down, it will be recalled, by W. V. D. Hodge.

I think it was on a dark Wednesday afternoon that Todd came to our house in Clarkson Road, together with John Cockcroft. So here were the two scientists I respected most in Cambridge bent on preventing my resignation from going through. They had been in discussion with the vice-chancellor, who said he felt the university would have second thoughts about the institute proposal. Provided it was scaled down to a suitable level, they would be glad to discuss it with me. I believe I had already told Todd that I thought the changes in the proposal wrought by the Royal Society were grandiose. Inevitably, therefore, this was a change I had to agree with. In so agreeing, I was being edged into recrossing the watershed I had traveled over during the past week. The weather in the Lake District had been turbulent, with rain, wind, breaking skies, and big high clouds. As I had gazed over opening distances, I had liked what I had seen, and I was in no way unhappy about the decision I had taken. But now I was suddenly turned back whence I had come, lacking the philosophy of life I was soon to learn from someone a decade older than myself, a man with whom I was destined to spend many days in far-flung regions of the Scottish Highlands.

When I came to meet him, Dick Cook had climbed all the Alpine peaks with heights greater than 4000 meters, except for the north wall of the Badile, a testing climb in the Swiss Alps, immediately to the southeast of the Maloja Pass. During the next few years, he would do that too, in five hours rather than the scheduled nine. Dick was a fast mover, both up and down. In those years, he was President of the Lake District Fell and Rock Climbing Club. Often over the fire in the evening, he would tell me of his mountaineering experiences. He was convinced that his survival, when so many of his early climbing companions had perished in a wide range of accidents, was due to his conviction that every mountain had a god who oversaw what happened on every crag and who could summon the most devastating storms from out of a clear sky. Dick was pure Greek. As well as the main mountain gods, he saw a disembodied spirit in every individual part of a mountain, a pantheon that must be appeased and made benign before any climb could be attempted. He would have understood Homer perfectly, if, in his youth, he had been taught to read Homer. He expressed himself more simply than Homer. Instead of Homer's attention-arresting opening sentence,

> The wrath of Achilles is my theme, that fatal anger which, in fulfilment of the will of Zeus, brought the Greeks so much suffering and sent so many gallant noblemen to Hell, leaving their bodies as carrion for the dogs and passing birds,

Dick would simply say things had to feel "right" before he would "go." Incidentally, when scanning the pallid programs offered by modern television, how one wishes a few Homers were among us today. An enterprising lecturer in some faculty of English could make a whole course of lectures out of Homer's first sentence, just as the well-known writer C. S. Lewis did in a course entitled "Some Difficult Words." It happened that I found myself sitting next to Lewis at a dinner in Magdalene College. When I asked him what was so difficult about his words, he replied: "They're not very difficult, actually, but, by God, they will be by the time I've finished with them."

Dick Cook's concepts of pantheism extended even to inanimate objects. In traversing the Obergabelhorn in Switzerland, he inadvertently dropped his ice axe. It slithered away down a steepening snowslope, gaining speed towards an almost vertical drop of several thousand feet. When it was almost at the brink, it ran across a rock that threw it up into the air. By a chance in ten thousand, it came down into the snow at the edge of the precipice with the point of the shaft driven hard into ice below the snow, so that it stood there upright on the edge. Dick's guide refused to hear of going down to fetch the axe, but he agreed, in return for a doubling of his fee, to let Dick down on a rope. So Dick went down to rescue his axe, just as he would have done to rescue a person.

Dick Cook died of Alzheimer's disease. From outside, it is hard to know what somebody else's disease feels like. But I know what one aspect of Alzheimer's disease feels like: terror, abject terror in the early stages. There came a day when Dick would forget which mountain he was on, or where he was supposed to be going. He was a man who had never shown fear, but he was frightened then, and, after a few such occasions, there came a stage when Dick wouldn't go out on the mountains anymore. There came a time, looking on from outside, when one couldn't feel one was looking at Dick Cook anymore.

Such experiences raise the question of what it is that constitutes a person. The rationalist would say the nervous system, the brain and its ability to transmit electrochemical impulses around the body. When that goes, as it does in Alzheimer's, the person goes too, even though the body remains. The problem is, of course, to know what is involved in "going." Is it simply a matter of ceasing to exist, or is it a kind of unhooking of an external "something" from the nervous system? The rationalist would argue that postulating anything external is an unnecessary complication. If it were not for subtleties, such as the ability of human consciousness to intervene at a deep level in the interpretation of quantum mechanics, it might seem so. But there is certainly a strong subjective feeling otherwise.

My parents died in the 1950s, my mother in 1953 and my father in 1955. My mother died quickly from a blood clot, following what should have been simple surgery for appendicitis. I was informed by phone that she had been rushed to a hospital and that the operation was a success, so there was no urgency for me to interrupt a course of lectures I was giving in Cambridge. I waited three days until the end of the week, by which time it was too late. My father's death, from lung cancer, was painful and slow. It was one of those cases in which the morality of the medical profession is to keep the sufferer alive to the last possible moment, even though there is not the slightest chance of a recovery. It was a poor reward for my father's having stood to his machine gun for three long years during the First World War, not the sort of thing a grateful country is supposed to bestow on its sons. Just before he died, my father's last words were: "How long is the journey, lad?" He was not a religious man; he always claimed to be a rationalist. So his question caused me to think outside rationalist lines more than any words I have heard from the pulpit, although with no profound conclusions.

Another remark that affected me was from a specialist-in-death, a thanatologist. She said: "Everybody dies peacefully." This is, of course, untrue. A person who takes strychnine does not die peacefully. But I knew what the woman meant—natural death. I believe I experienced what she meant only two months after my meeting with Todd and Cockroft. In early December, I had a minor operation that nearly wasn't. Owing to the incompetence of a general practitioner, not the surgeon, things went drastically wrong. It happened after a simple operation was supposed to be safely over. Several hours later, I vaguely remember the shriek of a ward nurse and lights around me—as I later discovered, there was panic to get hold of the surgeon and the anesthetist quickly for a second operation. Afterwards, the surgeon complimented me on my stoicism. But it wasn't stoicism. It was just what the specialist-in-death had spoken of: an overwhelming impression of peace. Mountaineers occasionally have an experience like Dick Cook's experience with his ice axe. They fall and begin the slide to an obvious death, when some freak aspect of the fall causes the slide to be checked. Such lucky cases report that, after the swift desperation of the last attempt to prevent the fall, what follows is a sensation of peace. It is as if, once the unhooking of "something" from the chemicals that constitute our bodies seems inevitable, the mind becomes indifferent. That this should be so for a superbly fit mountaineer answers whatever rational objections I can come up with at the moment.

With the medical crisis over, I recovered quickly and was home for Christmas. Early in the new year, I drove to the northwest, to spend three weeks convalescing at the Old Dungeon Ghyll Hotel in the English Lake District. Walking out on the hills modestly each day did something useful besides furthering the immediate objective of medical recovery. Since the early 1950s, I had suffered from a ligament problem in my left knee. Halfway through the day, typically, the ligaments would suddenly give way, and any steep descent would then become painfully slow. The problem arose, I think, from trying to do too much while still out of training. I would remember the way it had been in the 1930s, would try to do the same things, and would end up with the sore knee. However, now that I was obliged for other reasons to take things easier, the knee began to strengthen, and really, from then on, the trouble effectively disappeared. Indeed, the years of climbing all the Scottish mountains, about 300 of them, were still ahead. On some days, Dick Cook would come to Old Dungeon Ghyll from his home in Bowness on Lake Windermere, and we would go together on the snow-covered tops. Another friend, Norman Baggaley, came midweek from Blackpool, the three of us making an outing I had never done before: up the valley known as Far Easedale, then along the boundary fence to Sergeant Man and back to Grasmere, where the poet Wordsworth once lived, with a lot of snow and ice on the high ground. During that walk, the three of us agreed to go to Scotland in mid-March on what was to prove a magical trip, an experience that played a key role in permitting me to remain balanced in the irritatingly difficult mess that followed on from what I had agreed to in Cambridge with Lord Todd and John Cockcroft.

The agreement had been for a sum in the neighborhood of £750,000 to be provided, in approximately equal parts, by the Nuffield Foundation (of which Todd was chairman), the Wolfson Foundation (of which Cockcroft was a member), and the newly forming Science Research Council (SRC), with the university's contribution being limited to providing land for a building. Two days after the visit from Todd and Cockcroft, I had a visit from a high official of the SRC-to-be, whom I had known well in my association with the Department of Scientific and Industrial Research, the forerunner of the SRC. The plan seemed neat and rational: the Wolfson Foundation would provide the financing for a building, the Nuffield Foundation would support staff salaries, and the SRC-to-be would buy and finance a computer for the proposed institute. It had all sounded good, with the money falling into convenient packets. But the division would have the eventual consequence that accounts would need to be kept in several ways. Indeed, when the university was eventually obliged to enter the scheme, in ways still to be described, accounts would be required in three different ways—one for the university, another for the SRC, and a third for the foundations—like the Rat setting out pistols and swords for himself, the Mole, and the Badger in Kenneth Grahame's book *The Wind in the Willows*.

I returned to Cambridge in late January to find no letters or other documents that would serve to advance the plan. On inquiring into the delay, I was told that the money that I had understood to be promised had really to be applied for. Both the government and the foundations required petitions from me for what had already been promised. The implication was that I knew the system and shouldn't expect anything different. Precisely because I did know the system and had seen nothing to compare with the promises I had been given, I had indeed thought it would be different.

By now I was beginning to regret that I hadn't followed Dick Cook's second rule of safety: Never go back on your decisions. If Dick decided it was "right" to "go" for a mountain, he went, with 100 percent commitment. If he decided it wasn't "right," no amount of cajoling by others would make him change his mind. Although I learned, in the next few years, to follow much of Dick's philosophy, and eventually almost all of it, I hadn't followed it that one day in August 1970. The party was physically strong but inexperienced. Its members particularly wanted me to lead them to the top of what I knew to be a deceptively dangerous mountain. As it was quite often with Dick, the day didn't look "right" to me. But I foolishly succumbed to general pressure from the party, coming, in consequence, as near to tragedy in the mountains as I ever came. But for two very lucky breaks from the excruciatingly bad weather that developed, I think more than one member of the party would have been lost. The eventual outcome from my succumbing to persuasion in October 1964 was not to prove much different, although I hope the telling of what happened will at least prove interesting, and perhaps a little instructive.

Writing a proposal for the Nuffield Foundation was easy, because it could be very similar to what I had done already in 1960. To make a sensible proposal to the Wolfson Foundation, however, a tentative building design would be needed. Luckily, I had seen a building at the University of California's campus in La Jolla—the Institute of Geophysics and Planetary Sciences building, as it was called—that seemed to me ideal for the purposes of the Institute of Theoretical Astronomy. I had dimensions—office sizes, corridors, heights, widths, the dimensions of seminar rooms, and so on. The building had won an important architectural prize, and so I knew the subtle interrelations that are hard to judge in advance were well conceived and successfully executed. It remained to decide materials, which could not be the same in Cambridge as in California. I also had to decide the number of office units, which was conveniently adjustable to the eventual size of the proposed institute. For this and for the materials, I needed to know exactly where the building would be positioned. Recalling the university's promise of land, I visited the University Estate Management Office to consider just what site might be available.

No doubt the vice-chancellor and the senior officers who had promised the land had in mind what was called the West Cambridge Site, but this was tied up at that time in a melee of interlocking committees in which years of delay were more than likely. As well as this problem with the land, there had been an issue, in connection with the Nuffield application, on which I had sought advice from the university, explicitly from W. J. Sartain, the Secretary of the Faculties, as he was called. Sartain was always friendly towards me, for the reason that I was one of his rare visitors—perhaps indeed his only visitor—who was not seeking to get his hand into the university's "chest," a metaphorical money box. Sartain usually wore a dark suit. He had straight, black, slicked-back hair, and, despite being a little bulky, he had a generally satanic look.

Budgeting costs for academic salaries in my application to Nuffield had been no problem. It was on the matter of overhead that I needed advice. Sartain told me to add an overhead of 100 percent—that is, 50 percent of the eventual total. At my shocked response, he turned more gloomy than usual, saying there was no explaining why, but that was how it always turned out. It was a matter he had investigated carefully over many years, he went on, breaking suddenly to say: "You know, I've had three threatened suicides this week."

The suicide threats must surely have had to do with the New Museum affair that was going on at that time. The New Museum Site consists of a narrow strip of land near the Cambridge Market Square, enclosed by Corn Exchange Street, Bene't Street, Free School Lane, and Downing Street. Cambridge Town had the Bene't Street end, the university had the rest. This had been the site on which James Clerk Maxwell had built the first Cavendish Laboratory, to which additions were made in the 1930s. It was where the Arts School, in which Wolfgang Pauli had trod the squeaking board, was located. The university's first chemistry department had been built at the Downing Street end, and, over the years, other scientific departments, including the Biochemistry Department and the Mathematical Laboratory, had been squeezed piecemeal into the confined space. Before the war, the pressure had already become so great that Geography and sciences with lesser clout had been obliged to move away across Downing Street into what was called the Sedgewick Museum Site. When Alexander Todd had faced up to the need to move the Chemistry Department away from its constricted quarters of the 1920s and 1930s, he had made a leap across the Sedgewick Museum Site as far as Lensfield Road.

Between Lensfield Road and Bateman Street, on the one hand, and St. Andrew's Street and Trumpington Street, on the other, there is a great square of land known as New Town. Todd had arranged an option swap with Cambridge Town, whereby the university would relinquish its hold on the restricted New Museum Site in exchange for the right to the great square of New Town, thereby gaining for the university something like a sixfold increase of land available for building. Todd's colleagues in the university had spurned this chance because, although the area of the New Museum Site was greatly limited, its potential volume was not. It was thought that, by growing upwards, the volume could be extended essentially without limit. What happened then is perhaps best understood by thinking of a screeching crowd of gulls falling on the bits thrown out by fishermen as they trim their catch. Heads of departments would open bottles of champagne at home on days when they gained fifty cubic feet of volume, and they would wake screaming in the night when they lost fifty.

George Batchelor was in the thick of it. He secured University Committee agreement for a high tower for his new department of Applied Mathematics and Theoretical Physics, which eventually soared up to challenge King's College Chapel. At that point, it was sensibly turned down by the County Surveyor.

In my youth, a ditty that ran as follows was momentarily popular:

> The common cormorant or shag
> lays eggs inside a paper bag.
> But what these unobservant birds
> have never noticed is that herds
> of wandering bears may come with buns,
> and steal the bags to keep the crumbs.

Cambridge had many unobservant human birds. What those who constituted the General Board never noticed was that the New Museum Site was potentially explosive. The old buildings were full of wood, and all of it was tinder dry. In the daze in which most administrators seem to live, the General Board gave thumbs up to Alan Cottrell's Department of Metallurgy being accommodated there with its raging-hot furnaces. Pretty soon, fires were biting, and, because they tended to occur during the night when nobody was watching, they had a good hold by morning. George Batchelor had secured space for his new department across a kind of bridge in the middle of the site, biased towards the Free School Lane side, taking up a place underneath the eaves. But this was precisely where one of the fires took hold, and it was inevitably gutted. Thereafter, the department, over which so much had been said speciously at meetings of the Faculty Board of Mathematics, was consigned to gloomy quarters behind the old University Press building.

As the County Surveyor turned unrelenting thumbs down on the best-laid schemes of department heads, it gradually became apparent that something had to be done. Alexander Todd's scheme for acquiring New Town having been ineptly permitted to die, it was then essential to seek space on what was called the West Cambridge Site, which actually consisted of virgin fields most of the way to the outlying village of Coton, a site that was to ensure that future students of physics and chemistry were to be separated by two to three miles of heavy traffic.

In response to my inquiry about whether there might be a sliver of land somewhere else—my building being long and not very wide, I needed only a sliver—I was shown a few patches edged in red pencil, including a field in front of a house known as Madingley Rise. Early in the century, the house had been bought by H. F. Newall, a keen astronomer whose abilities candor would describe as consisting largely of his clubbability. The house was bought so as to be close to the Observatories under the direction of the Plumian professor. Newall then contributed, being a wealthy man, to the founding of a solar-physics observatory, and he also seems to have financed an ad hoc professorship for himself, a relaxed attitude about such matters being general in the early years of the century. The funds provided served to endow a Professorship of Astrophysics, to which F. J. M. Stratton succeeded after Newell's death. Stratton had much the same build as W. V. D. Hodge but was still more rubicund. Immensely popular with the astronomical establishment, Stratton obtained what were then considerable government funds for developing the Solar Physics Observatory, of which he became the director, with Eddington being the director of the observatory itself. When, in 1946, Harold Jeffreys turned his back on the traditional relation between the Plumian Professorship and the Observatories, Stratton's successor, R. O. Redman, became the boss of the entire outfit. It was thus into a field below Newall's house and bordering the Observatories that the University Estate Management Office now directed me, a field traditionally given over to the grazing of horses.

On a Sunday morning in February 1965, together with my son, I visited the proposed field, finding that the design of building just didn't seem to fit anywhere—certainly not in the middle, and hardly in any one of the corners. It was then that my son pointed out a strip of woodland running parallel to the Observatories boundary. It had just the right length and width. The deciduous trees around the periphery of the wood were in good condition, but the trees in the inner regions, which would have to be removed to make way for the building, were spindly larches in an inferior state. Essentially nothing that was good would have to go. The following morning, I returned, therefore, to the Estate Management Office to say it wasn't the valuable field I wanted but the inconsequential copse with the spindly larches in it. Instead of smiles, however, I was greeted with unrelieved gloom. The university didn't have the land after all. More accurately, they had it and they didn't have it. They had it in the sense that nobody else had it, but they couldn't use it, either the field or the copse, because Trinity College had a building lien on all the land around Newell's old house. Somehow, whatever you touched in Cambridge either let out a noise like a foghorn or collapsed into dust.

The next morning, I visited the Bursar of Trinity College. I explained the situation, pointing out that, since the deciduous trees on the periphery would be retained, only the spindly larches in the middle being lost, nothing amiss would be done to the environment. The redoubtable Bursar thought for a minute and then said: "I am sorry to inform you that our senior fellows have a special affection for those larch trees. Having regard to all the extra port that will be necessary to keep them happy, this exercise is going to cost you £10,000."

My next port of call was therefore the University Assistant Treasurer's Office. The Assistant Treasurer's name was Gardner. He had been a high official in Kenya, or maybe in Uganda. In appearance, he was a toughened-up version of Yul Brynner. After swearing more or less silently for several minutes, he decided that, since the university had agreed to provide the land, it should swallow the £10,000.

The completely straightforward application was the one to the Science Research Council. After determining what computer gave the best value for the sum of money that had been specified, it was simply a matter of making an application for the purchase of the computer in question on behalf of the proposed institute, giving the same arguments for the funding at the institute as in the other applications, the same arguments I had given years earlier in 1961–1963.

At the end of the Lent term at Cambridge, I was soon away to Scotland with Dick Cook and Norman Baggaley, our intention being to make for the sandstone mountains of the far north, basing ourselves at Kinlochewe. From Perth, we took the A9 to Inverness, emerging in the evening from dinner at a small hotel an hour south of Inverness to find snow falling. As we drove through Inverness, the snow thickened into a blizzard, but, along the ten miles open to the sea from Inverness to Beauly, it was something a good deal worse than a blizzard. Never have I been in snow so heavy. I doubt that there are many places in the world where the air can become so charged with the stuff as Scotland in early spring, a condition that adds an element of real danger to anyone caught out in distant country. For a while, mountaineering schools taught that, by carrying a survival bag weighing a couple of pounds and by burrowing into a self-made cave in snow, survival can be guaranteed. The idea is good, and it works ninety-five percent of the time. But snow can be so fine and so whipped by wind that no effective cave can be made, and the stuff somehow penetrates survival bags. Only after a seemingly well-equipped party was lost in just such a blizzard as we then encountered was it realized that this otherwise good technique is by no means a complete panacea. When we left Inverness, we were still thinking in terms of reaching Kinlochewe. Six or seven miles later, we had been reduced to hoping we could reach Beauly.

Generally speaking, snow falls more heavily in the eastern Highlands than in the west. Without having a clear idea of where we might go, we set off the following morning with the notion that things would be clearer to the west. In such situations, it is always a good idea to have an objective of some sort, even if you don't quite know what it is. The main road down the Great Highland Glen had been ploughed, making it relatively easy to drive southwest from Inverness to Invergarry, where we turned west onto the road for Kyle of Lochalsh and Skye, where we thought the Cuillin Hills (which are mountains, really) would be either comparatively free of snow or, at any rate, immensely spectacular. Any thought of reaching Skye that day disappeared, however, when an overturned lorry blocked the passage down Glen Shiel for five hours, half of which we spent eating lunch at the Cluanie Inn, and the other half of which we spent in stamping around in the snow, casting envious eyes at the chain of mountains that forms the Five Sisters ridge, the wall on the north side of Glen Shiel. After sliding and skidding down to Shiel Bridge, we edged along the side of Loch Duich as far as Ardelve, just beyond Dornie, where we found a hotel we liked so much that we stayed there for three days. There was little to be done for the first and second day, except more stamping around in the snow on the lower slopes, but, on the third day, we made an ascent that took me into a world I had not experienced before, a world I had not thought to exist in the sedate conditions of the British Isles.

Dick Cook wanted a mountain he hadn't climbed before. In view of the conditions, he chose the simplest to meet this requirement, some twenty miles to the north across Stromeferry, a mountain called Moruisg, which would have been a simple lump of a thing on a fine summer's day. With a strong wind in our backs, we went up easily enough, although it became exceedingly cold on the top 1000 feet, making me glad I was wearing a down jacket under my anorak. The descent, in the face of a furious icy wind, was something else. My spectacles iced up immediately. Removing them made me half blind. We were all half blind anyway from the fine particles of ice that the wind drove like a million fiery darts into our faces. To counter the pain, I bit on the cord that fastened the hood of my anorak, finding, by the time we reached the bottom, that I had bitten it clean through. It was the worst wind I ever encountered. Not in its strength, for in later years, when I lived in the English Lake District, there were days when shepherds could not go along the roads except by clinging to the stone walls. And I once took fifteen minutes in a raging gale to climb the top fifty feet of Ben More Assynt. But the cold of it and the driven particles of ice were quite exceptional.

In the next days, there developed a snow condition on the mountain tops that, although uncommon, does happen within the experience of most mountaineers who visit Scotland in the spring months. In the Lake District, it happened once in the fifteen years I lived there. It has to do with a heavy spring snowfall followed by extreme cold, which freezes the surface into a layer of ice about an inch thick, through which an ice axe cuts easily but through which it is impossible to penetrate even with heavy boots. When this happened in the Lake District, the local hospitals filled with fallen visitors.

With no wind on the day after our ascent of Moruisg, we climbed Aonach Meadhoin above Cluanie and then, for about two hours, traversed the ridge towards the Five Sisters. The icy surface continued all the way down from the summit ridge to the road 2000 feet below, which is just what an unchecked slip would have led to, a 2000-foot fall onto the projecting rocks of the Shiel Valley. To cut one's way down with an axe would be immensely fatiguing, and it would hardly be possible for a lone climber without the gift of powerful legs. One easy way is to carry crampons, which, in later years, I always did routinely from late November to the last of the spring snows. A second, even easier way, if you are lucky, and if the descent is down a southern slope on a clear day, is to depend on the warmth of the spring Sun. By about 2:00 P.M., the direct heat of the Sun may have melted the hard icy surface. What happens then is wonderful to a degree it is difficult to describe. All in a moment, you go from being totally helpless in the absence of ice axe and crampons to being totally in command of the situation. This happens in the moment when the boot bites through to softer snow beneath the upper crust, which still remains strong enough, however, to banish the possibility of avalanche.

Descending now needs no conscious downward movement. All you have to do is prance, and down you go, smoothly and at speed. But not silently, for the multitude of little ice chips you have set free go down with you in a cloud that makes a singing noise. Because this condition never occurs except on a clear blue day with the Sun bright towards the south, all the mountains around you are ablaze with light. There is just one snag, however: the descent seems over almost before it begins, so swiftly do you go, even down several thousand feet. The finest of scree slopes does not compare with it. In that week, we experienced singing snow twice, on the day above Cluanie and in Glen Lyon on our last day.

On a summer day, the traveler on the way to the Isle of Skye who turns west at Invergarry will see a striking mountain to the west that remains in view through much of the long ascent of the road towards Cluanie. This is Gairich, on the south side of Loch Quoich. The north ridge plunges to the loch as straight as a ruler. We made our way to the dam at the lower end of the loch, leaving the car there. The day was overcast and forbidding, so that we would hardly have thought of tackling Gairich had it not been for the exhilaration of the previous day. First, we had a six-mile slog over uneven, partially snow-covered ground to reach the starting point, with the prospect of a six-mile slog later in the day back to the car. We found a gully up to the north ridge and then a long, steeply rising ridge to the top. Ice was everywhere to a depth of a few inches, the snow cover having been blown away by the winds of previous days. Crampons made light of the ridge, however, both up and down. It was my first fairly lengthy experience in the use of crampons, and I was amazed at the security they conferred. In my twenties, I had come to fear ice. Now, at fifty, I lost some of the fear, thanks to the crampons. So even that apparently unpromising day turned out to be exhilarating in its own way. It also turned out remarkable in a respect none of us had quite experienced before. Arriving back at the car, we drank enormous quantities of tea, and we also consumed large quantities of a huge cake that Norman Baggaley had brought along. Then the unslakable thirst continued through an evening spent at the Lovat Arms in Fort Augustus. Why we had lost so much water on an apparently cold day, I never quite understood.

A day of fantastic wind followed. Since snow was blown up in the air to form diffuse white clouds several thousand feet above the mountain tops, it must surely have been even fiercer than the day on Moruisg. It needed only a glance at the conditions, otherwise sunlit again, to tell us nothing was possible except to drive along gently to the south. We did this, arriving eventually at the Breadalbane Arms in Kenmore at the lower end of Loch Tay, whence, on the morrow, we made our final foray into Glen Lyon, again being favored by singing snow.

On the day of icy wind on Moruisg, there was honey in the comb for tea at the hotel in Ardelve. After dinner in front of a blazing fire, Dick Cook produced the little suitcase in which he kept his stock of maps. He often did this in the evening, usually to take out a gardening periodical. On this evening, however, he took out a slim volume in which, after carefully leafing through a number of pages, he entered a tick from a pencil, a single tick mark. Curious to understand what was happening, I asked to see the volume. It had the title *Munro's Tables*, from an original compilation by Sir Hector Munro. Inquiring as to what this meant, Norman explained that a Munro was a separate peak above 3000 feet in height. *Munro's Tables* listed all there were in Scotland, about 280. Moruisg had been one that neither Dick nor Norman had "done" before, which was why Dick had now added a tick mark against it. When Norman asked how many of the 280 I had "done" myself, I began by thinking I had done quite a lot, but, after about an hour of study, I answered in a chagrined way: "Only about thirty." Norman had done eighty, and Dick had done about 160. Each night thereafter, I studied *Munro's Tables* with growing wonder. Pretty soon, after returning to Cambridge, I acquired a copy for myself, discovering, on Ordnance Survey maps, mountains by the score that, until then, I didn't know existed. If the five "Munros" we had just climbed could yield such splendid days, what might the 245 still unknown summits reveal? I decided that I would climb all 280, beginning in 1965 at the age of fifty. The long hard day on Gairich suggested that I had built up, since January, a fair measure of toughness. Although cramponing down a long icy slope is not difficult technically, it puts a strain on the legs and particularly on the knees. The welcome fact that my formerly weak left knee had shown no recurrence of the ligament strain suggested that my hitherto persistent trouble with my knee was behind me now. So why not? In September 1964, I had sought to fly away a free bird, only to be enticed back into the cage. Climbing all the Munros would be something of a substitute for being a free bird.

So to what did I return in Cambridge? To a reply from the Nuffield Foundation, which was exactly as Alexander Todd had said it would be; and to one from the Wolfson Foundation, which was as John Cockcroft had said it would be. I returned also to a tortuous situation at the newly forming Science Research Council that was to cause a further fifteen months of turmoil. Before the turmoil would be soothed away, I would have climbed my first hundred Munros. Or, to put it another way, I would have climbed the height from sea level to the summit of Mount Everest ten times over. There are few troubles in life that one can't learn something from. In this case, indeed, an external observer might have learned something important. It was that, in deciding in the mid-1960s to fund scientific research through a system of research councils, the government had laid a vast egg for itself. Personally, too, I had learned to distinguish the private sector from the public sector. Some think this to be a matter of economics, but it is more a matter of mental structure. In the private sector, the higher the position of a person who gives you a promise, the more reliable it will be. In the public sector, it is the opposite: "Put not your trust in princes," says the Psalmist, "nor in the son of man, in whom there is no help." A little vague, perhaps, but generally along the right lines, as far as the public sector is concerned.

Chapter 24

The Institute of Astronomy,
but Still in Slow Stages

IN MAY 1965, the Science Research Council responded in a surprising way. My grant application would be approved, except that it could not be for an institute in Cambridge. The SRC official who had visited me the previous October knew perfectly well that the project depended on the good will of three parties: the Nuffield Foundation, the Wolfson Foundation, and the SRC itself. The first two accepted Cambridge as the site and, indeed, probably required that Cambridge be the site. Knowing this from the beginning, why had the SRC made such a peculiar decision? Because it had referred the matter to a lower committee, two pegs down from the SRC itself. It was as if a bank manager had offered you a loan and then left the final decision to an assistant to the assistant manager. This is a technique that government servants use to evade responsibility. When things go wrong, it becomes almost impossible, in such an arrangement, to see who is to blame. The bank manager points to his assistant, and the assistant to his assistant, and so on, in all kinds of convolutions.

If the Institute of Astronomy was not to be in Cambridge, I asked, where was it to be? Anywhere, I was told; anywhere at all, so long as it wasn't in Cambridge. This put me in good spirits. Provided everybody now stuck to their guns, there would be an impasse not of my choosing, and I would be able to walk away, free and unscathed, from what appeared to be becoming, in American parlance, a can of worms. I had to visit the United States in connection with other work, and I went off in a relaxed frame of mind. The visit occupied about the right length of time to suggest that I had considered the SRC proposal in a sober frame of mind, permitting me to respond immediately upon my return. Yes, I would agree to abandoning Cambridge. I would also agree to resigning my Chair in Cambridge, a necessary step. My choice of a site would be the very far north of the British Isles, almost as far north as it is possible to get, short of plunging into the North Sea. I would go to Inverness.

There was little likelihood that such a proposal would be accepted by all parties, but, if it had been, I think I still would have done it. A staff of up to twenty-five would be self-sustaining. It was the correct hunting-party size. The town of Inverness had wanted a university a few years before, but it had been turned down by the government, probably because the Highlands did not provide a sufficient student catchment area for both Aberdeen and Inverness. This recent event suggested that, if the Institute of Astronomy were at a higher level than an undergraduate university, it would win support from the town, a town with perhaps the finest river in Britain flowing through it. Willy Fowler said that, at the time, he thought I was out of my wits. A few years later, when he came to know Inverness and the Highlands, he changed his opinion.

The SRC then went as far as consulting the Scottish Office, only to be told that they would not support the choice of Inverness but would very much like to have the institute in Edinburgh or Glasgow. So the SRC then asked me if the institute could be located in one of those two cities. I now became quite angry. First the SRC had promised Cambridge, then they had said anywhere. As far as I was concerned, they could now jump in the River Ness, which flows very rapidly out to sea.

Just when my escape seemed imminent, the SRC backtracked. Yes, they would support the establishment of the institute in Cambridge after all, but only on the condition that the university were to make a 20 percent financial contribution to the project. I visited the vice-chancellor, begging him to refuse. Providing land for the institute had so far required little beyond the cooperation of the Estate Management Office, and of the treasurer's agreement to pay £10,000 to Trinity College. But a 20 percent contribution would require consultations with the Faculty Board of Physics and Chemistry, with the Faculty Board of Mathematics, with a new piece of bureaucracy called the Council of the School of Physical Sciences (a school that, until then, everybody had thought defunct), and, ultimately, with the General Board. Assuming that all these hurdles could be jumped, there would still be the real possibility of a *non placet* in the Senate House.

One half of me knew all this perfectly well, but the other half was still the diffident young fellow who had arrived in Cambridge in the autumn of 1933. The diffident half had respect for a vice-chancellor. When the vice-chancellor said he thought it "right" that the university make a contribution such as the SRC had suggested, I became reluctantly convinced. And, in truth, with the vice-chancellor taking this line, to have resigned there and then would have seemed peculiar to everybody, including my own family.

A vice-chancellor's tenure of office was then a mere two years. In those days, one vice-chancellor succeeded another without much continuity. I came eventually to realize that it was a primary motivation of any vice-chancellor simply to get through the two-year term without a scandal breaking loose. And so I think it was in this case. Pretty soon, the vice-chancellor who had taken the high moral ground of thinking it "right" for the university to accept the responsibility of a 20 percent contribution would be gone, leaving a successor knowing little or nothing about the agreement.

What happened was rather worse than the other half of me had expected, for the General Board decided to hold what was called a Discussion on the matter. This was a meeting in the Senate House, before the vice-chancellor, that any member of the university could attend and before which any member could speak, with their remarks reported later in printed form. George Batchelor appeared to specify his conditions for not mounting a *non placet*. There were two main ones: the institute should not participate in university teaching, and the institute should not seek to poach members of his staff. So the university lost the fairly considerable amount of teaching we might have done, essentially because of the meddling of the SRC, whose charter required it to support education in the universities.

As 1965 passed into 1966, with the affair grinding relentlessly on from one committee to another, I gained support from Todd, Cockcroft, and even the SRC, on account of the university's intent to secure 100 percent control of the institute in exchange for 20 percent of the financing. Outside the faculties, this was seen as improper, and it became possible to sway the General Board to the extent that the university eventually got only 40 percent of the control. Indeed, the General Board accepted my suggestion that the institute be placed directly under its own aegis rather than under a faculty, and that it should operate under a management committee on which Todd and Cockcroft would sit, with Todd as the chairman. The university's 20 percent contribution devolved into paying rates on the institute building, providing assistance out of the university's secretarial pool, and paying fuel and electricity bills.

The climbing of 100 "Munros" in this period was a contact with sanity. There were also sane issues in the wider Universe. George Gamow's idea that the chemical elements could have been produced in the early moments of a big-bang Universe had fallen on stony ground. The building chain could not be carried beyond gaps at atomic masses 5 and 8, and, from the early 1950s onwards, there was overwhelming evidence to show that most of the synthesis of elements took place in stars. There remained a small residue of Gamow's idea, however. Notably, it would be possible to produce helium in an early Universe of the big-bang type, together with a small quantity of deuterium (^2H) and an even smaller quantity of lithium-7. But a similar quantity of lithium could be produced from the disruption of carbon and oxygen by cosmic rays, while it hardly seemed, in the 1950s, that helium—produced, we knew, in ordinary stars—was much of an issue. Thus, the idea of an early Universe producing chemical elements fell into disfavor, although two of Gamow's students, Ralph A. Alpher and Robert C. Herman, continued for a while to publish papers discussing a supposedly primordial synthesis of helium and possibly of other very light elements.

In returning over old ground in the mid-1950s, I became disturbed by the difficulty of producing enough helium in stars to explain the total quantity that was observed in galaxies. Calculations suggested that about 2 percent of helium would be produced relative to hydrogen, whereas observation revealed that the helium concentration in galaxies was about 30 percent. Tommy Gold, Hermann Bondi, and I published a letter in 1955 in the *Observatory* drawing attention to this problem. The letter suggested that hydrogen-to-helium conversion might take place in nonvisible stars located inside dense clouds, which clouds emitted radiation strongly in the far infrared—which turned out to be true, but not sufficiently true to explain a helium discrepancy between 2 percent production in stars and the observed 30 percent. In preparing a lecture course on cosmology in Cambridge in 1963–1964, I decided the problem was indeed so severe that a return to Gamow's position was inevitable. To this end, my colleague Roger Tayler and I reworked the old calculations with later data on the neutron half-life and with more than one form of neutrino. Our conclusion was that the matter in the Universe had to be born in a high-temperature radiation bath in which, for every proton or neutron, there had to be a thousand million or more radiation quanta. Because we could not be certain of the exact amount of primordial helium, we could not specify the ratio of quanta to particles precisely, but it had to be of this general order.

A distinguished cosmologist of the younger generation remarked the other day that the widely favored big-bang theory is not so much a theory as an illusion, a view from which I started in 1947–1948. Big-bang cosmology is an illusion because, after asserting that matter cannot be created, it proceeds to create the entire Universe. It does so outside both mathematics and physics, by metaphysical assertion. Thereafter, once created, the Universe expands like a large uniform cloud subject to its own gravitational pull, which tends to decelerate the expansion. If the initial metaphysical expansion is fast enough, the cloud will survive the outward motion and will disperse with a considerable final speed. If, on the other hand, the metaphysical expansion is too slow, gravitation will eventually slow down the outward motion to zero. Thereafter, the cloud will reverse itself and collapse back whence it came—metaphysical nothingness. Like a pencil balancing on its point, there is an intermediate case, in which the cloud just happens to slow down to zero as it expands infinitely apart. This pencil balanced on its point, or extremely near its point, is how the Universe is supposed to be, according to big-bang cosmology.

Such a view would easily have been accessible to Isaac Newton or his followers, with mathematical results no different (as E. A. Milne showed) from the equations obtained by the Russian mathematician Aleksandr Friedmann in 1922–1924, using Einstein's form of gravitation. A major reason for the popularity of the big-bang theory is undoubtedly that it is simple enough to place no burden on the mind. Undoubtedly, too, there are many who are attracted by its retreat into metaphysics. For myself, I find the retreat into nonexplanation unsatisfactory, contrasting so markedly with the exquisite subtlety of all science outside cosmology. Can the Universe really be so crude while all the rest is so refined?

Tommy Gold describes the standard big-bang theory like this. An enthusiastic young astronomer explains to a general audience how the motion of the Earth around the Sun prevents the Earth from being pulled, by solar gravitation, into the Sun. At the end, a little old lady comes up and says: "Young man, you have it all wrong. The reason the Earth does not fall into the Sun is that it is held up on the back of a large turtle." To this, the young astronomer replies with an easy smile: "And what holds up the turtle?" The old lady becomes confidential, replying: "An even larger turtle." Then, as the young astronomer's mouth opens once again, she continues: "It's no use, young man. It's turtles all the way down."

It was in November 1947 that I decided to try to push metaphysics out of cosmology, the occasion being a meeting in Birmingham of the Physical Society. Rudolf Peierls had asked me to speak on the synthesis of elements in stars. At the end, he remarked something as follows: "You have told us about the synthesis of nuclei from hydrogen. Next you'd better tell us where the hydrogen comes from." I had traveled to Birmingham, first by driving the open Singer two-seater, which I had bought from George Booth for £5, as far as Northampton, and had gone on from there by rail. The weather was icy cold, and it was during the return journey, at close on midnight, frozen silly in the open car, that I decided to see if I could discover where the hydrogen came from—in other words, to see if matter could be created. How this turned out I shall describe in my ultimate chapter, where I will attempt to come to grips with the continuing struggle between physics and metaphysics in cosmology.

In July 1966, the Senate at Cambridge passed a grace founding the Institute of Astronomy. So, at last, work could begin. It had been possible by now to refine earlier financial estimates that had originally been made in a heterodox way. The usual technique is to start with an artificially low estimate, to suck in the donors. Then, having got the donors committed, the estimates rise towards reality—not in a single move, because that would shake the confidence of the donors too much, but in a carefully escalated set of increments designed marginally to avoid suspended shock.

Perhaps because, in my youth, I really did know people who had to plan their financial affairs with accuracy in order to survive, I have always felt differently about cost estimates. I started from an estimate that I thought would genuinely meet the eventual cost. With a contingency added, this meant that, unless the estimate was poor, there could very well be a margin left over on the project. The Anglo-Australian Telescope was pretty well built in this way, and so was the U.K. Schmidt; but, on all the committees of which I was a member, I rarely ran into anything but the unrealistic, orthodox method of estimating costs.

Anyway, revised estimates for the building gave a reduction of £35,000. This so astonished officials at the Wolfson Foundation that they argued something as follows: If this chap can go down from £175,000 to £140,000, perhaps, if we look into it, the cost may come down further—say, to £125,000? Pretty soon, then, we had a bright, wide-awake official at a meeting of the University Building Committee. It was at this meeting that the treasurer of the university lost his temper, transformed all in an instant into a raging bull, so that, for one glorious moment, I thought the treasurer's mad charge at the Wolfson official would bring the entire project to a gory end.

For myself, the bright official reminded me of fresh air and oranges, the way all bright people on committees do. In my boyhood, there was always a fair on Whitmonday on rough ground above Shipley Glen. People came up from the valleys below for an outing and especially for a breath of fresh air. I must have been seven years old in the year a stall vendor held up big yellow oranges and shouted: "Oranges! Full of fresh air!" I have seen bright, eager people on committees all the way up to the Council of the Royal Society, and even on an occasion when Harold Wilson invited the Royal Society's officers to a Cabinet meeting. Invariably, they were full of fresh air.

As a Parthian shot, the Wolfson official told us that Sir Isaac had lost confidence in architects and wouldn't pay for them anymore. A tour of university and college buildings constructed in the 1960s will give modern visitors an inkling of why Sir Isaac felt as he did. Thus, my estimate of £140,000 would be net, without 15 percent added for architects' fees, which it would be up to the university to deal with. On the morrow, I visited the treasurer to make a deal: no added fees in exchange for the building committee going defunct. It was too good a bargain for the treasurer to miss, so that was the last of the building committee. Meetings thereafter were with the builder, Rattee and Kett, known in the university as Kett.

The building was started on 1 August 1966, and it was finished on 1 August 1967. Kett had become a subsidiary of J. Mowlem, and Mr. Birmingham, from Mowlem, converted my sketches into working drawings. Mr. J. Balser (from Kett) and I toured brickmakers in East Anglia, eventually to obtain the brick that was used. Balser had heard of big wooden-sashed windows that happened to be available at a good price, and these were the ones we used. On one count, I failed to overrule the triumvirate of Balser, Birmingham, and Poindexter, an elderly quantity surveyor from the University Estate Office. I wanted a wooden roof with simple canvas waterproofing for protection. They all wanted heavy concrete blocks, arguing there would be no leaks that way. In the short term, this proved to be true. But, a couple of decades down the road, the concrete did begin to leak, and then it cost a pretty penny to put right. Canvas might have needed replacing every ten years, but only at a comparatively low cost.

So, by 1 August 1967, there I was, the institute started, with R. O. Redman, the Professor of Astrophysics (Newell's legacy), on the one side and Sir Edward Bullard on the other. I always considered that Redman would have done best to have reached some arrangement with me already in 1958. My predecessors had been directors of the Observatories long before the professorship of astrophysics existed, and the circumstance that Harold Jeffreys had vacated the directorship should not have interfered with there being a long-term return to the traditional state of affairs. I would have been happy for Redman to have continued as director to his retirement in 1972, with the return to tradition deferred until then. Unfortunately, Redman made no move that way.

The problem with Redman was that he had an unnaturally intense sense of territory, which was exacerbated by Teddy Bullard, whom he regarded as a predator without scruples. Teddy was forever trying to get hold of Redman's land, on all manner of pretexts (some ingenious, others specious). At one point, circa 1967, he instructed his staff to park their cars on Redman's land, to which Redman immediately replied by erecting a chain of bollards. Redman came to see me about that, as he had begun to do more and more, speaking of Teddy as if he were a trickster out of Soho. Actually, if Redman hadn't defended his ground so fiercely, there would have been none available in recent times to accommodate the move of the Royal Greenwich Observatory to Cambridge. All of it would have been snatched by the Department of Geophysics.

Teddy Bullard had been a neighbor in Clarkson Road, as he was now at Madingley Rise. Although he had an offhand antiestablishment manner, this was really a pose. On every critical committee vote that I ever saw, he always voted pure establishment. Because he had a sense of humor and was always cheerful, I liked him, but, on a sensitive issue of business, I preferred Redman, gloom and all. I felt I could trust what Redman said, but I would have run a mile rather than rely on Teddy. He and I received instructions from the university administration on the same day, stating that, in the interest of fuel economy, we should change the heating systems of our respective buildings from gas-fired to oil-fired. Because I had just installed gas-fired boilers costing £10,000, I shredded the letter. Teddy, on the other hand, obeyed his letter and proceeded to erect the fifty-foot flue pipe required for oil-burning heating systems by town regulations—an establishment man at heart. If it had affected him, Redman would have called an emergency meeting of his Observatory Management Committee. Then, with the committee's support, he would have written a hot letter to the university administration, with the probable consequence that, after further correspondence, the instruction would have been canceled, by which time it would cost, in everybody's time, a hundred times more than the original fuel economy was meant to save. This small episode will explain how it came about that a profound difference existed between my opinion of myself as an administrator, which was good, and the opinion of others.

By the summer of 1967, computers had improved considerably since the spring of 1965, when I had prepared my application to SRC. Rather than £375,000 for the computer plus running costs over five years, I could now get the same results for £275,000. The simplest procedure would have been to upgrade the quality of the computer by £100,000; and, if I had done that, I would have ended up £150,000 better off. Because £275,000 would be adequate, however, I had no wish to spend public and foundation moneys unnecessarily. So I simply reported that £100,000 would be saved on the computer. In the original letter of agreement between the foundations, the university, and the SRC, it had been agreed that costs would be shared, with the foundations and the SRC each being responsible for 40 percent, and the university, the remaining 20 percent, without the nature of the expenditures being specified in detail. Clearly, I thought, the foundations and the SRC should each save £40,000, and the university should save £20,000. This was not what happened, however. The SRC began by taking the entire £100,000, thereby triggering a reduction of £100,000 from the foundations and £50,000 from the university. I mention this detail not out of a sense of grievance (although, even at this late date, the behavior of the SRC still strikes me as peculiar) but as a sort of alibi. When we come to the rather fraught ending of the next chapter, I would be happy to claim that, in accepting this financial prestidigitation, I established in some degree that I am not a person given to tantrums.

Chapter 25

The Thirty-ninth Step

ONE MORNING in October 1967, I came into my office at the new institute to find Jim Hosie sitting there. I had last seen him at lectures in 1935–1936. He was a mathematics graduate from Glasgow who had come to Cambridge for a further two years, going on from there to a post of high responsibility in the Indian civil service. Returning to the United Kingdom in the late 1940s, he had begun a second career, and now, some fifteen years later, he was the director of the Science Research Council's Division of Astronomy and Space. His business was to ask me to become the chairman of the astronomy section of his division—essentially the same committee as in Blackett's day at the old Department of Scientific and Industrial Research. Additionally, I would be a member of the combined Astronomy, Space and Radio Board and of the Science Research Council itself. It seemed a big change from the inconsequential position I had been in over the past two years.

The SRC appointments, in my case, were for five years, except for my membership in the Anglo-Australian Telescope project, which was to last for eight years. The AAT was a long-running cliffhanger, both technically and politically, that has been faithfully described in a recent volume: *The Creation of the Anglo-Australian Observatory*, by S.C.B. Gascoigne, K.M. Proust, and M.O. Robins (Cambridge University Press, 1990).

The five years at the SRC were to include successes, but, unfortunately, not as many of them as there might have been. There was the rebuilding of the Lovell Telescope at Jodrell Bank, the provision of a new large radiotelescope for Ryle, the building of the 48-inch Schmidt telescope in Australia, the beginning of infrared astronomy, and a general increase in the number and sizes of research grants to astronomy. The failures were to establish a Northern Hemisphere observatory with a likely site in the Canary Islands, to build a new, large, steerable radiotelescope for Jodrell Bank, and the provision of a first instrument for work in the rapidly developing field of millimeter-wave astronomy.

The problem with a new, large, steerable radiotelescope was the sheer scale of Bernard Lovell's conception, a paraboloid with a diameter of 400 feet. At first sight, 400 feet does not seem so much larger than the 250 feet of the existing instrument at Jodrell, but it was four times greater in weight, taking it close to the limit of what anyone believed to be possible. A projected 600-foot radiotelescope at Sugar Grove, West Virginia, collapsed literally in the twinkling of an eye. It went so quickly that I am not sure anybody actually saw it change from an ordered structure to a chaotic mass of twisted metal. After many years of operation, a partially steerable 300-foot radiotelescope at Green Bank, West Virginia, did the same, to the astonishment of astronomers everywhere. The more Lovell's concept was discussed, the more convinced I became that Taffy (E. G.) Bowen had hit about the optimum size for a steerable metal paraboloid, the 210-foot instrument at Parkes, New South Wales, Australia. Lovell, I think, had already gone somewhat beyond the optimum in the 250-foot radiotelescope at Jodrell Bank. Nothing of this could be said in 1967, of course. So much momentum had been built up for the 400-foot concept that it had to be treated as a project with high priority, coming in, according to Lovell's estimate, a little below £5 million, nearly the same as the British contribution to the Anglo-Australian Telescope (the latter being an optical instrument, not a radiotelescope).

Readers of Lovell's autobiography *Astronomer by Chance* will find Bernard's life to have been a succession of challenges, all brought to successful conclusions. There was the wartime challenge to produce much-needed radar equipment for night fighters, the challenge to produce radar maps of the ground, the challenge to adapt wartime equipment to the needs of astronomy, and, of course, the ultimate challenge of the 250-foot radiotelescope itself. A pattern existed in all these adventures, a pattern that would have been recognized by Ernest Rutherford and, indeed, by Lovell's own mentor, Patrick Blackett. It was a pattern of moving quickly into a project and then proceeding by increasing experience to a successful result. But not everything can be done in that dashing way. Those of us working for the Admiralty during the war could not proceed by making a beginning and going on from there by trial steps. Things had to be fully engineered from the start.

Over the years from 1967, the feeling developed generally among those of us who were uncommitted at the SRC that the 400-foot radiotelescope project must fall into this second class—less adventurous, perhaps, but essential, if both the physical disaster of Sugar Grove and serious cost overruns were to be avoided. It was a question of obtaining a complete design and of knowing its cost, a requirement the project was, unfortunately, never able to fulfil, even within rather wide financial constraints.

I think it was simply that the physics of materials was against it. I know Lovell felt abandoned, especially after his high hopes in the period 1965–1970. But, at the risk of giving offense where I would much like to avoid it, since I regard Lovell as an outstanding figure of my generation, I feel that he was lucky. Nothing could have been worse than to have run at last, after so much that was excellent, into an irreversible catastrophe.

I had the temerity, at one time, to try tempting Bernard out of his fixation with the 400-foot radiotelescope. For a space of five hours, I got him to contemplate abandoning the monster in favor of a major development in millimeter-wave astronomy. There would have been £3 million for it, an enormous sum circa 1970. If I could have kept Bernard awake all night instead of allowing him to sleep on it, there might have been a chance, and with marvelous consequences.

There was hope of a millimeter-wave project from another direction. At the review by the Science Research Council of the year's scientific achievement held in the spring of 1971, I spoke of American work on the detection of complex molecules in space. This fired the SRC's collective imagination sufficiently for me to be asked to formulate a project that would permit a quick entry into the field. With the United States well ahead, there was no hope of a successful, speedy entry, except with an American partner. I was fortunate to obtain the best, the Bell Telephone Laboratories in New Jersey, which agreed to provide, at no charge, the difficult, sensitive "front end" of the receiver. John Saxton, the director of the Radio Research Station at Slough, agreed to handle the "back end," also at no charge. Using a site on Siding Spring Mountain in New South Wales, close to the 48-inch Schmidt telescope, would again allow us to avoid the cost of infrastructure, leaving most of the SRC grant of £0.3 million for the aerial and tracking gear. The plan received the unanimous support of the SRC astronomy committee, but it was blocked on the Astronomy and Space Board by Martin Ryle, with the help of an electrical engineer called Eric Eastwood. The awkward feature of SRC policy was that no project could go through if it was opposed by anyone of senior status. I thought that, possibly, it was really to me that Ryle objected, so I handed the entire project to John Saxton. But Ryle and Eastwood still blackballed it, ostensibly on account of the foreign collaboration. Ryle claimed that he could handle the "front end" himself, keeping the Americans out; but, in fact, no British millimeter-wave facility became available thereafter until well into the 1980s, by which time the opportunity for pioneering work in the field had gone.

We also failed, in 1967–1972, to establish a Northern Hemisphere observatory in the Canary Islands, which we thought might provide a suitable site. A small committee was set up to report on the best mechanical design of a large telescope, since we thought the equatorial mounting of the Anglo-Australian Telescope, while good, was probably unnecessarily expensive. The committee reported in favor of a more compact altazimuth mounting, and this indeed has been the design chosen for most large telescopes in recent years. The problem came with obtaining a report of similar quality on the proposed observatory site. This had to be organized, the way things were, through the Royal Greenwich Observatory. But Richard Woolley, the Astronomer Royal, did nothing about it for more than a year. Faced with his persistent noncooperation, we declined to go into a proposal with the site unspecified, judging that, even if the SRC were to validate an incomplete proposal, the Treasury would not. Building a telescope on Spanish territory was difficult, in those days. Only a solid technical case for doing so could have succeeded. Thus, a favorable moment for obtaining the financial support needed for a Northern Hemisphere observatory was lost because of the very person who was supposed to stand foremost for British optical astronomy. Although I was not privy to the decision taken a while later to separate the office of Astronomer Royal from the Royal Greenwich Observatory, I would suspect that this affair had something to do with it. I should record, however, that Woolley did not always act uncooperatively; but, unless he was continuously flattered, he was quite unpredictable.

Those were the years in which the SRC's allocation of money, its "vote," as it was called, rose at an annual rate of 10 percent or more. Hosie was particularly good at getting his hands on the increasing flow of money, essentially through well-conceived projects being available to absorb whatever came along from the Treasury. The game was to be better prepared than the other divisions of the SRC. Where projects could not be made watertight, as with Lovell's 400-foot radiotelescope, the Northern Hemisphere observatory, and, to a lesser degree, the millimeter-wave initiative, the opportunity was lost.

Those were also the years in which the SRC committed sins, both of commission and of omission, that were to haunt it ever since. To increase its budget maximally, the SRC was avid to become the paymaster for Britain's international scientific commitments, especially of CERN, the international nuclear laboratory in Geneva, and ESRO (European Space Research Organization) in Paris. By doing so, the annual increment of 10 percent was calculated on such international subscriptions as well as on the domestic budget, a practice that some might describe as sharp—and not too bright, either. International subscriptions are paid in hard currency. With the value of sterling generally falling in relation to the hard currencies throughout the 1970s and 1980s, the SRC was forever being asked to make up currency declines from out of its own internal budget.

There were always piercing exclamations from the SRC's senior officers about the "unfairness" of such impositions, but, so far as I could see, it was perfectly fair, because it was the way the SRC had wanted it in the first place.

Those who planned the research-council system made a blunder with respect to the SRC. After it was decided that the SRC's main objective should be to foster research in the universities, it was also decided, as an addendum, that a collection of the government's own scientific establishments should be rounded up and placed under the SRC's administration. If the establishments in question had been serving a vital function, this would not, of course, have been feasible. Their "owners" (as, for instance, the Admiralty was the owner of the Royal Greenwich Observatory) would have objected too strenuously for any such change to have been possible. It was precisely because the institutions in question had all outlived their functions that their owners allowed them to go.

The raison d'être for the Royal Greenwich Observatory had been to keep the nation's clocks and to supply necessary data for ships at sea, in the form of a publication known as the *Nautical Almanac*. The scientific basis for this work lay in what is often called fundamental astronomy, which consists primarily in measuring the positions of stars in the sky. The entire scale of astronomical distances, ranging to the most distant galaxies, depends ultimately on such measurements. These measurements continue to be of importance, as shown by the commitment to the recent European satellite *Hipparchos*, which, unfortunately, was rendered partially inoperable on launch. Nothing of this should have been taken away from the RGO. But for the RGO to become engaged in the same forms of research as the universities—indeed, to copy from the universities— was quite wrong. The process had been started already by Richard Woolley's predecessor, Harold Spencer-Jones. When he appointed Tommy Gold as Chief Assistant, it had been a move in that direction—an apparently small move, but with the seed of devastation in it. Woolley had accelerated the trend, with the avowed intention of converting Herstmonceux Castle into a latter-day form of university.

In the 1920s, radio had been a mystique, and the government of the day had founded a laboratory for research in the subject—which, at the time, seemed sensible enough. But, by the 1930s, radio research was no longer the avant-garde activity it had been a few years before. Then, in the wartime years, the old-style research sponsored by the government was vastly outdistanced by new discoveries coming originally from the universities. All reason for having the radio laboratory had gone, and, as would happen in business, the laboratory should have been terminated. Instead, it grew ever larger, a phenomenon endemic in government. This was another odd lot that fell to the SRC, adding yet another straw to the camel's back.

The worst cases, however, were in nuclear physics. The 1939–1945 war ended with the belief, among physicists, that almost any modest accelerator would suffice to unlock the secrets of particle physics, and there had been no shortage of universities anxious to get into the business of building such accelerators. The prewar Cavendish Laboratory, with its high concentration of nuclear physicists, had by now disintegrated its complement into university departments throughout the country. Birmingham, London, Manchester, Glasgow, together with a residuum still in Cambridge, were all totally certain of the situation, all seeking to stress their claims at meetings of the Royal Society, on committees of the DSIR, and in demands leveled at the weak Attlee administration. Pretty soon, however, in a myriad of accelerator projects, the spreading disaster began to show itself, and wiser counsel at last prevailed. Attention to available resources forced through the concept of a national laboratory with a defined duty to serve the needs of everyone. It was to be called the Rutherford Laboratory, and it was to be sited near the government's Harwell establishment some ten miles south of Oxford.

Having regard to the British penchant for naming its worst white elephants after its greatest men, the Rutherford Laboratory was doomed before even a sod was turned. The initial misjudgment was repeated. The problems of particle physics were not one but two orders of magnitude larger than had been thought in 1945. Instead of being staggered by the double magnitude of their misjudgment, nuclear physicists lost nothing of their confidence. Indeed, as they saw it, their subject and, with it, themselves were important in proportion to their demands for public money— doubly so, now that the nearly bottomless pit of high-energy physics was becoming apparent. The government was urged into joining the European consortium located in Geneva, CERN. This was actually a correct decision. What was not correct was to continue with the Rutherford Laboratory, which had, paradoxically, become both too big and too small. It did not have the function of a university in education, nor was its scale sufficient to make it effective in research. The correct decision would have been to abandon it, but, instead of that, it was passed on to the SRC.

The nuclear physicists basked in the new situation. To understand what it means to bask, it is necessary to observe sunflowers in a field, to watch how thousands upon thousands of them keep their faces constantly towards the Sun as the morning progresses. This was the way it was with the nuclear physicists and the SRC. CERN was an absolute, of course; but only a little lower on the scale were Rutherford and grants to a wide range of universities. With vaulting ambition, nuclear physicists would even have a second national laboratory, an electron accelerator sited at Daresbury in Cheshire.

All this did astronomy some good. The nuclear physicists were our main competitors, and, since everybody could see they were playing a game, we tended to get the lion's share of whatever large sums were going. Two chemists, Ron Nyholm from London and Ewart Jones from Oxford, wrote a diatribe against nuclear physics. It was Ewart Jones who had pressed the Research Council for a millimeter-wave project as a halfway house between astronomy and chemistry, the project eventually killed by Ryle. It will perhaps be understood now why any disagreements within one's own ranks led instantly to money that could have come to astronomy being immediately diverted elsewhere.

The drollest moment in my association with the SRC occurred at the beginning in October 1967. Brian Flowers, later Lord Flowers, had not long been Chairman of the Science Research Council when, on a visit to Cambridge, he asked to see me. In a conversation at my home in Clarkson Road, he raised the topic of the institutions that the SRC had so unfortunately inherited. Would I be willing, he asked, to consider cutting back the activities of the Royal Greenwich Observatory? The institutions were consuming too much of the SRC's resources, he explained. Within four years, Flowers would be riding hard in the opposite direction, leaving me to cope with the situation he thus created. The cause of Flowers's retreat from Moscow was a device called a high-flux neutron beam reactor (HFBR). Having failed to stem union pressure at Rutherford and Daresbury, Flowers saw the problem, especially at Rutherford, as finding something useful for the staff there to do. The HFBR was brought forward as the solution. The case for it was peculiar.

Witness after witness came to give evidence to the SRC, explaining what wonderful things such an instrument would mean for their research. It was actually all so trite that Flowers (who, in my opinion, was pushing the project to get the unions off his back) was reduced to arguing that, although no single project could be judged to have first-class importance, there were so many of a smaller nature that they added together to a formidable requirement. The interesting thing was to watch how the various council members behaved in the face of this taradiddle.

Understandable support came from chemists, who, to this point, had not had any large project that was relevant to them. The less understandable support came from council members whom Flowers wheeled in from time to time. There were three industrialists, experienced enough to see that the capital cost of the HFBR plus its operating costs would consume all the free money available to the SRC for close on a decade. They spoke against the project, or, at any rate, they spoke about it doubtfully. But, when it came to the vote, one went with Flowers and the other two absented themselves from the part of the meeting where the vote was taken. It was this behavior of the industrialists that taught me you can't tell whether a fellow committee member is strong or weak or middling simply by listening to what he says. You must watch how he votes or doesn't vote.

Lovell was naturally appalled at the thought of the SRC's free money all disappearing down one huge hole, with inevitable consequences for radioastronomy. It took three months for this simple message to get through to the nuclear physicists. When eventually it did get past the static that always seemed to buzz inside their heads, they too were appalled. I knew there was a neutron flux reactor at the Brookhaven National Laboratory on Long Island, New York, where an old friend of my research student days, Maurice Goldhaber, was the director. On a trip to the United States, I made a point of visiting Maurice to find out if there was any great pressure on his neutron reactor. I found that it was seriously underused and that none of the witnesses who had come before the SRC with tales of how important it would be to them to have a reactor at the Rutherford Laboratory had ever applied to use the facility at Brookhaven. When I reported the Brookhaven situation to the council, my story produced two minutes of embarrassment, after which they were at it again. Eventually, the minister for science killed the thing. But, unfortunately, bureaucracy has immense longevity. Years after I retired from the council, and with the skeptical minister also gone, the SRC (SERC by then) revived it, spent an immense sum of money on it, and proceeded to howl to the government for more.

My general experience was that, the higher I penetrated into the system, the more irresponsible it became. At one point, we were asked by the government for advice on a matter in which the return for money was poor. On this, nobody disagreed. But, for diverse reasons, it suited most that the waste should continue. The arguments from all corners of the table took the form they always do the higher you go: "Yes, we agree the situation is somewhat bad, but. . . ." After establishing, in their opening remarks, why they should vote *against*, they then equivocally and finally proceed to vote *for*, which covers all the possibilities.

On that occasion, the chairman decided that remarks made by Lovell and me were too awkward to be minuted, lest they come to ministerial attention. This effective denial of free speech led Lovell and me to discuss the fraught subject of resignation. Before arriving at a decision, we agreed to take "soundings." Through the good offices of a friend, I was able to have dinner with the minister for state, only one step down from the minister of science himself. Within fifteen minutes, I could see Shirley Williams did not understand the issue, nor did she apparently care much about it. Her interest was in the educational system, with the eventual outcome that those of us from relatively underprivileged backgrounds would, in the future, find it more difficult to enjoy the same advantages as those from more privileged backgrounds. Lovell consulted the well-known scientific guru Victor Rothschild, who offered advice that ran as follows: "Resign, and you lose all influence. Swallow the situation, and you can go on fighting." This was true. In the late 1950s, Peter Thorneycroft, a rare chancellor who knew something about industry, resigned over inflationary policies insisted on by Prime Minister Harold Macmillan. His financial secretary, the famous classical scholar Enoch Powell, also resigned. Thorneycroft and Powell were correct. Macmillan's policy opened the floodgates of postwar inflation. But the eventual vindication of their position did neither Thorneycroft nor Powell much political good. Neither rose to positions of so-called power again, although perhaps they gained some reward in ease of conscience. What their case showed was that it is more important to a hunting group not to break ranks than it is to be correct or courageous.

Unlike Thorneycroft and Powell, we decided to follow Victor Rothschild's advice, to accept that we could be muzzled by not having our views minuted. But the thought was now active in my mind: Was this a hunting group to which I really wished to belong? By the summer of 1971, the same question would be coming from other quarters. First, however, I must come to the story of the Royal Greenwich Observatory and of what ultimately became of the Flowers resolution of October 1967.

Before we got around to the RGO in 1969, in 1968 there was what Hosie called a Southern Hemisphere Review, the review committee consisting of Bob Wilson from University College, London, Hosie, and me. We visited all the Southern Hemisphere personnel the SRC had inherited from the Admiralty, together with the staff at the Radcliffe Observatory in Pretoria under A. D. Thackeray, who had been at the Vatican meeting of 1957. It was the latter who had done almost more than anyone else to establish the link in the distance scale associated with what are called RR Lyrae stars, oscillating stars analogous to Cepheid variables, work that was then being taken over by his successor, M. E. Feast. Because the review committee had nobody on it who was self-interested, the work was easy and good-tempered.

The most important long-term effect of the committee's report was the building of a 48-inch Schmidt telescope to complement in the south the 48-inch Schmidt telescope on Palomar Mountain. In its aim to complete the mapping of the region of the sky around the southern celestial pole, it was entirely successful. For two decades thereafter, it has continued at the forefront of research, especially in relation to quasar surveys and to connections with radioastronomy. In view of this eventual success, it may seem surprising that the proposal to build it ran initially into a great deal of opposition, to a point where I began to despair that the British would ever do anything sensible in optical astronomy. Even Geoffrey Burbidge expressed doubts, pointing out that the European Southern Observatory was ahead of us in the construction of a Schmidt for completing the mapping of the sky. Moreover, the ESO site in northern Chile was astronomically superior to the proposed British site alongside the Anglo-Australian Telescope on Siding Spring Mountain in New South Wales. These were formidable practical objections that would have depressed me greatly, but for an experience with ESO in the early 1960s.

Viscount Hailsham was then the minister for science. He was under pressure from the Foreign Office to override the Astronomer Royal, Richard Woolley, on siting a large telescope in Australia. Because of its interest in entering Europe economically, the Foreign Office was eager instead for British astronomers to join the European Southern Observatory in Chile. With the Commonwealth Office pulling in the opposite (Australian) direction, the issue was seen by Hailsham as complex. The occasion was a two-day meeting in Paris at which the ten or so ESO countries were to sign their protocol. Because ESO wanted Britain "in," we were invited to send representatives. Hailsham chose Roger Quirk (his assistant), Richard Woolley, and me. Quirk was a Dickensian character, not quite Mr. Pickwick, but with that general kind of image. It is relevant to the story that he spoke French fluently with an aggressively English accent. We flew to Paris, landing there at about 5:00 P.M. In a taxi, Quirk said grandly to the driver, "The Quai d'Orsay, my man." Thereupon he produced and waved a large white card, continuing to us in English, "We have an invitation for drinks at the Foreign Office." When Woolley and I had digested this information, he added, "I know the place." It must have been 6:00 P.M. when the taxi stopped beside the guards at the big iron gates that protect the French Foreign Office. Quirk waved the big white card again and shouted, "The reception!" At this, the guards bowed us through, and, a couple of minutes later, Quirk paid off the taxi, repeating his previous assertion: "I know this place."

A man emerged from a big lift just as we reached it. "The reception, if you please?" Quirk boomed at him, waving the card yet again. The man shrugged in Gallic fashion and grunted something. "He says we should try the sixth floor," Quirk explained, as we started up the lift for ourselves. It was in this way that we penetrated the inner sanctum of the French Foreign Office.

A few lights were burning in offices, although, by this time, most of those who worked there had gone for the day. We burst in on the remaining occupants, always with Quirk brandishing his card. We tried another floor. We started to look into offices with the thought the reception might be there. It would surely have been a strange reception that took place in any of the offices I saw. The desks were snowed under with papers; there were racks with pigeonholes stuffed with papers; the shelves were a foot deep in papers. There were empty coffee cups and countless discarded cigarette ends. Marching along the corridors, Quirk took to shouting, "Where is the reception?" Woolley, meanwhile, said nothing. Those were the days in which a style of tobacco pipe with a large bowl of shallow depth was much in vogue. Woolley smoked vigorously, so that the pipe glowed in the dark corridors. "I am getting quite annoyed," Quirk shouted.

It went on longer than it takes to tell; but, at last, Quirk took a literally close look at the white card, holding it up to his eye in the dim light. "The reception is not here," he said. "Definitely *not* here. It is at the other side of Paris," he announced at last. So now an undersecretary from the Ministry of Science, the Plumian professor, and the Astronomer Royal will be arrested for espionage, and it will be the case of the century. We have absolutely no business here. The card says absolutely nothing about the Quai d'Orsay. Nobody is going to believe our story—I certainly wouldn't, if you told it to me.

I give Roger Quirk full credit for having got us in, but I give myself credit for getting us out. In most sticky situations, the saving tactic is to walk correctly, in some cases to slouch, for which I have some talent. Quirk wasn't really quite right for slouching, but I thought he might get by simply by making a sort of rotating motion with his head. Woolley was clearly wrong for slouching, and it surely would have been disastrous for him to have tried it. So we set him to walk behind us, as if to give the impression that he was keeping the two of us under surveillance.

At the other side of Paris, we arrived as all the others were leaving. There was just time to gulp down a random collection of drinks. There was no time for elegant conversation. The time for elegant conversation was at the protocol meeting, which took place the following morning. As I watched lawyers from the ten countries arguing niceties of meaning (I watched in fascination as a German lawyer looked up the meaning of a French expression in an enormous dictionary), I thought an observatory organized in this way is not going to work. It will be slow, ponderous, and ineffective. This was why I wasn't too greatly worried at Burbidge's criticism of the 48-inch Schmidt telescope project. Despite their considerable start on us, I didn't believe ESO would be first to the post in mapping the southern sky—nor was it. The saving grace for ESO came only later, when it abandoned separatism, when representatives from the many countries came to reside together and to work together on a day-by-day basis, as they had done from the beginning at CERN in Geneva.

It was with some such thought about the Royal Greenwich Observatory that Hosie began what he called the Northern Hemisphere Review, the year following our review for the south. But now self-interested parties were to be on the review committee, and the position was instantly more difficult. I objected, but my objection was overruled by Flowers, which should have told me to cut myself adrift at that point. Instead, I accepted the compromise of being allowed to nominate two British astronomers from abroad as counterweights to Richard Woolley and Hermann Brück from Edinburgh. For the past three years or so, Jim Hosie had been busily building up the University Observatory at Edinburgh into what was fast becoming yet another SRC establishment. In fact, the University Observatory I had known in the 1950s was renamed the Edinburgh Royal Observatory, and it did indeed become an SRC establishment, which was not what Flowers had said he wanted it to become on the occasion of his visit to Cambridge in October 1967. It was another indication of how the SRC's yawing ship was being blown along by the winds of expediency.

The Northern Hemisphere Review Committee consisted of Bernard Lovell, J. M. Cassels, a professor of physics at Liverpool, the space scientist Bob Boyd from University College, London, Geoffrey Burbidge from the University of California, Wallace Sargent from the California Institute of Technology, Richard Woolley, Hermann Brück, Jim Hosie, and me as chairman—nine of us in all. As secretary, Hosie called a really immense number of meetings, despite which Burbidge and Sargent managed to attend all those that were relevant. We visited Herstmonceux on several occasions, Edinburgh too, and every British university with an astronomy department. We took depositions from every British astronomer of appreciable seniority. Our report was divided into two parts, a hardware part, recommending the establishment of a Northern Hemisphere observatory, and an organizational part. It was because Woolley objected to the indignity of being involved in the organizational part that he refused to cooperate on the hardware part, an attitude that only worked against him. In the beginning, the Royal observatories probably had a majority of the uncommitted members on their side, but they ended with none. It was impossible not to notice the difference between the university astronomers, who thought access to telescopes was a privilege, and those at the Royal observatories, who insisted on being paid overtime for it. In the previous year's Southern Hemisphere Review, I had been astonished at the emphasis given by everyone we interviewed, young and old alike, on their pension rights. In more than thirty years at Cambridge, nobody had talked to me about their pension rights. More and more, we came to appreciate the psychological gulf between those engaged in research in institutions run by the government and those so engaged in the universities, the majority of us not being favorably impressed by the situation in government institutions.

But there were signs and portents that should have been noticed—for instance, in the Hawaiian Island of Maui. Jim Hosie and I were returning from a meeting of the Anglo-Australian Telescope Board held in Canberra in mid-February 1971. The AAT was similar in structural form to the 4-meter Mayall Telescope at Kitt Peak, Arizona, for which reason Hosie had expressed a wish to visit the Kitt Peak offices in Tucson. We flew from Sydney to Honolulu and thence by a local airline to Maui. Qantas had arranged for the local flight, a hotel, and a rental car. After the 10,000-mile flight, we thought to take things easy for a day or two, with Hosie occupying himself pleasurably in putting the finishing touches on the report of the Northern Hemisphere Review Committee—a hard job well done, we both thought, Hosie explaining to me the subtleties of his finishing touches. He was much taken by Maui: the scenery, the food, the sun, and the sky. With the ennui of jet lag, we were early to bed and early to sleep—until 2:00 A.M., when the drumming started. Hosie and I were in different parts of the hotel, and the drumming was concentrated immediately outside his door, whereas, for me, it was only like distant thunder. It continued until dawn. It seemed impossible for such a phenomenon to happen twice, but it did. Hosie should have seen it as a warning that his time in astronomy was coming to an end—and, so far as British astronomy was concerned, so should I.

So what did the report recommend that was to prove just as disturbing in some quarters as those Hawaiian drums were in ours? It recommended that a board be formed consisting of a representative from each Royal observatory, a representative from each of three or four of the universities strongest in astronomy, and a representative, on a rotating basis, from two or three universities with small astronomy departments. The board would seek to coordinate research in optical astronomy for both Royal observatories and universities, the bite being that both would be funded from the same pool, instead of the Royal observatories having first access to SRC money, with the universities coming second. Woolley and Brück wrote a minority report of their own, and the other seven of us signed the main report.

The report came before the SRC's astronomy committee in October 1971, where it was supported by a 13-to-1 vote with 1 abstention, the opposing vote being from a member of Woolley's staff. The member from Brück's staff at Edinburgh voted for the report. But Hosie now had suddenly lost his bearings. He was riding in the opposite direction from everything he had urged for the past three years, as if every plus sign in his head had suddenly turned into a minus sign. As best I could tell, it was because the SRC had at last realized what, in the fever of Woolley's opposition, had been overlooked: that, by having a board ordering priorities, the SRC would lose something of its iron control over British astronomy.

The meeting of the SRC's astronomy committee in October 1971 was important to me in a different respect. The first term of life for the Institute of Astronomy began on 1 August 1967, and it would end on 31 July 1972. A grant application for continuation should have been made already in 1970, but it had unfortunately been delayed for a compelling reason. By October 1971, things were getting late, and I was anxious, therefore, to arrive at a decision, yes or no, about the institute's future. But, since there were indications that the cause of delay would shortly be removed, Hosie argued that it would be best administratively to wait until the end of January 1972, when, I was promised, a decision really would be reached.

At the January meeting, as it turned out, Hosie was galloping back in full flight in the direction whence he had come, as if his horse had just encountered a ghastly apparition. It seemed his career depended on preventing the Northern Hemisphere Review from making further progress. When he eventually came up with a sensible reason for delay, I temporized. Woolley retired at the end of 1971. He had been replaced as the director of the Royal Greenwich Observatory by Margaret Burbidge. Should we not wait (Hosie asked, in some desperation) until Margaret had the opportunity to comment on the report? I was able to agree to this as a reasonable suggestion. In the event, later in the year, Margaret Burbidge was asked for her comments. Unlike Woolley, she was favorable to the report and would have been glad to take the RGO to a position some way outside the influence of the SRC. This left SRC officers with the option either of proceeding to implement the report or of freezing it essentially illegally. They chose to freeze it. So the matter Flowers had paid a special visit to my house to discuss in 1967, over which many scores of hours of meetings had been held, on which Hosie had triumphantly put the finishing touches in Maui, which had been passed 7 to 2 by the Review Committee itself, passed 13 to 1 by the astronomy committee, and, at last, had been approved by the director of the Royal Greenwich Observatory, was cast into limbo by the arbitrary diktat of the officers of the organization that now held sway over all of the British physical sciences. It was not a situation to hold out much confidence for the future.

Hosie's pet project was called PILOT. The idea behind it was good, but the execution was misconceived. It was apparent from our experiences on the Anglo-Australian Telescope Board that, while great sums were being spent in astronomy to acquire the maximum flux of light from an astronomical object, only much smaller sums were being spent on developing instruments to use the light effectively. This was not a peculiarity of the AATB; it applied the world over. Hosie decided, therefore, to make available a considerable sum to be divided into packets to be awarded to a number of projects for the development in universities of astronomical instrumentation. I was against this maneuver, attractive as it might seem superficially, because, in my view, only people who are experienced in actually using large telescopes can judge what new instruments will really be useful. Hosie's idea was right in principle, but its execution was cart-before-horse. All he would succeed in doing, I felt, was to bore more holes down which public money would be poured. The only good thing to come out of his initiative that I ever saw happened when Alec Boksenberg's work on a charge-coupled device was used in a practical program together with Wallace Sargent on the 200-inch telescope at Palomar Mountain.

I am verging now on my own affairs. If I have not said much about the Institute of Astronomy, founded with such gestational pain in 1966–1967, it is because the institute was a quick success, and success makes only dull reading. By the end of the first year, in the summer of 1968, the institute had built to pretty well its full complement, about 35 people, including assistant staff. We had only a handful of summer visitors from abroad in the summer of 1967, but, by 1971, we had more than thirty drawn from the world over. The publications of the institute for the period fill three large volumes. For the purposes of those who generously made the grants that founded the institute, I had to keep two nonmiscible sets of accounts. When I last saw them, in July 1972, there was a surplus of about £10,000 in one and a surplus of £35,000 in the other.

After bearing with these successes, let me turn to events, in November 1971, that led out of the blue to disaster. Circa early 1970, I had a visit from Peter Swinnerton-Dyer and Brian Pippard, both members of the General Board of the University of Cambridge and both with a great taste for management. The bad news they brought was that the university would be likely to shut down the Observatories following the retirement of R. O. Redman in September 1972 unless I agreed to an amalgamation of the Institute of Astronomy with the Observatories. There were two clear-cut objections to this seemingly seductive idea. One was that the university was motivated to save money. Since there was little overlap in the activities of the institute and the Observatories, it was hard to see that anything but a loss to both would result from such a money-saving conjunction. More funds would be required from the SRC merely to stand still. Yet, with real telescopes soon to be available to British astronomers, the possibility existed, it seemed to me on reflection, for a strongly based observational program to be built up on the Observatories side, provided an experienced user of real telescopes could be attracted to the Chair of Astrophysics following Redman's retirement. Otherwise, it would surely be better for me to respond negatively to the General Board's emissaries, which was exactly what Alexander Todd advised me to do. The delicate arrangements that had been hammered out between October 1964 and July 1966 would have to go. Todd said he doubted I had a spoon long enough to sup with the denizens of the lower regions of the university. In this, he was to prove right, except it wouldn't be the lower regions of the university that would cause the trouble.

The General Board set up a small committee to make recommendations on the future of the Observatories, with Pippard as chairman. The committee had discussions with Hosie, certainly, and possibly also with Flowers. When I was called before it, I found it had been sold on Hosie's instrumentation ideas as a future way of life for the Observatories. I told the committee it was time now to stop preparing to do astronomy and really to do some, that the Observatories should seek to make maximum use of the large telescopes shortly to become available, with a view to making contributions to the rapidly expanding subject of cosmology.

The issue dragged on until the summer of 1971, by which time it had become urgent to get a grant application to the SRC on behalf of the Institute of Astronomy, which could not be done, I had been told, until the future shape of the Observatories had been settled. Then the situation seemed to clear itself. In July, Pippard and I at last appeared to agree, even to the extent of approving a short list of names for the professorship at the Observatories. We also agreed to press the Registry to call a meeting of electors to the professorship, and first papers to this effect were sent out in August. Nothing beyond what might normally be expected of Cambridge procedures had happened to this point. Expectation also suggested that, with five of the eight electors resident in Cambridge, a meeting might reasonably be expected by September, in which case the new organization's grant application to the SRC could be cleared in October.

It was a disappointment when the Registry reported that the earliest date at which a meeting of electors could be called was 26 November. Starting from August, this might seem absurd, especially in view of the urgency of the grant application at the SRC. But absurd things happened in the university every day of the week. The administration was being pelted by *non placets*, and there were the threatened suicides with which it fell to the unfortunate W. J. Sartain to cope. He was a saturnine-looking man, and the job seemed to suit him.

I had to be abroad in the second half of October, and, in the first part of November, I returned to find papers from the Registry that astonished me. My understanding of July with Pippard was gone, and Hosie's grand plan for instrumentation was back. Also gone from the list of electors was an exceptionally distinguished scientist, an elector of many years standing on the board. In his place was a person, while now external to Cambridge, who had many years of association with the Cavendish Laboratory. There had been three electors already in the Cavendish group—Pippard, Ryle, and Mott. Now this gave four. Because of the Northern Hemisphere Review, I could depend on Richard Woolley voting at every opportunity in a direction opposite to mine. Even supposing I could persuade the rest, this gave a 5-to-3 majority against me. So this was where my agreement to "rescue" the Observatories had led? My spoon was one vote short in its length. A 4-to-4 tie would have won, I believed, because, in a tie, the side with a poor policy always loses eventually. Nothing as ridiculous as this instrumentation thing could win except on a rushed vote, and it was some consolation to think that a 5-to-3 division of opinion had always, in my experience, guaranteed slow progress towards an election.

I rang the distinguished external elector who had been replaced to find out whether, for some personal reason, he had asked the Registry to terminate his appointment to the electoral board. While an elector might be replaced legally in some cases, precedent was very much otherwise. Especially an external elector should not have been replaced by the university after papers had begun to circulate, as they had done in August. The external elector in question told me that he had not himself asked to be replaced. He had received papers in August, but nothing since, not even a letter of thanks for his past service. This instantly put the issue in a different light. It was no longer the usual kind of Cambridge squabble. It had become something quite nasty, raising acutely whether these were people I wanted to be associated with any longer. Since Hosie's instrumentation concept was being used, I rang him with the clear statement that, if the electoral meeting of 26 November went the way it seemed to have been planned, I would resign. Although Hosie obviously didn't believe me, I thought that, since Jim was a fairly compulsive talker, my call should be sufficient to let the Cavendish group know what the score had become. I think the news must have reached them, for afterwards I heard that Teddy Bullard had said: "Fred won't resign. Nobody resigns a Cambridge Chair."

The main surprise at the meeting of 26 November was that the vice-chancellor accepted a motion to decide the issue there and then, an immediate decision on a 5-to-3 margin. The two votes on my side were the physical secretary of the Royal Society, Harrie Massey (the Massey of the Massey Conference of 1951) and (surprise?) William V. D. Hodge. For 30 years or more, Hodge had been the chairman of the Observatories Management Committee, and he could see perfectly well that the Cavendish group were out to emasculate the Observatories. The votes of Massey and Hodge were independent of mine, whereas the four votes of the Cavendish group were unacceptably correlated. Woolley's vote went against the interests of his old observatory. Woolley had held a job at the Observatories before becoming the director at Mount Stromlo in Canberra. He had every reason to wish the Observatories a high measure of success in real observational astronomy. His malice vote on account of the Flowers–Hosie Northern Hemisphere Review was the vote of a second-rate player. First-rate players fight hard where their interests are concerned, but they do not carry personal resentment across to other issues. Woolley was the reason why the meeting of 26 November could not be postponed for further reflection, as it should have been. Woolley was only six weeks from retirement at Herstmonceux. He had set up a job for himself in South Africa and would soon be gone from the British scene. The meeting had to be held after the distinguished external elector could be got rid of on 30 September and it had to be held and decided before Woolley left for South Africa. Otherwise, the 5–3 balance might only too readily have gone back to 4–4, with the likely consequence that things would eventually have been reversed.

But now I really did want to be done with it. Had the Institute of Astronomy grant for the period after 31 July 1972 gone through in October 1971, I would have been free to have acted immediately. A few words of resignation to the vice-chancellor, and I could have walked out into the free air, leaving the majority to explain the situation as best they could. But I was hamstringed by the delay in the institute grant. I had to play the matter gently until the end of January 1972, when I was able to make a trade with Jim Hosie. If he would pass the institute grant, I would agree to delay the Northern Hemisphere Report until the arrival of Margaret Burbidge. Hosie saw to it that the grant went through. I waited ten days for the minutes to appear in writing, and then I resigned.

Again, to give the Institute of Astronomy the best chance of survival, I made no comment in the ensuing babble. I was told that Flowers came rushing to Cambridge and that he and the vice-chancellor had hard words. If so, SRC involvement might have been deeper than I thought. My feeling remains, however, that the SRC was sucked into a clever plan.

It was a part of the plan, I believe, also to suck in the vice-chancellor, who was given to believe that, in pressing the vote, he was acting to please the SRC and so to secure money for the university. The letter the vice-chancellor wrote subsequently to the press showed ample evidence of a naive mind. Pippard also wrote to the press, referring to me as a petulant person. Pippard was the outstanding player in the Cavendish group, yet he failed to appreciate that what was only a game to him was not a game to me.

At the Institute of Astronomy, there were even more visitors than usual in the summer of 1972. We held our research meetings, discussions, and lectures in just the same way as before. We all managed to keep cheerful until the evening of 31 July, the last day of the institute in my regime. Several of the visitors and staff hired a boat, and we made a trip down the river Cam to Ely. At the end, Sverre Aarseth, who had come many years earlier from Norway as a graduate student and who was our best chess player, presented me with the Visitors' Book. Then it was over, my time in Cambridge, almost exactly thirty-nine years.

I stayed on for two weeks more, until the last of the visitors had gone. On Friday, 19 August, I walked out of my office late in the afternoon. The office was at the end of a long corridor down which I had to go before reaching the main outer doors. My intention had been to leave immediately, but, on impulse, I decided to take a last look around. In the lecture room, there was a wide expanse of blackboards. During 1966–1967, I had squeezed the construction budget to afford blackboards of the best quality. They were covered now by a maze of words, mathematical symbols, and general squiggles. I studied them for a moment, wondering what the squiggles had been intended to mean. I could guess some of the patterns but not others—as in life itself.

The main doors were big and heavy—lighter ones had not been possible for architectural reasons. They swung slowly shut behind me as I set off to walk for the last time from the institute to my home. The third period of my life had ended.

Chapter 26

The Bay of the Birds

AS SOON AS my tenure at the Institute of Astronomy was over, my wife and I coupled up a caravan to our car and set off early one morning in August 1972 for Bedruthan Steps on the north coast of Cornwall, about midway between the towns of Newquay and Padstow. The route was via Oxford and Swindon to Sparkford and Exeter, where, in those days, you still drove through the central streets. With about 150 miles of the journey done, we took a lengthy stop for lunch halfway between Swindon and Sparkford. It was only then that an overwhelming sense of relief from the past months and years of petty squabbling swept over me.

We parked the caravan not far from the spot where the Window problem of the wartime years had been solved, twenty-eight years earlier almost to the day. It was close to where the smugglers' track known as Pentire Steps goes down to the Bedruthan beaches. Down you go, with a scrambling twenty or thirty feet at the bottom, to find yourself on a splendid yellow beach with firm sand, excellent for bathing. No beach I have seen anywhere in the world is its superior. Some beaches in Australia come close, and one I found on the western side of the South Island of New Zealand was perhaps its equal. On the western coast of Ireland, there are beaches to beat almost anything in the world, but, unfortunately, they are outside the arm of the Gulf Stream, which gives such remarkable warmth to the waters around Cornwall and, indeed, to beaches extending even to the extreme north of Scotland. After the warmth of the Gulf Stream, the icy waters offshore in southern California come as both a physical and a mental shock.

Pentire Bay is covered by the sea for about half the tide. If you hit it with the tide half down, you therefore have about six hours before, willy-nilly, you must be out. You can't get through the narrow defile between Diggory's Island (immediately offshore) and the mainland, however, until maybe an hour and a half before low tide. Even then, you may get a bit

377

wet in the process. After passing the defile, you now have three hours before you must be out. Ahead is the largest of the Bedruthan bays. Ahead and to the right, as it was then, is a big rock shaped like an Elizabethan lady striding across the sand in crinolines, with her head seen in profile, a rock known as Queen Bess until, in recent years, she unfortunately lost her head and the name became meaningless. Perhaps her name should now be changed to Mary, Queen of Scots.

The way around to the third bay is passable dry-shod only within about an hour of low water, seemingly worse today than it was in 1972. There is, however, a hole through the bluff that separates the second bay from the third. You go across a pile of boulders, then through the hole, and finally across easy rock ledges into the third bay, so arriving at the bottom of the old descent from Rundle's excellent tearoom, all traces of which have now disappeared, in keeping with the laws of thermodynamics, according to which the bad drives out the good. Here also is the bottom of the new, unpleasant steps constructed by the National Trust. In the old days, we knew of two further bays to the south. To reach the first of them, you had to pass over slippery, seaweed-covered rock filling a defile between the mainland cliff and yet another rocky island. I recall, from the wartime years, the ebullient Hermann Bondi, racing in bare feet with his trousers pulled up, at what seemed a suicidal speed across those slippery rocks, as if he had lost all control of his mental processes.

To reach the fifth bay, you must go through the Hole of the Civil Servants. The hole is so-called by me because, in a narrow-grooved passage, it contains a three-foot-deep pool of icy water, into which (because the peculiar geometry of the place is such that no shaft of light shines from its surface) all but those who know it intimately inevitably plunge—likely enough, face downwards.

So we came to the fifth bay, which, compared with Pentire and the second bay, is a pretty dull place. In my younger days, I had been there for the sake of something to do and little else. But we were now in possession of information I had never heard before. The local National Trust farm, which we had first known in prewar times as Rundle's Farm, was now known as Littlefields' Farm, after its current tenant. Mr. Littlefields' boys were great explorers of the cliffs and the beaches. We learned from them that there was a way through to yet another beach, one normally accessible only through a hole within half an hour of a low spring tide. But right now in September was the lowest spring tide of the year, giving us maybe an hour in which the water line would be below the hole. So that was why we were there in the fifth bay, a dull place we would otherwise have no interest in visiting. Sure enough, there was the hole uncovered by the exceptionally low tide. We were quickly through it, into a bay the like of which I had never seen before. It was occupied by a myriad of sea birds, which rose and fell like white clouds in the sky.

We made our way along the sand to the farthest point, at which it felt as if one could put out a hand and touch the very last of the cliffs before the coastline turns east to the village of Trenance and the sands of Mawganporth. It seemed as if, with just one more success in negotiating obstacles along the shore, we would indeed have been round to Mawganporth. I cannot say whether this would, in fact, be possible, for we had no time to explore further. Every minute there was borrowed from the sea. Standing offshore was another of the large humps that occur all the way to this point from Diggory's Island, once parts of the land that the sea had gradually infiltrated and split into the small pieces that now form these islands. The remarkable thing here, not found near any of the other islands, was that a sandbar had been uncovered on the seaward side of the island. In places, the sandbar was being washed, even at this extremely low tide, to a depth of three or four feet; but, being of the usually firm yellow sand derived from the long-running breakdown of the cliffs, it was easily negotiable. Meanwhile, there was a long-running bedlam in the sky, with screeching, ever-wheeling hosts of birds, the entire scene never still for a moment, and with the Sun shining brightly on the green water.

When you reach some remarkable and essentially unique place, how long should you choose to stay? The likelihood that you will never be there again might suggest that you should stay a considerable while, to "drink it in." But it has never worked that way for me—just the opposite. The more remarkable the place, the shorter the time I stay. It is as if the very fact of being in such a place means that I don't belong there, that I am an intruder, to be tolerated only for a fleeting moment. Afterwards I have always found that I had the wit to have taken a mental snapshot of the place that remains everlastingly vivid, which I imagine would not have been so had I lounged about for hours on end. There was no such choice, however, in the Bay of the Birds. At best, we had an hour there, which, for safety's sake, we cut to something less than forty-five minutes.

I came away from that experience with what I realized later was a wrong idea. It was the sort of deduction leading to a mistaken identification of cause and effect that often occurs in science. The number of birds in the one bay alone had surely far exceeded those in all the other bays at Bedruthan. It seemed natural to suppose the difference had something to do with the relative frequency of human penetration, the commonness of it in most bays and the extreme rarity of it in the Bay of the Birds. But there was never any evidence of human interference with birds, nor on the high cliffs everywhere could there have been. Other birds are surely more important predators than humans these days. What matters is a sufficient availability of nesting sites to support a colony, the idea of the colony being for birds of a particular species to support each other in warding off egg-raiders of other species. In most of the Bedruthan bays, the cliffs are rather smooth, without the necessary profusion of ledges. In the Bay of the Birds, however, the cliffs were seamed by a maze of ledges, making it ideal for supporting a large colony. The nature of the rock also allowed humans to descend the cliffs in some places but not in others, the situation being anticorrelated: where humans could get down, there were few sites for birds; where there were many sites for birds, humans could not get down. Thus, what at first sight seemed like a direct connection between birds and our presence on the beaches was actually due to something else that controlled both.

Shortly after that trip to Cornwall, we quit the house we had built fifteen years earlier in Clarkson Road. Cambridge had too many equivocal memories for us to have made staying there an attractive possibility. But where else might we go? Instead of making an immediate decision, we began by considering four general areas that appealed to us: the Lake District of northwest England, Cornwall, northern Yorkshire, and central Wales. (I would also have liked the Scottish Highlands, but my wife immediately vetoed that on grounds of remoteness.) Then we moved around those general areas and kept a lookout on each. It happened that it was in the Lake District that the first sufficiently outrageous possibility turned up. It was indeed so preposterous that we bought the place immediately. The valley running more or less east-to-west from Penrith to Keswick is well known to the millions of visitors who come to the Lake District each year. So too is the deep trench of Ullswater, also running generally east-to-west from Pooley Bridge to Patterdale. What is not so well known is a high valley running from a gap in the east between Great Meall Fell and Little Meall Fell to the summit of Great Dodd in the west. The three more or less parallel valleys are geologically distinct. Ullswater is cut through dramatic Borrowdale volcanics at its upper end, giving place to the pretty Silurian outcrops in the east, the low Penrith–Keswick valley goes through darker Skiddaw slates, while the high intermediate valley is geologically nondescript. It has an aureole of rock at its upper end, partially cooked stuff, and it has wind-blown conglomerate at its Great Meall Fell end. Probably because of the comparative fertility of the conglomerate, it has as many farms as the larger, better-known Patterdale Valley below. It was there at the very top end, in the highest house of this geologically unremarkable valley, that we chose to live. Except for the Kirkstone Inn, I never found any habitation in the Lake District that was higher, or more susceptible to winter snows. Indeed, the slope of the Dodds to our west was a popular winter place in the Lake District for skiing.

The view from the house to the north was cut off by a local hillock called Cockley Moor, that to the west by the rising bulk of the Dodds, and that to the southwest by nondescript hillocks that rise eventually, bit by bit, to the high summit of Helvellyn. In fact, you could walk over hills all the way from Helvellyn to the house by the shortest route in about an hour and a half. The southwest was the direction from which the biggest storms came, the winds being beyond expectation in their fury. All the way from the southwest to the northeast, however, the view had a very long range, extending in the south and southeast up to ten miles towards the high mountain route along which Roman armies marched rather than risk the forests, bogs, and marauding Celts of the lower valleys.

The house itself was nearly linear, with the main rooms and all the bedrooms facing the long-range view to the southeast. It had a big stone patio outside also facing that direction. Every day, when the sky was clear, we had the Sun from dawn until about 4:00 P.M. Although the Lake District is supposed to be a place of poorish weather, I had more the impression of living in sunshine in that house than in any other house in my experience. Even in places with a strong sun, as in California, the impression of sunshine is weaker because, once the Sun rises high in the sky, people there take steps to avoid it. At Cockley Moor, we deliberately lived in it.

The house had about three acres of ground, one of garden (to which topsoil had been brought in), one of trees, and one of grassy paddock. Because everything was well watered and because the soil was nutritious peat, all plants that favor high acidity grew luxuriously, almost madly, especially alpine gentians and other mountain flora. After we had planted a sheltering belt of sycamores, birches, and beeches in the paddock, as replacements for big firs that high winds kept taking out, birds came in profusion. None ever seemed to notice us. Nor did the moles, the badgers, or the hares. It was like living continuously in a hide (what Americans refer to as a blind). Except you didn't need to hide in any literal sense. You just went about your business and the animals went about theirs. You thought you owned the land, but the animals were sure they did. The paddock had been allowed to go to wrack and ruin, so we bought one of those tough ride-on mowers. After the immense struggle that was required to cut through the big tussocks of rough grass for the first mowing, it became easier. With the grass cut down to about three inches (the ground was too uneven for anything finer) once a month, the better grasses gained ground and wildflowers began to grow in profusion, adding still more to the variety and number of birds. There was a wet place in the paddock that I found impossible to mow, a place, I suppose, that was kept wet by some broken old land drain. Once beeches and birches were planted there, it became less dank, grasses grew longer, and, all of a sudden in the summer, you found the grass seeds being stripped at a great pace by goldfinches— but not as fast as I once saw the red berries stripped from the rowan trees that grew down the bed of the stream that ran through the wood and paddock. The tiniest birds were the goldcrests, living in the tops of the pines, but they were very hard to see. Most years, in the autumn, we would have a migrating flock of fieldfares descend on us for a day or two. I once saw what must have been thousands of fieldfares fall onto the rowans. In less than three minutes, the trees changed, as if by magic, from being bright red with berries to being spring green again. The different species of birds normally kept out of each other's way. This was especially so when the fieldfares landed. All the rest simply went silent and waited until the visitors had gone. Of particular delight were the warblers with their yellow breasts. Despite watching warblers on scores of occasions, we never learned to distinguish the different types. Indeed, I never felt any interest in identifying different species the way most bird watchers do. I was always more interested in seeing what birds did.

Although those of different species normally ignored each other, they cooperated when an owl was around, some dive-bombing it, others twittering unceasingly in the owl's ear. Lumbering crows would take off to disturb the hovering of predatory hawks that nested in crags a couple of miles away. The biggest problem for us was the pair of flycatchers that nested five yards outside the front door. The first year they came, we watched the eggs being laid, being hatched, and the young growing and fledging. Then one morning the nest was a wreck. Feathers were strewn over a wide area. An owl had waited until the brood was almost ready to fly and had then taken the lot. When the flycatchers returned the following year and started up the process again, we had no taste for watching it through to the same inevitable end. So, as amateurs in the nest-construction business, we set about adding a superstructure that we hoped would prove impenetrable to an owl but that would still permit the coming and going of the flycatchers. It was a fraught several weeks as the young appeared and grew towards the danger point. But this time they got away, and I felt much the same sense of satisfaction as I did when, at about the same time, the determined Vince Reddish, the director of the Edinburgh Royal Observatory, showed me the first successful photographic plate taken with the 48-inch Schmidt telescope on Siding Spring Mountain in New South Wales.

Over a few days in the autumn, the bird population fell steeply away as the migratory species took off for the south, leaving only the winter regulars. Blackbirds coped, but we were too high for thrushes. Wrens came and went, surviving in the warmer winters but unfortunately dying in the colder ones. Although tits had been amply visible in the summer, it was now that their skill and dexterity really became prominent. The tits reminded me of a poem written by my great-grandfather, Ben Preston. It concerned a working man of Victorian times, but it had none of the Dickensian sense of a Bob Cratchit in toil and agony, so beloved of the intelligentsia of the nineteenth century. For this particular working man, absolutely everything in life goes right. The chap expresses his cheerful philosophy of life in the same opening and final verse. In the local vernacular, it reads:

> I lewk for a dlimmer whenivver it's dark,
> I whissel when t'winter wind ovver me blaws,
> Then it's pleasant i' summer to hearken to t'lark
> Yit I like a tomtit 'cos he sings when it snaws.

Which was exactly what they did.

My situation was similar to what it had been when I was eight years old. I had gone roaming the woods and fields in revolt against the school system. Now I was in the Lake District in revolt against events in Cambridge, taking note, as before, of what was happening around me in a way that was impossible either from the classrooms at Mornington Road School or from the committee rooms in State House, then the headquarters of the Science Research Council. Just as eventually I had been obliged by the objective need to attend school, so now I was obliged to consider the financial future. I was without pension, something of overriding importance to everybody I had met in government employ. I was 57 years old, not of an age ever to gain access to a lucrative pension anywhere else. Such rights as I had would become due on my sixtieth birthday, in 1975, when I would receive a delayed handshake from Cambridge of approximately £25,000, not a sum to set the world on fire. Thanks to my wife's foresight in building the house in Clarkson Road, we had gained about £10,000 on the exchange of houses, most of which would have to go on necessary repairs. Thus, provided I could earn the equivalent of my salary in Cambridge, our position financially was not much changed. I had the good will of friends in the United States, essentially none of whom could believe what had happened. In American universities, the year is mostly divided into quarters, with scientists usually working three of the quarters—autumn, winter, and spring, taking the summer off—and they are paid, pro rata, three-quarters of what they would receive for working the whole year round. I planned to take an American appointment (that had been generously offered to me) for two quarters rather than three—autumn and winter—returning to Cockley Moor for the other two. This would equal my Cambridge salary with a little to spare. I would thus be working half the year in a really first-class scientific environment rather than being under constant pressure to create such an environment.

But this seductive plan could not begin for another six months. There would be meetings of the Royal Society until November and of the Royal Astronomical Society until February 1973, of which I was a vice-president and the president, respectively. For these, I would either have to remain in the United Kingdom or commute from the United States. At the risk of an overload of commitments, I had held these positions at the time of the critical Cambridge meeting on 26 November 1971. In some degree, they explained my lack of preparation for that meeting. No serpentine tactics on my part had seemed necessary, for I simply could not believe that Cambridge would prefer merely to build ancillary instruments rather than concentrate its interests on active observations of the Universe, using the excellent telescopes that would soon be coming into operation, and would do so against the advice of the current president of the Royal Astronomical Society, of the current physical secretary of the Royal Society (Harrie Massey), and of the recent physical secretary of the Royal Society (William Hodge). There had to be some explanation, I suppose, but, within the realms of rational thought, I have never been able to find it.

Conscious of its need for my vote against an Australian plan to seize control of the Anglo-Australian Telescope, the Science Research Council had activated the string with which it pulled the minister for science—at that time, Margaret Thatcher—to move her to write requesting that I be retained as a member of the Anglo-Australian Telescope Board for an extra year. This had happened before the Cambridge debate. Because I had agreed to stay on the board for another year, I felt committed in the autumn of 1972—not particularly for any moral reason, but because sticking to what you say you will do is a simplification in life.

By September 1972, the struggle to prevent the Australian National University's attempt to take over the Anglo-Australian Telescope had become tense. Although the board chairman, Taffy Bowen, had yet to declare his position openly, it became apparent at meetings in Canberra in February 1972 that, on a formal vote, he would be likely to side with the British position, giving us a 4-to-2 vote on the board. This caused the university to make representations to Malcolm Fraser, then the Australian Minister for Education and Science. In Canberra, Fraser called us to a meeting in Government House, telling us bluntly that we must comply with the Australian plan to steal the telescope. As the official representative of the British government, Jim Hosie did most of the talking on our side. Basically, he simply told Fraser that we did not intend to move from our position. At that, Fraser hinted broadly that he would be discussing, with his opposite number in London, the question of British board members being dismissed. A similar idea had already entered his mind with respect to Taffy Bowen. It was then, when the question of removing Taffy was examined, that the Australian government had its first shock—or rather a double shock. In the intergovernmental Agreement Act of 1970, the British and Australian governments had both signed away the ownership of the AAT to the six-member Telescope Board, of which I was a member, while, in a separate Act, the Australian government had even signed away the right to dismiss its own board members.

In this acrimonious atmosphere, the board had sought to get the pungent air out of its lungs by holding a meeting in La Jolla, California, where Margaret Burbidge was resident. This was at the end of April 1972. Now, at last, the issue of control and operation of the telescope appeared to be settled on a 4-to-2 vote. Control and operation would remain with the board; it would not be passed to the Australian National University. Then, in August 1972, at about the time we were coupling up our caravan for the trip to Cornwall, Malcolm Fraser and Margaret Thatcher had a meeting in Canberra to discuss the legal meaning of the Agreement Act of 1970, with the outcome that it meant what we had taken it to mean, which was actually what it said.

The position would probably have stayed that way if the Australian government had not changed at the end of 1972. The new Labour administration—under Gough Whitlam as prime minister, with W. L. Morrison as minister for education and science—was immediately put under pressure to renew the struggle by Sir John Crawford, the vice-chancellor of the Australian National University. Perhaps it was to demonstrate the superiority of his government to the previous one that the cudgels were taken up immediately and willingly by Whitlam. In view of the legalities that could not be changed, the only new move available was to get rid of Taffy Bowen. Because this could not be done directly, it had to be done by stealth. Retired now, Taffy was offered a job in Washington, D.C. The job was highly congenial, it took Taffy among wartime friends, and it was well paid. Once he had accepted it, he was then told that his duties would require his resignation from the Anglo-Australian Telescope Board. Faced with the choice between what was unpaid and becoming increasingly unpleasant, on the one hand, and what was paid and pleasant, on the other, Taffy, in February 1973, chose the latter. This, by a long-standing agreement between the two sides, made me the board chairman, which was precisely why the SRC had wanted me to continue as a board member even after I had left Cambridge. Quite how the Science Research Council, which could surely have prevented the situation in Cambridge, now had the nerve to ask for my help was, I thought, like the peace of God—it passeth all understanding. Anyway, I accepted the chairmanship as an interim measure until my board membership should lapse at the end of August 1973, thinking innocently that, after the Thatcher–Fraser meeting of the previous August, the 4-to-2 decision in favor of the board retaining control had been accepted in good faith and was now behind us.

But not so, apparently. Once I had assumed the chairmanship, Hugh Ennor, the administrative head of the Australian Department of Education and Science, as it then was, told me to prepare for storm clouds ahead. Beyond that, however, he told me nothing. In any case, I knew there were storm clouds ahead, but I thought they would be of a nonpolitical nature. I knew Margaret Burbidge would be departing from the board. Her place on the board was taken eventually by Vince Reddish, who was just as staunch as she had been, but, in the spring of 1973, there was no way of knowing who would be appointed. Our project manager had resigned just as the construction phase of the telescope was about to begin. Instead of the building of the telescope being the responsibility of a single contractor, the way it usually is with telescopes, the board had used an array of contractors, with the consequence that the glass for the mirror came from the United States, but with the figuring done in the United Kingdom, yet with secondary optics made in Australia, the main mounting made in Japan, the main gear wheels made in Switzerland, the electronic drive from Japan, the telescope tube from the United Kingdom, the TV system and the computers from the United States, the spectrographs from the United States and the United Kingdom, the gear box from Japan. . . . To many with the experience of building telescopes through a single contractor, it had seemed a recipe for disaster.

With all this in the front of my mind, I had no difficulty in agreeing with Hugh Ennor's prognostication that storm clouds lay ahead. Luckily, perhaps, Ennor didn't tell me of the particular cloud he had in mind, and the situation did not come out until many years later. I thought the Australian replacement for Taffy Bowen was to be Paul Wild, Bowen's successor at CSIRO. But it seems that Whitlam instructed his minister that the replacement should be a very senior member of the Australian National University, and the name apparently being whispered was that of the legendary Nugget Coombs. I never met Nugget Coombs, but I had heard of him through Joe Jennings, my red-bearded friend who became an expert on the Australian outback. Joe was currently in the ANU School of Pacific Studies. In the course of his exploratory work in New Guinea and in Australia itself, he made discoveries that were eventually to be rewarded with the Gold Medal of the Royal Geographical Society. Long before the AAT was ever mooted, Joe had told me of the redoubtable Coombs, who, it seems, dominated every scene he entered—enough to curdle my blood, had the Whitlam plot been known.

History would surely have been different had the plot gone ahead. The imagination runs riot as to what might have happened. It was not my style of chairmanship to give way to anything except reason. I think British members would have stood firm and that, in this, they would have been supported by the British minister and, if necessary, by higher authority still. The British members would have stood by the 4-to-2 La Jolla decision, with the consequence that the board would have undergone irreversible fission, the logical consequence of which would have been for the Australian government to have bought us out. The price of objections by Bondi and Woolley to the Northern Hemisphere proposal from Redman, Ryle, and me in the spring of 1967 would then have been eight years in time, although relatively little in money. The price for Australian astronomers might well have been worse. Because of the resignation of our project manager, allied to the multicontractual manner of the telescope's manufacture, and because opinion outside of the ANU would have become seriously inflamed, it is easy to see that the assembly of the telescope might have proved the disaster that many experienced astronomers in the United States expected it to be. That the telescope was actually an immense success as an instrument (although it was always to suffer from the limitations of its site) and that the United Kingdom's 48-inch Schmidt project was also a major success (again, against a fair measure of outside opinion) was to form the rock on which British confidence in astronomy in later years was to be built. Throughout the 1950s and 1960s, it had always been thought that radioastronomy would be the British standard bearer. But major projects in many other countries were to overtake and surpass British radioastronomy, and it was through the Cinderella of optical astronomy that the victory would be won. Jim Hosie was forever urging Cambridge and Jodrell Bank to unite into a national observatory for radioastronomy. Had this been done, the outcome might have been different there, as it might have been if the radioastronomy establishments had taken up developments in the infrared and in millimeter-wave astronomy.

The day was saved by popular opinion among Australian astronomers. Before acceding to Whitlam's instruction, Morrison decided, perhaps on Hugh Ennor's advice, to hold a meeting, in June 1973, of Australian astronomers generally to discover their views. Finding it heavily against the Australian National University, the Labour administration, being populist in its protestations, had little alternative but to decide against the attempted takeover of the telescope, thereby ending what might have been quite a bolide impact with only a whimper. The best thing then was for the Australian government to forget about the telescope, which it did until the inauguration day on 16 October 1974, by which time the real storm clouds mentioned already had also cleared away, with the telescope clearly en route to the success it eventually became.

My personal plans worked well until the demise of the Heath government in 1974. In its first budget, the second Wilson government introduced punitive taxation against British citizens working abroad, except for those who were out of the country essentially for the whole year. I was now angry with the British system. Previously, I had tolerated it with a measure of amusement; but now I felt like the old fellow in a remote Wyoming town who spent a small fortune on television sets because of the frequency with which he emptied his six-shooter into the screen whenever a politician appeared. Anyway, I now had either to live permanently abroad, leaving my wife alone in the snows of winter, or we had to sell the house at Cockley Moor, both of us moving abroad, since the house, like a small child, was not one that could be left alone, except perhaps for a short time in calm weather. At any moment, a wind might come, stripping away the slate roof and dumping the contents of the house in the racing river a quarter of a mile away. Or the electricity might be cut off, and the contents of our two big deep freezers would go up in microorganisms. Or the automatic boiler would not come on, there would be a freeze-up, and we would return to find the entire house covered in icicles, as in a German children's story. Or the sheep would get in and mow the garden flat, especially when the stone walls around the property would "flush" after a sharp frost—which is to say, fall down. Or the central heating would boil, and a neighbor, seeing steam coming out of the windows, would call in Harvey the red-haired plumber. Harvey would go up into the loft, which had a vast area but a height from ceiling to roof of only two feet or so in places. When away, we would have visions of Harvey getting stuck up there, and of us returning months (or, according to the Wilson government, years) later to find his whitened bones. So it went on, with things happening that neither Labour politicians nor Treasury officials had conceived of in designing their soak-the-rich budget of 1974. Nothing is more absurd than the perennial Labourite notion of soaking the rich. In their very nature, the rich cannot be soaked. That is precisely what richness is for, to avoid being soaked. It is the fellow in the middle who always gets drenched, just as it was with me in 1974.

I was angry as well with Americans for changing their immigration laws a few years earlier so as to discriminate against Europeans, discriminating most against those who had contributed most to their success. So the United States was not to be considered. My wife and I had always thought France to be the best country to live in. Unfortunately, I was not sufficiently fluent in French for this to be a serious possibility. In the resulting vacuum, the idea of Australia occurred to us. It was indeed for this reason that my wife came to the inauguration of the Anglo-Australian Telescope in October 1974. Let me describe the state of affairs.

The SRC was keen to keep the chairmanship of the AAT Board in British hands, which it would do if I stayed. On the other hand, it would be difficult, almost to the point of impossibility, for the chairman to be resident anywhere except in Australia. Once the telescope came into operation, there would be many small problems to be settled pretty well on a day-to-day basis, requiring immediate local contact with the chairman. My own interests dictated that I should be near the staff responsible for maintaining the telescope and also in a position to talk with visiting astronomers using it, when there would be many questions of scientific relevance to discuss. Three places for locating the staff had been mentioned—Canberra, Sydney, and Coonabarabran, a small town 25 miles from the telescope site. There was now one of the best telescopes in the world nearby, together with the United Kingdom's 48-inch Schmidt telescope, which was well advanced in its sky survey. In three hours of easy driving to the south, there was the CSIRO 210-foot radiotelescope (with John Bolton as director), and three hours to the north, there were the CSIRO solar installations. My wife took half a day to reject the Canberra possibility. Because of its larger size, she took a day and a half to reject Sydney. After ten days looking carefully over the Coonabarabran area, talking both to local people and to people from the United Kingdom involved with the 48-inch Schmidt, she said: "Well, if this is where you want to come, I'm willing." Aside from the possibility of France, it was the only time she ever expressed a willingness to live outside the United Kingdom.

So it came down now to the views of board members. In December 1974, we held a meeting in San Francisco, halfway between London and Canberra, at which the issue of the siting of the AAT staff was one of the main items on the agenda. On this item, the Board split 3-to-3; all the British members—M. O. Robins from the Science Research Council, who had replaced Jim Hosie, Vince Reddish, who had replaced Margaret Burbidge, and I—preferring Coonabarabran. The Australian members were divided between Canberra and Sydney, but all were against Coonabarabran. Mac Robins made an appeal to the Australian side on the grounds that British users of the telescope would need to travel halfway around the world to reach the telescope and would be greatly helped if the staff could be on the spot, close to the telescope. This appeal was rejected by the Australian side on the grounds that they knew Australia better than we did. Actually, I think they knew very little about Australia outside the main cities.

By 1974, I had been many times into the remote Australian bush with my friend Joe Jennings, the chap who was awarded the Geographical Society's Gold Medal for his explorations in Australia. Three months earlier, we had been in camp in the Grampians of southwestern central Victoria. We had taken eight hours to climb what in Scotland would have taken forty-five minutes, with Joe saying at the end: "Aye, that was a bit of moderately thick bush." Joe had a big red beard, he dressed as an archetypal Australian bushman, and he knew mineral deposits everywhere but would never tell would-be exploiters where they were. Joe was a great talker. When camping in Scotland, we would go to some local hotel for dinner. Joe had an acute inhibition about giving formal lectures, but, in a crowded hotel lounge over coffee, he would talk with the greatest of ease. It was always the same. It would begin quietly enough with his explaining to me how some local landform had come into existence. The people nearby would start to listen and, bit by bit, the rest of the lounge would fall silent except when somebody asked Joe a question. For two to three hours, they would sit spellbound, hearing things they had never dreamed of about the country they had driven through that day with unseeing eyes. I used to beg Joe to commit these talks to writing. He always said he might but never did.

It was the same with the small farmers we sometimes visited in the Australian bush. Pretty soon it would emerge that Joe knew more about the bush than they did, about the landforms, the plants, and the animals. He was the only mountaineer I have been out with who, in a thick mist, had a clear three-dimensional picture of the terrain in his mind, which made him far more accurate in keeping to a route than any of the rest of us. It was essentially the same thing in finding a route through thickish bush, about the same as a thick Scottish mist for difficulty. Anyway, Joe would go on holding a cockie's family spellbound, the wife having cooked us a meal fit to bust. Eventually, they could hold their curiosity no longer. It is a mistake to think all Australians are blunt. In the cities, they are brash enough, but, in the deep country, they have a proper respect for your affairs. So they would hold themselves in until the pressure of curiosity became so ungovernable that it burst forth at last. It was, of course, Joe's accent. He came from Pudsey—about seven miles from Bingley, where I come from—and he made no more attempt to conceal his place of origin than I do. At last they would ask where he came from. At this, Joe would roar with laughter, shaking his red beard, and boom out: "I'm a bloody Pom." I have never really understood why referring to Englishmen as

Poms seems excruciatingly funny to the Australian mind. Maybe it is a corruption of the French word for potato.

When the Australian Board members rejected the plea from Mac Robins, I thought to myself how very few were the examples I had come across in the higher flights of science administration where anybody had considered the interests of anyone other than himself. Maybe then it would be a good idea to be out of it totally, rather than half in and half out, as I had been since leaving Cambridge. The only provision we had for a 3-to-3 split was to refer the matter back to the British and Australian governments. With a likely 1-to-1 split there, that would be no panacea, although, to be sure, it would have lifted the problem off my back. A better solution seemed to me to award an effective casting vote to the Observatory Director. Joe Wampler had come from the University of California at Berkeley on a two-year leave of absence. Wampler was getting the AAT off to a fast start, in return for which it seemed reasonable that he should have a say in the matter. Wampler preferred Sydney. I suggested to the board, therefore, that we make a trial period of one year in Sydney, after which the matter would be reviewed.

Of course, this was equivalent to a decision. In the coming year, the board would accumulate staff with a bias towards Sydney, and they would be keen to stay there. If it had happened the other way round, the board would have accumulated staff with a bias the other way. With the thought that it would not be my problem in a year's time, I told the Science Research Council that 31 August 1975 would be my last day as a board member. The limit had been reached to events that began when Jim Hosie first visited me in October 1967. In the intervening eight years, the AAT and the 48-inch Schmidt had both been completed, and both were fine telescopes, as different from the Isaac Newton Telescope as it was possible to be.

Now it was necessary to cope with the tax laws of the new Wilson government, a government whose policies were to lead nearly to national bankruptcy in 1976 and whose activities generally were to be among the drollest of the century. I would have to remain entirely outside the United Kingdom for a complete number of government financial years, until a sufficient monetary cushion had been built up for the future, thus, in the words of the novelist Jack Priestley, giving the tax man his ultimate victory. To avoid the prodding and prying of Priestley's tax man, I thought it prudent to resign all remaining connections with British clubs and committees. Otherwise, I could see the Inland Revenue demanding signed and sworn statements from every such association to the effect that I had not attended any of their meetings. This is the way things always seem to go.

Thus, my severing of connections with the British scientific scene was not a matter of choice. It was forced by the Wilson government of 1974.

By the spring of 1977, after two years, the nomadic phase of my life ended, however. By then, it seemed that the modest financial cushion I had been working towards had been achieved. The intervening years were spent partly in writing, partly in research in institutions in the United States, and partly in giving lectures. My wife came over to the United States for a section of each year. In September–October of 1976, we made a great trip, thinking at the time that we could repeat it if we chose. Unfortunately, however, we never have. It was the American bicentennial year, but then, in the autumn, most of the vacation traffic had gone.

Our bicentennial tour started at 5:00 A.M. in Pasadena. Before the morning rush in the Los Angeles area began, we were across the San Gabriel Mountains, stopping for breakfast at Lancaster in the Mojave Desert. Then north up the Owens Valley, ever north past Mono Lake and into Nevada, stopping for the night just south of Carson City. The following day, we crossed back into California via the Donner Pass, nearly a quarter of a century after I had left Otto Struve in Berkeley to get across the Donner before a blizzard struck. There was business with a publisher in San Francisco. Then across the Golden Gate Bridge, but not on and on to the north by the main highway. Instead, we turned west for the sea, which we followed for more than a day. Then we turned back inland again to the Redwood Highway, looping west gradually to the sea again at Eureka, near the northern frontier of California. There had been heavy rain, leaving a great deal of mud on the road as we crossed into Oregon and then over Grant's Pass. After a night in a remote valley, we pressed on through Portland as far as Olympia in the state of Washington, with Mount Rainier in the distance. Then we drove up the Olympic Peninsula to Port Angeles, where a ferry goes over to Vancouver Island in Canada. Then north and ever north, up the Fraser Canyon as far as Prince George, British Columbia, at the verge of the Alaska Highway. My memory there is mostly of three days spent in a log cabin on the northern bank of the upper Fraser River. From Prince George, we cut east through the Rockies at Mount Robson. In the early morning, the mountain was clear, its immense ice fields stretching upwards to seemingly impossible heights. We crossed into Alberta, then headed southeast from Jasper to Banff by the same route we had followed in 1953, but now it was a fast smooth road, rather than the unsurfaced track it had been a quarter of a century before (which was not an improvement, we thought). From Banff we went on to Calgary, and thence to Regina, Saskatchewan, and Winnipeg, Manitoba, with lakes produced by glacial action being the main interest in an otherwise featureless section.

East of Winnipeg, we reached the basin of Lake of the Woods in southwestern Ontario and northern Minnesota. In its bright fall colors, the forest was a revelation, and it became clear just why many outdoorsy Americans from Minneapolis prefer to take their vacation on the Canadian side of the border. Timing is important at Lake of the Woods, unless you are impervious to the bites of black flies. We hit things with the fall colors at their best and with the black flies disappearing before the onset of winter. Three weeks after we hit the Great Lakes region on the Canadian side, the temperature plunged some 50°F. Likewise, the road along the northern side of Lake Superior was an unexpected revelation, as was the peculiar Archean geology of the entire region. Then we drove eastward through Sault Sainte Marie to Sudbury and Ottawa (with me quite failing to remember just where I had been at the end of 1944 when I had been almost frozen solid on a December day while climbing a 50-foot vertical steel ladder) and then on to Montreal and Quebec. In Quebec, I concentrated on driving uphill and ended exactly where I wanted to be, on the Plains of Abraham. I had often wondered how the American War of Independence would have gone if the English general James Wolfe hadn't died there in battle with French forces under Montcalm. Maybe George Washington wouldn't have been able to beat him. Maybe, with Wolfe in command of the British forces, there would have been no rupture. Maybe he would have sided with Washington. Who knows? At all events, Wolfe was not the sort of man to have ordered the British navy to sail up the Bronx River, which is what the Admiralty in London ordered the local Admiral-in-Command to do. It isn't easy to think of many cases where one man might really have changed a major course of history. Wolfe was such a person, which was why I wanted to visit the Plains of Abraham.

The road south from Quebec took us across low mountains into Maine, New Hampshire, Massachusetts, and Connecticut, where we stayed with friends for a while. Then we drove past New York to Washington and up on to the Blue Ridge, which we followed, summit after summit, as far as Asheville, North Carolina, and thence northwest through Tennessee and into Kentucky, where we again stayed with friends. We crossed the Mississippi River into St. Louis, Missouri, and Mark Twain country. The fall colors were particularly superb all the way to Oklahoma City, whence we headed directly west to Santa Fe, New Mexico, the opposite to the day I had driven in 1953 with two young sailors who had just come back from the Korean War. From Santa Fe, we drove northwest to Grand Junction, Colorado, and thence by the Colorado River to the Grand Canyon, and ultimately back to Pasadena, where our journey had begun. It had been a different world from the tortuous windings of the past decade, which I was glad now to have put behind me.

The major casualty of the decade for me had been a relentless step-by-step decline from scientific competence, which those who are tempted into higher administration invariably experience. Most go dead mutton after five years. Many express the wish to "get back," but none I had seen had ever done so, any more than an athlete who has retired is likely to succeed in making a comeback. This was to be the ultimate challenge as I returned to Cockley Moor in the spring of 1977.

Chapter 27

Climbing the Last Munro

OVER THE DECADE from 1975 to 1985, I became interested, together with my colleague and former student Chandra Wickramasinghe, in the big problem of the origin of life. We came to think that life is a cosmic phenomenon and not the outcome of a number of highly improbable events that took place locally here on the Earth. It will seem strange to the nonscientist that this clearly interesting question should have provoked a high degree of opposition and resentment, but very certainly it did so. Curiously, it is perfectly all right to say that life exists elsewhere in the Universe, provided it is considered to have arisen separately in each place of its existence, even though to have arisen separately would require a repetition of the same highly improbable events as on the Earth. What must not be done is to regard life in many places as manifestations of the same cosmic process. This, it seems, offends deeply against the scientific culture.

Although—as yet, at any rate—we have not received any plaudits for a decade's quite hard work on this question, at least on one score we were successful. It was a consequence of our views that much of the material on the outside of our planetary system would have to be of an organic character, and also much of the solid material in interstellar space, a prediction that is now acknowledged fairly generally to have been correct.

A highly exuberant project from the late 1960s, which I completed in these years, was the climbing of all the Scottish mountains with heights above 3000 feet, some 280 of them, the so-called Munros. Ben Dearg, some 3550 feet in height and about ten miles southeast of Ullapool, was my penultimate Munro. I had been lucky to get through a forestry gate at Inverlael, gaining a couple of miles by car along a rough road up the valley of the river Lael. Ben Dearg was the second of two Munros that day. There was an interesting corrie (a cup-shaped depression) between them that I crossed in gathering cloud and rain, aware that, because of deteriorating weather, it would be a tricky matter to find the way out again—or, at any rate, to find it with the precision required to take me directly back down to my car in the Lael Valley. It was just the kind of situation Joe Jennings reveled in. Once he had oriented himself with map and compass, Joe would stride along without ever needing to reorient himself again. Never in the many days we spent together did he emerge anywhere but at our intended destination. When a heavy mist is down, I, in contrast, have to crawl along, reorienting myself every half minute. I was much aided that day by an excellent compass that I strapped on my wrist like a watch, making it easy to take a look at it every half minute or so. By this technique, I had found that, provided you have read the map correctly and have kept fixedly on the right compass bearing, there is nowhere you can be, in the long run, except in the right place. Even so, I was much pleased when I reached the track down the Lael Valley, arriving at my car in plenty of time to get back to my somewhat distant hotel in time for dinner—or so I thought. But, just as I stopped the car at the forestry gate near the main road, there was an ominous click from the gearbox, which turned out to be jammed. Luckily, it was the sort of jam that could be rocked out, but I wasn't strong enough to do it alone, and, by the time I had found help (the day was Sunday), I was indeed late for dinner.

Clearly, this was not major trouble, but, had I been Dick Cook, I would have taken it as an omen: this was in September 1971, two months before the Cambridge disaster. There was another omen not long afterwards. I used to keep the excellent compass in the flap of my rucksack, so that it was always there should I need it. Otherwise, I simply forgot about it, the way I always fail to think about small things that seem to be in order. Shortly after climbing Ben Dearg, I checked my rucksack after a plane flight to find that the glass on the compass had been smashed by baggage handlers. I never found another as good, a further indication of how things do not necessarily improve with time.

My last Munro was Blaven, a distant extension of the Cuillin Hills on the Isle of Skye. I had planned with friends to make an event out of it in May 1972, carrying up bottles of champagne and a generally sumptuous repast. But the events in Cambridge destroyed this plan, which required joie de vivre as well as good food and champagne. Nor was the wish there from 1972 to 1975, nor was there the opportunity from 1975 through early 1977. Then, at last, after returning to Cockley Moor in the spring of 1977, I thought again of Blaven. It made little sense, however, to revive the idea of a party. I would go there quietly, alone, and finish what I had begun so zestfully twelve years earlier. Thereafter, I made several trips to Skye, but, on each occasion, the weather turned indescribably foul, to a point where I began to think that I was destined never to climb Blaven. Dick Cook, with whom I had done the first Munro in 1965, Moruisg, attributed personalities to mountains. If a mountain, even a very difficult one, took a liking to you, it would let you through to its summit with little trouble. But, if you got yourself, for some reason, into the bad books of even an easy mountain, there was no way you were going to climb it without trouble. It was a philosophy I found curious, but, since it had carried Dick safely up all the classic climbs in the Alps and to the summit of a 21,000-foot Himalayan peak when he had reached the advanced age of 57 years, I had to respect it.

I had begun to think I was surely in Blaven's bad books when, on 23 October 1980, I awoke in a hotel in Broadford on the Isle of Skye to a cloudless morning. The intention in 1972 had been to make the rock-climbing traverse from the nearby mountain of Clach Glas, but now, done at the end of October after a cold, clear night, I judged this to be out of the question. Cold autumn nights can lead to conditions as bad as real winter. The rocks first become wet after rain and the water then freezes to give a thin layer of ice (*verglas*, as Alpine mountaineers call it) everywhere. Old-style clinker-nailed boots help by cutting through the thin ice, but modern rubber-soled boots leave you stumbling helplessly.

Blaven is a humpbacked mountain with two tops differing in height by a mere eleven feet, the northern top being the higher. There is an easy route up the northern top on its eastern side and another up a stony gully between the two, but my climbers' guidebook praised the long southern ridge from Camasunary, the name given to a croft about three miles to the north of the village of Elgol. Since this would give superb views over the main Cuillin summits to the west, I decided to take it, the other routes being dull slogs. So I drove from Broadford to Elgol and made my way along a track to an enclosure of fields at Camasunary. The southern ridge was delightful, just as the guidebook said it would be, but, the higher I climbed, the more I began to feel that Blaven really had me in its bad books, with consequences that Dick Cook would have predicted. Somewhere ticking in my mind I remembered it was a rock climb between the two tops of Blaven, that it wasn't just an easy rise of eleven feet, as my guidebook seemed to suggest. I arrived on the southern top at noon to find it was so. There was a steep plunge of 150 feet to a gap beyond, and the rock was indeed slippery with verglas, out of the question for a lone man of 65, never too much of a rock climber anyway. Trusting the guidebook was the same flaw of temperament that had made me trust precedent in Cambridge.

The crags fell away steeply on both sides of the ridge. Because of the icy state of the rocks, it seemed pointless to examine them in detail. Indeed, the rocks to the west were just as intractable for me as the ridge itself, while there seemed to be an essentially vertical precipice on the eastern side. But you can never be quite certain in the Cuillins; things that appear to be impossible may turn out otherwise, like problems in life. The weathering of this remarkable 75-million-year-old volcano has produced an amazing variety of freak rock formations. Descending a little to the east, I saw a gully going steeply down to screes now perhaps 200 feet below. The gully was evidently straightforward for perhaps 100 feet, where it was blocked by a big boulder, beyond which I could see only the lower screes. Without too much hope, I made my way gingerly down to the boulder to find that, on its outer, eastern side, the cliff did indeed fall, as I had expected, vertically to the screes. But on its upper side, amazingly, there was a groove containing a trap dike of firm, ice-free gabbro rock that went down to the bottom like a steep flight of steps. The boulder was so shaped, moreover, that I could squeeze between it and the upper side of the gully so as to reach the trap dike. Fifteen minutes later, I was on the northern top. It had been my persistence, Dick Cook would have said, that had caused the mountain to relent, a point of view well understood by Greek authors before their culture was overwhelmed by the Romans and the Christians, the latter with a preference for devilry and witchcraft over the spirits of the groves, the rivers, and the mountains.

In 1971, we had planned for champagne and a goose (or a turkey, at least) for the summit party. Now, it was a solitary flask of tea and a packet of Mr. Kipling's apple pies. As I ate the pies, I looked across at the main ridge of the Cuillin Hills, stretching tortuously from Garsbheinn to Sgurr nan Gillean, which I had first visited in 1936 at the threshold of my research career.

As I descended the mountain by the way I had come, it was no longer so cloudless; but the sky was still softly suffused with blue, and—immediately at my feet, as it seemed—the islands to the south of Skye were dotted in a richly colored sea. I had enjoyed many beautiful days in my youthful wanderings on the Cuillin Hills, but none to surpass the beauty of that late October day on which I climbed my last Munro.

Chapter 28

A Lucky Ending

I WAS A FEW months over the age of 65, near the canonical age of retirement, when, in October 1980, I stood on the summit of Blaven. More than my project of climbing all the Scottish Munros seemed to be finished. By then, most of my work, excepting that in cosmology, had been decided for better or for worse. Mostly for better, fortunately. The cosmology was otherwise, however, and it is this lingering topic I am going to discuss here in my final chapter.

I will be immodest enough to claim a reasonably substantial role in the rise of big-bang cosmology, which took place in the 1960s, and which is still quoted as the foundation of belief in this theory. In 1963–1964, there was work with Roger Tayler that had the effect of refocussing attention on the achievements, around 1950, of George Gamow and his colleagues. Then, in 1966–1967, there was work done in collaboration with William A. Fowler and Robert V. Wagoner that suggested that the element lithium (or, at any rate, one of its isotopes) was the product of the big bang. Indeed, the latter investigation has served as a springboard for the development of the concept of nonbaryonic "missing mass," which, by 1980, had become all the rage. This was amply sufficient for me to have become one of the dozen or so influential torchbearers for the big-bang theory, after whom the astronomical world in general follows. But it seemed to me, again for better or for worse, that the big-bang theory did not develop, as the years passed, as a successful theory is supposed to do. A successful theory should progress as a river grows, from trickles on numerous hillsides, to a moderate number of tributaries, and then, by amalgamation, into a broad stream. This has not happened with the big bang, which has almost no success to offer in return for a great deal of effort over the past two decades.

Cosmology is the study of the whole Universe. We can all agree that such a study is highly ambitious. To claim, however, as many supporters of big-bang cosmology do, to have arrived at *the* correct theory verges, it seems to me, on arrogance. If I have ever fallen into this trap myself, it has been in short spells of hubris, inevitably to be followed by nemesis. Our efforts to understand the whole Universe should, it seems to me, be viewed in the way, some seventy years ago, we village boys set about raiding in September the ripening orchards of the larger houses of my home district. If we could get away with it, well and good. But, if we were caught and punished, we had only ourselves to blame.

It is the crowning achievement of science that its entire spectrum can be shown to be derivable from just one broad stream of argument, a stream that began as a trickle in the early nineteenth century due to the work of the French mathematician Joseph Louis Lagrange. The method is concerned with calculations involving a finite volume of space-time— but *any* volume you care to choose. This generality is crucial, for it gives universality to the laws of physics that emerge from the resulting calculations. It is this essential breadth of vision that is lost in the big bang, and it is lost in a very flagrant way. Einstein's famous work on the theory of relativity was founded on the principle that the laws of physics should not depend on how the points of space-time are catalogued by means of what is called a coordinate system. Yet this is just what the big bang does. It catalogues space-time in a particular way and then says that, with respect to its particular choice of coordinates, the broad stream of physics holds, or does not hold, according to a particular division in the catalogue, a division that depends on its own arbitrary mode of construction. Stated more conventionally, in a particular coordinate system, there is a time before which the laws of physics do not hold and after which they do.

It was this artificiality that led Hermann Bondi, Tommy Gold, and me to begin thinking, in 1946–1947, about alternatives. We saw that alternatives would require the creation of matter, and we were at first repelled by the idea. However, the events of the cold winter of 1946–1947 fell out such that I found myself returning to it. I saw that, to make mathematical sense of it, I had to suppose the existence of a so-called scalar field, which turned out to have properties that greatly surprised me. Much of my surprise was due to not really thinking through what happens in everyday physics when some local process releases energy—say, the burning of a pile of wood. We are well used to the loss of energy that occurs as the wood becomes oxidized, the balance of the energy being emitted as light and heat. The light and heat may be absorbed by our surroundings, but, to the extent that they are, our surroundings then reemit the energy. Perhaps, after several such absorptions and reemissions, the energy ultimately goes out into space, emitted by the Earth along with the much larger amount of energy that the Earth receives from the Sun. And it is there that we normally think the sequence of processes ends, as it would if the Earth were alone in flat space-time. But, in the actual Universe, the energy is ultimately absorbed by the Universe itself. It goes into making a slight change in the rate of expansion of the Universe.

The same thing happened when I investigated the properties of the scalar field, which I called the C-field because I was interested in the possibility of the field serving to create new matter. Creation required energy. It took in energy—the opposite to a wood fire, which gives out energy. The resulting C-field went out from the region of creation, eventually to lose itself in the expansion of the Universe, energy being balanced at all stages.

What puzzled me at first in my investigations of January 1948 was the sign of the effect. It shouldn't have done so, since I had supposedly learned all about signs in standard relativity theory as long ago as my final undergraduate year of 1936. But it may be that, like me, quite a few scientists have the sign wrong in their heads. Thinking of the case of the wood fire, does the positive energy going out from the fire have the effect of slightly speeding up the expansion of the Universe, or does it slightly slow down the Universe? I will just bet that many would say that the fire speeds up the expansion, which is wrong. Positive energy going out with a positive pressure slows the Universe, no doubt about it. This is what Einstein's theory says. We tend to get the sign of the effect wrong through thinking of the explosion of a local cloud under everyday conditions. Internal heat and light do go into speeding the expansion in that case. For the Universe, it is the opposite; and, with the C-field carrying *negative* pressure effects to compensate for the positive energy of newly created particles, the reaction on the Universe was to cause expansion, not to check it. To get the situation right, think of expansion as a *negative* form of energy; then everything follows the right way round.

In 1948, I had no physical ideas about the details of creation. Such ideas were only to come slowly over succeeding years. So, in place of physics, I made a mathematical hypothesis. I assumed matter to be created everywhere at some slow rate, which I visualized as adjustable, as one might adjust the flow of water from a tap. Subject to this hypothesis, I was able to prove what hit me as a remarkable result—namely, that the rate of expansion of the Universe always came into balance with the flow from the tap. Turn up the tap, and the Universe speeds up; turn down the tap, and the Universe slows. Now I felt I understood *why* the Universe was observed to be expanding: it was because matter was being created everywhere.

Let me insert a bit of a diatribe against the big bang at this point. Why, in the big-bang view, does the Universe expand? Think first of a large local cloud of gas set (by some unspecified agency) in outward motion. If the motion is too weak, the self-gravitation of the cloud will cause it to slow down until it comes eventually to rest and then falls back on itself. If the initial motion is too strong, the self-gravitation of the cloud will not be able to slow it much and the cloud will disperse completely, with all its particles still moving outwards at high speed. Like balancing a pencil on its point, there is an intermediate situation with the cloud neither falling back on itself nor dispersing at high speed. The initial motion is so finely adjusted that the cloud just—and only just—manages to disperse. A little less initial energy, and the cloud would fall back; a little more, and it would disperse too rapidly. This picture of a local cloud generalizes to the big-bang view of the Universe, in which the entire Universe originates like the pencil balanced on its point. The balancing must be ultrafine. The density range in the cloud from the first moment contemplated in the theory (10^{-43} seconds after the supposed origin of the entire Universe) until the present is so vast that the pencil has to be balanced to an accuracy of about 1 part in 10^{60}. Written out in full, 1 part in 1,000,000,000,000, 000,000,000,000,000,000,000,000,000,000,000,000,000,000,000. How is this incredible balance achieved? There is no answer from the big-bang supporters, except with the implication of divine adjustment.

So, with the C-field, I now had an escape from this dilemma. I had a mathematical law that forced a balance between the rate of expansion of the Universe and the density of the matter that it contained. There was no need for an initial situation balanced to infinitesimal accuracy. The situation could start quite out of balance, and, as time went on, it would come increasingly into balance. And I didn't need any appeal to a moment of origin for the Universe. The Universe could be as old as you like.

It was just such a result as a young man on a small salary and with a family to support dreams about, and you can safely assume that I set about writing a paper on it with the utmost dispatch. But I had a heavy teaching load in February 1948, with the university working a six-day week. On good days, I would be finished in time to catch the 4:00 P.M. school bus from Cambridge to Great Abington—these were the days when we were living at Ivy Lodge, the house without electric power. Central heating in a cold winter was not even a vision for the future. The house had a pleasant drawing room with an open fire. With coal supplies minimal, this was the one fire we allowed ourselves, managing elsewhere with portable kerosene stoves. The work went in a rush towards the end, and one night I told my wife that I was going to stay up after the family were abed until the paper was finished, which it was at about 2:30 A.M.

I am unhappy to think that I must have proselytized for my paper. Otherwise, I fail to see how I could have given a seminar on it already by the first of March. The seminar was in the Cavendish Laboratory, with Paul Dirac in the front row. So, too, somewhat surprisingly, was Werner Heisenberg. I happened to be in Germany the following year, and I was flattered to be told that, on his return, after spending six months at St. John's College, Heisenberg had said that the C-field interaction between the creation of matter and the expansion of the Universe, the negative-pressure effect, was the most interesting idea he had heard during his stay in England.

Hermann Bondi and Tommy Gold were there too at the Cavendish seminar, and they immediately came up with an interesting further idea. The model I had described was for the creation tap to be turned on uniformly everywhere, giving a Universe that, on a large scale, was the same at all times. Bondi and Gold argued that, by observing a relatively small sample of such a Universe, one would know what the entire Universe was like everywhere, not just from place to place but also from time to time. Hence, astronomical observations made at a great distance, at the limiting range of very large telescopes, should (red-shifting effects apart) reveal the same state of affairs as very much more local observations. And, since the latter could be done with relatively modest equipment, it followed that a great deal could be learned about the distant Universe without the need to make huge investments in immense equipment, such as the 200-inch telescope that was coming into commission on Palomar Mountain at just about that time. The position, as Bondi and Gold advanced the argument, was wonderfully hopeful for astronomical observers. It was wholly different from the big-bang theory, in which the observer has no hope at all of penetrating to the supposed origin of the Universe, even if the entire productivity of the human species were expended on it.

But, between us, we had given too many hostages to fortune. To see my way through the mathematics, I had assumed a uniform creation tap, while Bondi and Gold had contracted to explain everything, no matter how far distant, in terms of what was known for galaxies and other astronomical objects within the nearest few tens of millions of light-years. This was a citadel it was going to prove very hard to defend.

Actually, we did quite well for five years or more. Attempts by observers to carry the day against the theory ended for a while in defeat for the observers. But one or two remarks, made to me privately, raised apprehensions that at least had the effect of preventing hubris from appearing onstage. The theory required new galaxies to be formed and at such a rate as to renew the entire distribution of galaxies in a time of the order of 10 thousand million (that is, 10 billion) years. Big-bang cosmology does the same. But big-bang cosmology contrives, as it does in every aspect of astronomy and astrophysics, to make the process of the origin of galaxies unobservable, whereas our theory required the origin to be very observable. But it wasn't observable, Walter Baade told me; and, in the mid-1950s, my friend Allan Sandage insisted that distant galaxies were slowing down in their expansion rate, whereas, according to our theory, they should have been accelerating.

Walter Baade was kind enough not to make his criticism very public. Nevertheless, I knew it was seriously meant, and it has not been until recent times that I have seen how to answer it in a satisfactory way. A reply to Sandage's criticism was already apparent at the time, however. It was that, in comparing galaxies at different distances, it was possible that like was not being compared with like. This issue has never really been settled to everybody's satisfaction. Only two years ago (1992), at a meeting of the German Astronomical Society, I found cosmologists who take much the position I took in 1955. Meanwhile, I have swung towards Sandage's old view, only to find him less sure about it today than he was twenty-five years ago.

The troubles for the theory that began in the mid-1950s, and that became more severe as time went on, permitted Bondi, Gold, and me to blame each other. I could say, and did, that they had made the scale of their representative local region too small. I thought it should have been 1000 million light-years instead of a hundred million. Actually, it should have been much bigger still. But the worst mistake was my own: my obstinacy over the creation tap, which I kept always open and at a slow trickle. This has turned out to be as wide of the mark as it could be. The correct position is that the tap is nearly always closed; but, on the rare occasions when it opens, a flood emerges. I had plenty of warning. In 1950, my wife and I toured Switzerland in our papier-mâché DKW. I gave a talk at the Technische Hochschule in Zürich. Wolfgang Pauli was there, and, at dinner with him that night, he raised just this question, saying: "If you could understand the physics of how creation happens, it would be much better." Unfortunately, it was to be a long, long time before I came to grips at last with this central issue of the theory, not until I returned to cosmology in 1985, after many years away from the subject.

Because the C-field is a so-called scalar, a quantum description requires it to be made up of bosons, particles with properties analogous to the quanta of an ordinary radiation field. From this, it is to be expected that interaction processes with matter, including the creation process, will be analogous to radiation processes. The latter, being well known, provide a useful catalogue of reference for thinking about the creation process from the quantum point of view, as Wolfgang Pauli was suggesting. The nearest analogue is the process that leads to the strong emission of radiation from a laser, which process leads to conclusions that are very much in agreement with a wide range of astrophysical observations, all those observations that relate to what is often referred to as high-energy astrophysics—strong radio emission, x-ray emission, active galactic nuclei, and quasars.

A laser is a device in which a number of light quanta combine to induce the downward transitions of atoms, with more quanta coming out than go in, at the expense of the downward transitions of the atoms constituting the material of the laser, which atoms must first be "pumped up" to a higher level before the laser can be triggered. But, for the C-field, there is an important inversion of sign to consider: the inversion of sign that produces expansion of the Universe through a negative pressure, an inversion that causes matter to go up in energy at the expense of the negative effects in the C-field. Technically, this inversion is not hard to achieve. In terms of the method developed by Richard Feynman for describing mathematical interactions of an ordinary radiation field, a function written as δ_+ appears in the quantum amplitude for the process in question, while a time-inverted function δ_- appears in the complex conjugate of the quantum amplitude. For the C-field, the two functions δ_+ and δ_- are simply interchanged. My trouble, in 1948, was that I didn't know all this, for the understandable reason that, in 1948, it hadn't yet been done—nor had lasers yet been invented.

A laser is triggered by fine tuning between the frequency of the quanta responsible for the firing and the frequency associated with the transitions of the pumped-up atoms constituting the laser material. In a like manner, the creation of matter is triggered by a fine tuning of the C-field quanta and the frequency associated with the mass of the newly created particles. Now even if such a fine tuning existed initially, it would be lost very soon indeed, when viewed on a cosmological time scale, because the frequency associated with specific particles remains unchanged, whereas C-field quanta in extragalactic space have their frequencies progressively lowered by the expansion of the Universe. So it can be seen now that my guess of 1948—a guess made for mathematical convenience rather than from physics—was a mistake. The tap cannot dribble everywhere at a constant rate. In extragalactic space, it cannot dribble at all. The C-field quanta there must be too low in frequency.

But light quanta gain frequency if they fall into a gravitational field, and so it must also be for C-field quanta. And, if the gravitational field is sufficiently strong, such a rise (for those C-field quanta that happen to fall into it) will become sufficient to open the creation tap. The flood of particle creation generated by the analogy to laser action is then triggered. The existence of the C-field is hypothetical, essentially forced in order to understand how matter came into being. It is essential in order to avoid the breakdown of physics that otherwise occurs at the supposed big bang. Granted the existence of the C-field, its physical properties can then be formulated in the manner I have just outlined, leading to the view that, whenever sufficiently condensed aggregates of matter are formed, the creation tap is opened. Large negative pressures are generated with the same expansionary effects ensuing as occur cosmologically. A falling together of matter, towards what is called a black hole, is the process that triggers matter creation, and it is a process that causes any aggregate that approaches a black hole too closely to self-destruct, to blow itself apart, the energy of the explosion being compensated by an expansionary push on the Universe.

At this point, observation takes over—or, more properly speaking, should take over. In recent years, what might be called a black-hole establishment has arisen, composed of individuals who talk to each other in positive language, as if black holes were as certain of existence as tomorrow's sunrise. Yet there is not a scintilla of observational evidence to support their position. What there certainly is evidence of are highly condensed aggregates of matter producing very strong gravitational fields. There is a great volume of evidence of violent activity associated with such aggregates, but the evidence is all of outbursts, never of the continuing infalling motion that would lead to the formation of a black hole. The evidence is all of a process that self-destructs. Tommy Gold has described the psychological process whereby people meeting often enough among themselves can become swayed towards quite erroneous conclusions, in a way that the solitary thinker would be proof against. The process is the same as the one whereby whole societies become sucked into amazing misjudgments, as happened in extreme form in the South Sea Bubble and in the astonishing episode of Dutch Tulip Mania. It is a process whereby assurances from others take on the appearance of reality. Once established and running, the process even overrides sharply contrary evidence—for a while, at any rate.

The picture that now opens out is greatly changed. Instead of being locked into a never-ending sameness, as it had been with my mathematically imposed condition of 1948, there is now a wonderful freedom, with the highly condensed aggregates of matter, taken on a wide scale, connected to the rate of expansion of the Universe. Let there be a high average density of aggregates, all blasting away, and the Universe expands rapidly. Let the aggregates disperse through self-destructive explosions, and the Universe slows under its own self-gravitation, perhaps even falling back in on itself.

Now it is a property of the C-field that, quite apart from its relation to the creation of matter, it cannot be compressed indefinitely. The negative pressure within it eventually rises more rapidly than self-gravitation, with the consequence that an imploding Universe, or an imploding local object, is eventually halted in its inward motion. Thereafter, the motion is reversed, and expansion is repeated. That is to say, without creation, the C-field acts to produce an oscillatory motion. Oppositely, with a great deal of creation, the oscillatory tendency is overwhelmed by expansion. And with a modest degree of creation, the Universe oscillates, but it becomes a little expanded at each oscillation. Broadly speaking, this, I believe, is the path to be followed—at any rate, it is the model best suited to the general time range in which we happen to live. To put it in numbers, each oscillation takes about 40 billion years to complete, 20 billion years from minimum to maximum and 20 billion years back from maximum to minimum. And, becoming a little larger at each oscillation, the Universe doubles its scale in about 20 oscillations—which is to say, in about 800 billion years. The situation is freed from the claustrophobia of the old mathematically constrained model that I suggested in 1948, and it is immensely farther-ranging than the big bang.

Much as I have always disliked the big bang, there is a sense in which one of its concepts survives in the properties of each individual particle of matter during the fleeting moment of its creation. A Machian formulation of general relativity, which Jayant Narlikar and I proposed in 1964, with its scale-invariance and its proof that gravitation must always be attractive between particles of matter, shows that newly created particles have the special property that their Compton and gravitational radii are comparable with each other. Such particles were first conceived of by Max Planck, and so are conveniently referred to as Planck particles. They are much more massive than ordinary particles, about 5,000,000,000,000,000,000 times more massive than a hydrogen atom. Their investigation represents the ultimate aim of the experimental physicist. But such a goal lies far beyond the capability of any equipment that can at present be contemplated as a practical construct. Planck particles are believed to decay into subsidiary particles in a time of about 10^{-40} second, with the subsidiary particles then decaying in a progressive cascade that ultimately yields a shower of the much less massive particles with which we are more familiar. This is in a time scale ranging from 10^{-24} second to 10^{-10} second. It is because the big-bang Universe is believed to have begun as a sea of Planck particles that physicists have become interested in cosmology. But the same is the case for the theory discussed here. Each particle, at creation, is like a big bang in itself, its physical properties as it decays having similarities to the earliest moments of the supposed big bang. Thus, whatever attraction the big bang may have for physicists is paralleled here. The situation is indeed improved from the physicist's point of view, since the particles he seeks are no more distant in the Universe than the nearest astrophysical high-energy source, whether an active galactic nucleus, a quasar, or a radio source. The supposed big bang, on the other hand, is beyond the range of observation, and, as the young Paul Dirac used to tell us students in the 1930s, "That which is not observable does not exist."

The difference between the present picture and the situation of the 1950s is that the theory is no longer tied to an immensely fast creation rate. It now takes 800 billion years for the Universe to double in scale, not the 10–20 billion years proposed in the 1950s, which cuts the needed creation rate by about 50. The rate at which nearby galaxies are required to form is also cut by 50, which is why Walter Baade insisted that observation simply would not permit the excessively fast rate that seemed then to be required. There is also the difference that, when the astronomer measures the expansion rates of the galaxies, it is the properties of the oscillations that occupy his attention, because the oscillations are fast, masking the slower overall expansion of the Universe. And, if one lives in the expanding half of an oscillation, as we are required to do, the Universe will decelerate as it comes towards maximum phase. It was this deceleration, it seems to me, that Allan Sandage found in the mid-1950s.

But, if the oscillations are averaged away over a very long time scale, it is the slow overall expansion of the Universe that remains, and this has the properties that Hermann Bondi, Tommy Gold, and I claimed it to have—except that everything happens far more slowly than we supposed. Ages are also greatly affected. Our galaxy is more like 300 billion years old than the usually claimed 10 billion years, which explains a lot of persistent trouble that astronomers have had over the estimation of ages. It also explains where the so-called "missing" dark material has gone and why galaxies, especially of the elliptical kind (which are the oldest), are exceedingly inefficient emitters of starlight. Stars like the Sun, which are reasonably efficient producers of light, are of recent formation, belonging indeed to the last 10 billion years or thereabouts. The stars of 300 billion years ago have long since gone, if they were at all efficient. Their products are essentially dead, as far as the production of light is concerned. Otherwise, the remaining stars of 300 billion years ago are faint so-called brown dwarfs, which are barely perceptible because they emit so little light.

The tall, charismatic Martin Ryle, from the Cavendish Laboratory, was the first to state openly that the steady-state theory of Bondi, Gold, and Hoyle was wrong. This was in the second half of the 1950s, at a time when nobody knew anything definite about the nature of the majority of radio sources. Ryle's idea was to count radio sources in the following way, much as Edwin Hubble had done for galaxies twenty years before. A flux level, S, was defined, and then, over a specified fraction (about a third) of the sky, an attempt was made to count the number, N, of sources with flux levels greater than S. The process was repeated for a number of values of S, leading to the determination of $N(S)$ as a function of S, a function that was then plotted logarithmically in what was called a log N–log S diagram. It can be shown that, statistical fluctuations aside, such a counting process carried out for any standard kind of object in flat Euclidean space-time results in a straight line of slope -1.5 in the log N–log S diagram. The theory of Bondi, Gold, and Hoyle led to the same slope of -1.5, provided the objects were not very far away. They could be outside our galaxy, but not a thousand million light-years away or more; otherwise, the slope would be less: -1.4, say, or -1.3.

Between 1955 and 1960, Ryle carried out three such investigations. There was what he called the 2C survey (C for Cambridge), which, he claimed, yielded a slope of -3 or more, and which, therefore (according to him), ruled out our theory. But the 2C survey had so few sources in it that it was seriously affected by statistical fluctuations. The next survey, the 3C, had about 250 sources, and it was said to give a slope of -2 or somewhat steeper than that, and, once again, Ryle claimed that his result ruled out our theory. We, in our turn, argued that, if the slope could change, through statistical fluctuations or observational inaccuracies, from being -3 to -2 in only a year or so, then, in a further year or two, it might become -1.5 or even less. Ryle's third shot was called the 3CR (R for revised) and it was said to give a slope of -1.8.

The negative slope meant that, as S decreased, N increased, and the steeper the slope, the more N increased. Thus, a slope of -1.8 had a steeper rise of N as S decreased than a slope of -1.5. Now, in such a situation, a steeper slope could mean either an excess of N at smaller S or a deficit of N at larger S. If the latter, the deficit of sources of large S in Ryle's 3CR survey needed to be only a handful, again making his claim statistically dubious.

The stage was thus set when, in early 1961, I had a telephone call from the headquarters of the Mullard Company. The Mullard Company had made extensive donations to radioastronomy at the Cavendish, so much so that the radiotelescopes at Lord's Bridge on the Barton Road had become known as the Mullard Observatories. A polite voice informed me that, during the coming week, Professor Ryle would be announcing new, hitherto undisclosed results that I might find of interest and asked if my wife and I would care to accept an invitation to be present. So it came about that, in the afternoon a few days later, I turned up with my wife at the Mullard headquarters in London. A smartly dressed Mullardman of about my own age led us into a modest-sized hall in which a number of media representatives were assembled. We were escorted by our host to the front row, where my wife was bowed into a seat. Then I was led on to a raised dais and bowed into a chair, not so comfortable as the one my wife had just been given. The smartly dressed man then withdrew, leaving me to gaze down on the media representatives. The rest of the stage decor consisted of a blackboard on an easel, a lowered screen for slides, and, I believe, a lectern.

So what was I to think about as I sat there under the bright lights? It needed no great gift of prophecy to foretell that what I was about to hear would have something to do with the log N–log S business. But was I being uncharitable in thinking that the new results Ryle would shortly be announcing were adverse to my position? Surely, if they were adverse, I would hardly have been set up so blatantly. Surely, it must mean that Ryle was about to announce results in consonance with the steady-state theory, ending with a handsome apology for his previously misleading reports. So, I set about composing an equally handsome reply in my mind.

A curtain parted, and Ryle entered. The Mullardman made a short introduction, and, pretty soon, Ryle had launched not into the promised statement but into a lecture. I was well used to its form, so I sat there, hardly listening, becoming more and more convinced that, incredible as it might seem, I really had been set up. The results involved the sources of what were now called the 4C survey. The 4C survey contained more sources than before, Ryle explained, greatly reducing statistical fluctuations. Yet the slope of –1.8 had been maintaining, showing that the steady-state theory was wrong, and would Professor Hoyle care to comment? The media leant forward in anticipation.

David Bates, with whom I had worked on the Earth's ionized layers circa 1948, had a story of a serious-minded academic sent by the Foreign Office to Germany, after the Second World War, to deliver elevating lectures to the troops, both British and American. An American C.O., having by then had enough of elevating lectures, introduced the serious chap to his men with a wide sweep of the arm, exclaiming, "He's the British Bob Hope! Boys, he'll slay ya." I know just what the academic must have felt, because that was the way I felt that day.

Ryle's supposed demolition of the steady-state theory was the lead story on the front pages of the London evening papers that night. For the next week, my children were ragged about it at school. The telephone rang incessantly. I just let it ring, but my wife, fearing something had happened to the children, always answered, fending off the callers.

The publicity had one good effect, however. Thereafter, the problem passed into hands more competent than those in the Cavendish Laboratory, at first with Bernard Mills, under the aegis of the ferocious Harry Messel in Australia. Knowing Harry, I think that, had he been in my position at the Mullard place, he would have been likely to have turned up with a hunting rifle and would have been prepared to use it liberally. What was eventually found was that the sources of the 4C and later fainter surveys have a slope in the log N–log S diagram that is little different from -1.5, except over a quite small range of S, a factor in S of about 3, where indeed there is a mysterious steepening of the radio-source count.

If the radio sources had turned out to be not very far away, the problem would probably have stopped at that point, since comparatively nearby sources behave as if they were in flat Euclidean space for which the log N–log S slope should indeed be -1.5. The discovery of a steeper slope over a small range of S would have been seen simply as a fluctuation of no particular cosmological consequence, and that would have been the end of it. But, as the years passed by, and explicit identifications of radio sources with explicit galaxies became more and more available, it gradually appeared that very large distances were actually involved, for which the Euclidean behavior in the log N–log S diagram apparently should not apply. Yet it did, more and more, as counts were made to lower and lower flux values. The counts made it look as if space-time were flat Euclidean— at any rate, to flux values, S, that were only a tenth as large as those that were available at the Mullard meeting.

The big-bang theory explains this remarkable result as a matter of happenstance. Radio sources change intrinsically with time and they just happen to do so in a way that compensates essentially precisely for the effect of the expansion of the Universe, so that, in combination, the situation mocks a Euclidean state of affairs. And nobody knows why. This would be bad, but what makes it worse is that the same thing has to happen for the apparent angular sizes of the radio sources. By happenstance, and without any apparent connection to fluxes, the sizes of radio sources must change with time in such a way as to mask the effect of the expansion of the Universe, again mocking a Euclidean structure for space-time. These are two examples of the large number of hypotheses that the big-bang theory must make in order to save itself. Instead of the theory serving to make deductions that can be compared with observations, as in normal science, the "theory" is really a catalogue of hypotheses, like a gardener's catalogue.

The explanation of the Euclidean puzzle turns out to be simple, in the theory stated above. Going backwards from oscillation to oscillation, radio sources that are uniformly distributed and intrinsically similar to each other can easily be proved to increase in N as S falls with increasing distance, just as in the Euclidean case, as $S^{-3/2}$. It is only when the cycles go back far enough, some ten cycles or more, with the slow overall expansion of the Universe becoming important, that the slope falls off, with N then increasing less rapidly than $S^{-3/2}$. The steep slope over a small range of S turns out to be caused by a discontinuous jump from the present half cycle to the previous cycle. It turns out to be all very straightforward, without any element of happenstance being required. The ultimately remarkable outcome, from the point of view of the radioastronomer, is how amazingly far back in time the log N–log S observations reach. Back to the eventual fall-off of slope, some 15 cycles of oscillation are involved. With 40 billion years occupied per cycle, this is a look-back age of about 600 billion years, much farther back than is achievable by any of the usual forms of astronomy, and, of course, vastly further back than the hypothetical big bang permits the Universe to exist.

In the late 1950s, it was found in the laboratory that a curious kind of particle forms when metallic vapors are cooled slowly, a particle shaped like a long, very fine needle. Typically, the lengths are a fraction, perhaps a third, of a millimeter and the diameters are about a hundred-thousandth of a millimeter. The first thing I did, when I returned to cosmology in 1985, after an absence of 15 years, was to calculate the electromagnetic properties of such particles: I found that they have immensely strong absorption and emission properties for radiation with wavelengths in the so-called microwave background, a background that had been discovered by Arno Penzias and Robert Wilson in 1965 and that is supposed by many to prove the correctness of big-bang cosmology.

Chandra Wickramasinghe checked and improved my calculations by high-speed computer, and it was the circumstance that particles known to exist in the laboratory have the ability to produce microwave radiation in a perfectly normal, everyday way that first made me acutely suspicious of the claim that only through the origin of the Universe in a big bang can the microwave background be explained. Add, too, that the laboratory condition of the slow cooling of metallic vapors is closely imitated in the cooling of metals produced by supernovae, that supernovae tend to occur in associations of massive stars, which, in concert, over relatively short periods, expel their material at high speeds, not only into their parent galaxies but out of them altogether into extragalactic space, and it was apparent that the means for producing the microwave background by straightforward astronomical means were all in place.

Even so, until I began collaborating again with Jayant Narlikar and my old friend Geoffrey Burbidge, I didn't appreciate just how easy and straightforward it was going to be. It was merely a matter of using intergalactic iron needles, of which no great density is needed, to degrade star-light into microwaves—which is to say, radiation with wavelengths generally in the region of one millimeter. To put it in numbers, we know that, over about 30 percent of an oscillatory cycle of the Universe, the cycle in which we are presently living, stars have produced about 2.10^{-14} erg of energy for each cubic centimeter of extragalactic space. Hence, over some 15 similar cycles—the time for the slow doubling of the scale of the Universe—the energy produced would be about 10^{-12} erg for each cubic centimeter of extragalactic space. Or it would be, if the slow expansion of the Universe were not taking place. The latter spreads the energy somewhat, reducing the concentration to about 4×10^{-13} erg for each cubic centimeter of extragalactic space, and this, after being absorbed and reemitted many times by the iron needles, can readily be shown to generate a black-body radiation distribution with a temperature of 2.7 kelvins, which is just the observed value for the microwave background. It is also easy to show that the resulting black-body distribution would be exceedingly homogeneous and isotropic, with temperature fluctuations from place to place, and also from one direction to another in the sky, of no more than a few parts in a million, arising from irregularities in the distribution of the sources of the starlight. All the requirements of observation are thus well met, and really without any significant hypotheses having to be made. The stars exist. The starlight is there, and its intensity is known from observation. Iron needles exist in the laboratory. Supernovae as sources of iron exist. All the components needed for understanding the origin of the microwave background quite outside big-bang cosmology are therefore known with certainty, and only relatively easy calculations are needed to put them all together. In astronomy and astrophysics, one cannot do better than that, especially inasmuch as the answer turns out to agree with observation more or less exactly.

How, in big-bang cosmology, is the microwave background explained? Despite what supporters of big-bang cosmology claim, it is not explained. The supposed explanation is nothing but an entry in the gardener's catalogue of hypotheses that constitutes the theory. Had observation given 27 kelvins instead of 2.7 kelvins for the temperature, then 27 kelvins would have been entered in the catalogue. Or 0.27 kelvin. Or anything at all.

Big-bang cosmology is a form of religious fundamentalism, as is the furor over black holes, and this is why these peculiar states of mind have flourished so strongly over the past quarter century. It is in the nature of fundamentalism that it should contain a powerful streak of irrationality and that it should not relate, in a verifiable, practical way, to the everyday world. It is also necessary for a fundamentalist belief that it should permit the emergence of gurus, whose pronouncements can be widely reported and pondered on endlessly—endlessly for the reason that they contain nothing of substance, so that it would take an eternity of time to distil even one drop of sense from them. Big-bang cosmology refers to an epoch that cannot be reached by any form of astronomy, and, in more than two decades, it has not produced a single successful prediction.

A state of zero progress also suits the world's scientific establishments. A scientific revolution, such as the one that occurred with the arrival of quantum mechanics in 1925, sweeps away the pillars of scientific establishments. Within only two or three years, they are gone, to be replaced by a new generation of young people in their twenties. Today, in a state of zero progress—indeed, with the stately galleon of science even blown backwards—there is little hope for young people in their twenties. They must do what aging gurus tell them to do, which is nothing.

A state of zero progress also favors demands on governments for larger and larger sums of money. Because, on the average, science has repaid society well over the years, scientists as a body hold a kind of blackmail over governments, blackmail that becomes stronger when things are going badly. For then it is possible to blame the becalmed state of science on inadequate financial support. More and more extravagant expenditures on larger and larger machines and instruments are demanded in order to escape from the impasse, whereas, in good times, no such blackmail needs to be exercised. The worse things get, the more scientists meet together internationally in the interest (supposedly) of progress. But, as Tommy Gold points out, perpetually meeting together locks people's beliefs together into a fixed pattern, and, if the pattern is not yielding progress, the situation soon becomes moribund. These considerations provide ample motivation for attempts to preserve the status quo in cosmology: religion, the reputations of the aging, and money. Always in such situations in the past, however, the crack has eventually come. The Universe eventually has its way over the prejudices of men, and I optimistically predict it will be so again.

Religion is a hard problem, one I will attempt to discuss a little by way of ending this book. The crude denial of religion that became prevalent among so-called rationalists in the late nineteenth century was, in my opinion, a response to the social and economic conditions of the time. It had no real intellectual value. And, setting aside also the ancient religions, the first modern statement on religion that I find of interest was that made by James Jeans in the 1920s, that God is a mathematician, although, in this, he was plainly following in the steps of a French mathematician of the early nineteenth century, Pierre Simon de Laplace.

Nowadays, however, I would not accord to mathematics the prestige that I once did. Mathematics consists *only* in saying that quantities are the same—as, for instance, 5, 2 + 3, and 7 - 2 are the same. The latter are what we call "obvious," and making the pronouncement of their equality would confer no prestige on a person who attempted to proclaim them from the housetops. Likewise, all mathematical statements, I believe, would seem obvious at an adequate level of intellect. They seem otherwise to us only because we are not good at mathematical thinking. We have to carry mathematical equalities—theorems, we call them—around in our heads as crutches to our thinking because, collectively, as a species, we are somewhat dim-witted and so cannot instantly perceive what is true, as a superior intellect might do. During the First World War, a young Indian named Srinivasa Ramanujan Ivengar traveled to England to work with the foremost mathematicians at Trinity College, Cambridge. Ramanujan was almost entirely self-taught; yet he had an uncanny ability to say that two entirely different-looking and ferociously complex mathematical expressions were the same. Often, he had no proof, and only occasionally was he wrong. I think it correct that, even to this day, no formally trained mathematician really understands where his perceptions came from. Ramanujan was a hint of what a superior intellect might be able to do. His case was also a hint that genetic possibilities for doing much better than we have yet done may lay, still unexpressed, in the human genome. If indeed a sufficiently superior intellect were able to reduce mathematics to a set of trivia, then James Jeans's description of God as a mathematician hardly seems adequately complimentary. It is as if a crowd of dogs were to attribute divinity to the ability to see that 2 + 3 and 7 - 2 are the same.

Theoretical physics is a different matter. As I now see things, theoretical physics stands intellectually above mathematics. The great discovery of Paul Dirac of the wave equation of the electron will serve again here. A certain operation acting on a spinor field led to a second spinor field. A similar operation on the second field led to a third, and so on, ad infinitum. There is nothing much, I think, that even a superior intellect could do further to this process, so far as mathematics is concerned. But what Dirac did was indeed something further. It was to say that the odd spinors in such a sequence are all the same, as are the even-numbered spinors—the second, the fourth, and so on. Mathematically, this need not be true. What physics does is to make it true, thereby *defining* the Universe. Thus, the Universe is a set of restrictions on mathematical quantities of the kind discovered by Dirac. God—if we are to follow James Jeans—is not a mathematician but rather the chap who thought up the restrictions.

I purposely don't want to make God too remote—nothing like the awesome God of the ancient Hebrews—because I don't believe that concept is right, impressive as it may be. I can explain the difference in terms of an old Spanish story. God, in disguise, falls in with a peasant walking on the road and asks, "Where are you going?" To this, the peasant replies, "To Saragossa," without adding the obligatory medieval addendum *Deus voluit* ("God willing"). For this disrespect, God turns the peasant into a frog and flicks him into the nearest puddle. After watching the frog flounder for a while, God reverses the process and says to the peasant, when restored to human shape, "And now where are you going?" To this the peasant replies, "To Saragossa. Or, into the puddle." The angry gods of the ancient world would have had the peasant straight back into the puddle. My God, in contrast, would make quite certain he got to Saragossa. A mistake of all fundamentalist religions is that their gods have no sense of humor. This is because fundamentalist religions are maintained over long periods by ritual, and ritual, in its very nature, has no sense of humor.

By returning to Paul Dirac's discovery, we can see other attributes of God. With the behavior of an electron determined by the equality of the odd spinors (and that of the even spinors), turn now to the solution of the simplest practical problem one can think of, a single electron bound electrically to a proton to form a hydrogen atom. It is natural to human thought—at least to my way of thinking—to imagine the way to choose the constraints that represent the laws of physics would be to ensure that their consequences were as simply derived as possible. But the opposite is the case. The apparently simple problem of the hydrogen atom is one that none of the multitude of textbooks of which I am aware that seek to inform us about the mysteries of quantum mechanics manages to solve fully—except one.* A skilled lecturer would need a whole term of lectures to cover it fully, which is why almost no student knows it. They rush into more complex and seemingly more interesting problems without bothering to make a proper job of the simplest.

The clear lesson, therefore, is that the laws of physics are specified so as to make for exquisite complexity, emphatically not the reverse. The amazing fact to contemplate is that the seemingly simple discovery of Dirac, taken with an elegant principle of symmetry in group theory, determines the whole of chemistry and, ultimately, of biochemistry, ranging up to the astonishing catalytic properties of proteins, on which life depends. It is all in there in the restrictions on the spinors. It is because of

---

*Claude Itzykson and Jean-Bernard Zuber, *Quantum Field Theory* (McGraw-Hill, New York, 1985).

this incredible chain of subtlety that I doubt the nineteenth-century denial of a purposive Universe, and also why I doubt the crude breaking of the physical laws that occurs in big-bang cosmology.

In the days in which I was involved in British science politics, a well-known chemist visited me with the aim of securing my support for a £10 million project, the money to come from the government, as always. What he wanted was to acquire a supercomputer that would work out all of chemistry. Instead of brusquely sending him packing, I had the tact to use universityspeak, which, being very soft, is supposed to turn away wrath but sometimes doesn't. No computer that humans will ever build will work out all of chemistry. Indeed, no computer that humans will ever build will do much more than can already be done with a desktop job costing less than £1000. If you ask what is needed to work out the full consequences of the laws of physics (chemistry, with biochemistry, being a fair slice of it), the answer is, Nothing less than the whole Universe. It is not too much of a guess to say that that is just what the Universe is: the calculation of the effects of the physical laws. This explains a problem that has puzzled theologians, philosophers, and scientists alike: Why is there a Universe at all? The theologian, with his belief in an all-powerful God, wonders why God didn't simply perceive the Universe. Why bother actually to have it? The answer is that the Universe *is* the simplest way of perceiving it. Any attempt to calculate the Universe, as my chemist acquaintance wanted to calculate all of chemistry, would end up with something a great deal more cumbersome. What really can be done in a compass less than the Universe is to perceive highlights, to manage meaningful argument by taking immense shortcuts, but with the loss of a great deal of detail.

In earlier chapters, I mentioned climbing nearly 300 of the Scottish mountains known informally as the Munros. About a half I did alone, as I did with the last one I mentioned in some detail, Blaven. The Munros are not dangerous or difficult, in the technical sense that an Alpine peak may be dangerous, although, if you are of a mind to it, you can find plenty of hard-to-climb rocks, especially on the Isle of Skye. The dangers come, rather, from sudden storms with ice-cold rain, or heavy snow, or unbelievably high winds, combined often with big distances to be made to safety. Inevitably, in being abroad on the mountains so often, I would run into bad conditions, particularly because the most spectacular situations occur in winter, in which I most enjoyed being out-of-doors. Inevitably, too, it became essential to learn survival techniques, of which by far the most important is anticipation. One problem that was difficult to anticipate, however, was the sudden descent of heavy mist. Being suddenly surrounded by what amounted to a thick cloud, in unknown country, far from one's destination, is in many ways analogous to doing scientific research. There are no straightforward paths to follow, and the number of mistakes one can make is legion. The best I could ever achieve was to proceed slowly (crabwise, as it were) with map and compass—for which, in scientific research, you can substitute observation and calculation. Working your way out of an awkward position takes what seems an age, and this too happens in scientific research. When I consider cosmology, as I started out in cosmology in the cold February of 1948, it seems no very great step to the position I hold today. I am perpetually astonished at the slowness with which perceptions come, just as one can scarcely credit the ineffectiveness of one's stumblings on a mountainside in the moment when the mist clears.

A good fraction of those Munros I didn't climb alone were done with my friend Dick Cook, whose ideas for coping with difficulties were much the same as mine—in fact, I learned most of my survival methods from Dick. Occasionally, however, mountains would be climbed in considerable parties—ten to twenty persons, perhaps. Down, on some of those occasions, would come a thick mist, and what happened then? Dick Cook and I would take off our rucksacks, pull out maps and compasses, obtain bearings, and start arguing about what was to be done. Among twenty persons, there was always somebody who knew better, however, someone who would stride away boldly into the mist, proclaiming that he knew to a jot where he was and to a tittle exactly how to proceed. Then the amazing thing happened. The rest of the party always—always, I swear—followed our hero, reminiscent of a song, much favored in my youth, about a dasher who carried a banner with a strange device, a device that, on examination, read nothing but "Excelsior!" I puzzled a lot about that device, but I never arrived at a satisfactory explanation of it.

Dick Cook and I would usually succeed in extricating ourselves from the mist-covered hills and would arrive back at our hotel in time for a bath before dinner. The others would stagger in, hollow-eyed, at later hours running into the early morning, or even into more distant time. Well, in my opinion, this is exactly the way it has been in cosmology during the quarter century or more that has elapsed since the discovery of the microwave background in 1965. And this is just the way it has been in the religions of the world since time immemorial.

The rule, then, is to proceed crabwise, and my problem now is how to proceed crabwise from where I stand today. Sir James Jeans, responsible for the concept of God the mathematician, has been unlucky posthumously. From a wrong mathematical analysis of the stability of stars, he was led to a mistaken view about the properties of stars, and so Jeans has appeared to leave little impression on astronomical development. But he foresaw the two distinct time scales discussed above, the time scale of order 10 billion years associated with stars and the time scale of 1000 billion years associated with the dynamics of galaxies and with the slow expansion of the Universe in the preceding interpretation. More remarkable still, he was responsible for the following sentence:

> The type of conjecture which presents itself somewhat insistently is that the centres of galaxies are of the nature of "singular points" at which matter is poured into our universe from some other, and entirely extraneous dimension, so that, to a denizen of our universe, they appear as points at which matter is being continuously created.

There now seems to be so much that is correct about Jeans's speculation that, far out as it might seem, we might wonder if meaning can be attached to the concept of "pouring" in from an extraneous dimension, thereby breaking the restriction of the entire Universe to the four dimensions of space and time. There is certainly something unattractive about regarding particle trajectories as beginning in the way that the creation of matter seems to require, so that it would surely be better if we could think of particles as having come from somewhere else.

It is worth considering, for a moment, the phenomenon in physics of pair creation. What happens is that a radiation quantum of high energy disappears in this process, giving rise to an electron plus a positron. To avoid the need for the electron and positron to have a beginning, Paul Dirac conceived of what came to be called "the vacuum," an infinite sea of electrons occupying negative energy states. It is reported that Wolfgang Pauli had a hearty laugh when he heard this notion, claiming Dirac had at last flipped his lid. The trouble for Pauli was that Dirac succeeded in calculating results that agreed with experiments. Dirac added the strange notion of "the vacuum" being unobservable. What the radiation quantum then did was to lift one of the electrons in the infinite sea from its negative state into a positive energy state, where it became observable as an ordinary electron. And the gap in the infinite sea was the positron. This was the way I learned it as a student in 1935–1936; but, in 1949, Richard Feynman gave an interesting alternative way of looking at the matter that is equivalent mathematically to Dirac's result.

With his characteristic originality of mind, Feynman made the positron and electron into essentially one particle, differing only in the sense in which they moved with respect to time. A positron was an electron moving backwards in time, future to past instead of past to future. Or an electron was a positron moving backwards in time. Either way would do.

What a radiation quantum then did when a positron–electron pair appeared was simply to invert the time sense of the positron. Thus, the positron is traveling backwards in time when it encounters the radiation quantum, and the interaction between them turns the particle path around and causes it to move forward with respect to time, when it becomes an electron. We are familiar with particles moving from left to right in space being turned round by a field so as to move from right to left. According to Feynman, it was the same with respect to time. And why not? Time should behave like space.

With this background, we can see a little of the way towards developing Jeans's idea. A particle moving in some extraneous dimension simply has its path turned into our world by the C-field I discussed above. Over many years, I did not take this idea seriously because I was in difficulty to understand why we couldn't see into the extraneous dimension. The mathematical description of the propagation of all fields seems to require that they are able to spread into any dimension that is available, by means of a so-called wave equation that includes all the available dimensions. So for this idea to have any chance of viability, a quite unusual mathematical restriction is needed. All our fields of physics have to be confined to what Jeans refers to as "our universe." It must be the fields rather than the particles that fail to penetrate outside. This is a very strange idea, but, since we have come so far already, it is worth thinking a little further.

In *Walden*, Thoreau wrote: "Our life is frittered away by detail . . . simplify, simplify." Do so by thinking in ordinary three-dimensional space. Collapse down the usual space-time into just two dimensions—say, the horizontal plane—and call it Π. Then the vertical direction is that of the particle paths when they are outside "our universe." And all our usual fields—electromagnetic, nuclear, gravitation—are in Π, where they can only act on particles moving in the vertical dimension in the brief moment when the particle paths cut through Π. The interaction that occurs in this brief moment may, in analogy to the radiation quantum interacting with the positron–electron path, have a probability of turning the particle paths from being perpendicular to Π to being in Π itself. Calculating such a quantum probability would be a formidable problem, but it would be conceivable.

Inevitably, in such a calculation, there would, with the particle in Π and so subject to the usual physical interactions, be a continuing possibility of the path being turned back out of Π, and, therefore, of the particle being returned to the "extraneous dimension." But, if we now play the card of the particle being of the kind mentioned above, a Planck particle subject to decay into a vast number of ordinary particles, such particles can become trapped into "our universe," provided decay can be got in before the particle slips back into the extraneous dimension. In principle, such a reversal could happen, but, in practice, it never does. So there is the beginning of an idea that can possibly be made to work—provided ordinary physical fields can be confined to Π, that is to say. Additionally, too, one could suspect that it is this trick of the decay of Planck particles that gives a time sense to our Universe, a time sense that is ultimately responsible for all those phenomena that lead to the degradation of ordered structures—as, for instance, hair that goes gray and lines that appear and deepen on the face—and that, in the end, lead to a visit from Shakespeare's fell sergeant, who is strict in his arrest.

Although, at present, this is only a crabwise move, it leads to a position from which I can perhaps explain my principal objections to the religious concept of an all-powerful God. Religions with an all-powerful God make no sense unless you also believe that God is pretty evil, or at least wholly indifferent to bad things that happen. If you believe in an all-powerful God, you have to ascribe to God a morality inferior to that of humans, which is quite a measure of condemnation. But the real point is that God is not all-powerful, God cannot overcome the evils of decay because the issue is not one that is open to choice. If you have the Universe, then you must have decay. If you have no decay, you have no Universe. Take your pick.

I have no particular liking for early Italian religious art. But, by the time religious art attained the standard of Raphael, it assumed an unintended degree of interest, for, when you look at Raphael's madonnas, you know what his girl friends were like. So, too, with Botticelli. And if today you combed the piazzas of Florence, it is a racing certainty that you will find girls essentially undistinguishable from the ones that Raphael and Botticelli painted. The phenomenon of life is an immensely clever way to beat decay, and, if a further solution to the problem of "I am me" can be found, the solution is complete.

Today we have the extremes of atheistic and fundamentalist views, and it is, in my opinion, a case of a plague on all their houses. The atheistic view that the Universe just happens to be here without purpose and yet with exquisite logical structure appears to me to be obtuse, whereas the perpetual quarrelings of fundamentalist groups is worse than that. Not all the religious quarrels I ever saw or read about is worth the death of a single child.

One can conceive of various Universes defined by different forms of mathematical restrictions. What I suspect is that the restrictions defining "our universe" are not just any old restrictions. The restrictions are optimized for their consequences. Or, to put it another way, God is doing His best, and to load off onto Him the all-powerful concept is a gross insult, an insult by people who do not merit the great trouble that has been taken on their behalf.

To return to crabwise possibilities: If Jeans was right in his postulate of an extradimensionality outside Π, then it is another racing certainty that there must be the possibility of communication existing in some way between Π and what is outside Π. By what has been said, the communication cannot be through the familiar fields of physics. It would need to be something else. This is a platform for the religious person to jump on. This is how to communicate with God, it might be said. Perhaps by prayer. It could be so, but I would prefer something less exposed to the accusation of self-deception. Better, it seems to me, to go for something we can all agree about. It could be consciousness. Consciousness is an experimental fact. More strongly than that, even; without consciousness, there would be no experimental facts. There would be no science, nothing but a nightmare puppet world. It is surely strange that science, utterly dependent on consciousness, should have little or nothing to say about it. With consciousness going outside Π and with science hitherto confined to Π, we can perhaps at last see why.

It is the nature of interactions that they have two ends. Our bodies, our brains and nervous systems, are clearly one end. But what is the other? At a crabwise guess, I suggest a grand information center. To be biblical again, it is said that God sees every sparrow that falls. Well, if He does, it is because we report it. It is because each of us, and every other animal in some degree, and every suitable aggregate of hydrogen, carbon, nitrogen, oxygen, and so on, anywhere over the whole Universe, is an agent, reporting back what we see and experience. Only in this way can I seem to make the beginnings of sense of what is going on. At present, it is but a speculation, but, come another thousand years of understanding, it may not be.

To continue to make sense, I have to suppose that something of a trick or illusion is practiced on all of us. We all have the impression that it is our individual end of the two-way channel that matters, whereas, surely, if the idea is correct, it must be the other end. The trick I can conceive of as being necessary to keep order among an immense number of channels (rather as, in a telephone system, individual conversations must be kept separate from each other). But we have clear clues that our incessant concern with self, with *me*, isn't right. I have remarked before that, after the extraction of a tooth, I have never felt the slightest warmth of affection for the thing. Most dentists know it, and they don't even bother to show the tooth to the patient. I suspect that the same holds for the rest of our bodies, which, like the tooth, are fine so long as they continue to work. And I suspect that, like the tooth, which certainly isn't *me*, neither are our bodies. Our impression otherwise is the trick, the illusion necessary, I suppose, to keep the two-way link operative. But it surely must be the other end that is somehow associated with *me*. It might seem a big plus mark to see one's way through this problem, but I suspect that things might not work too well if we could. After a lifetime of crabwise thinking, I have gradually become aware of the towering intellectual structure of the world. One article of faith I have about it is that, whatever the end may be for each of us, it cannot be a bad one.

Index